XI Meeting on Recent Advances in the Physics of Fluids and their Applications 2010

(FLUIDOS 2010)

Journal of Physics: Conference Series Volume 296

Colonia del Sacramento, Uruguay
3 – 5 November 2010

Editors:

Italo Bove
Arturo C. Marti
Cecilia Cabeza
Gustavo Sarasua

ISBN: 978-1-61782-919-2
ISSN: 1742-6588

XI Meeting on Recent Advances in the Physics of Fluids and their Applications 2010

(FLUIDOS 2010)

Journal of Physics: Conference Series Volume 296

Colonia del Sacramento, Uruguay
3 – 5 November 2010

ISBN: 978-1-61782-919-2
ISSN: 1742-6588

IOP Institute of Physics

Printed from e-media with permission by:

Curran Associates, Inc.
57 Morehouse Lane
Red Hook, NY 12571

Some format issues inherent in the e-media version may also appear in this print version.

Printed by Curran Associates, Inc. (2011)

For permission requests, please contact the Institute of Physics at the address below.

Institute of Physics
Dirac House, Temple Back
Bristol BS1 6BE UK

Phone: 44 1 17 929 7481
Fax: 44 1 17 920 0979

techtracking@iop.org

Additional copies of this publication are available from:

Curran Associates, Inc.
57 Morehouse Lane
Red Hook, NY 12571 USA
Phone: 845-758-0400
Fax: 845-758-2634
Email: curran@proceedings.com
Web: www.proceedings.com

TABLE OF CONTENTS

PREFACE

The papers published in this volume of the *Journal of Physics: Conference Series* were selected from the manuscripts submitted to the **XI Meeting on Recent Advances in the Physics of Fluids and their Applications (FLUIDOS2010)**, which was held in Colonia del Sacramento, Uruguay, 3–5 November 2010. FLUIDOS takes place every two years, usually in November, with the aim of gathering together researchers from all areas of the Physics of Fluids, to update themselves on the latest technical developments and applications, share knowledge and stimulate new ideas. This 11th meeting continues the successful experience of the previous ones which were held in different Argentinian cities. For the first time, the meeting was celebrated in Uruguay, more specifically, in the peaceful town of Colonia del Sacramento, designated a World Heritage Site by UNESCO.

The conference presented an outstanding program of papers covering the most recent advances in Physics of Fluids in the following areas:

- General Fluid Dynamics
- General and non-Newtonian Flows
- Magnetohydrodynamics
- Electrohydrodynamics and Plasmas
- Hydraulics, Thermohydraulics and Multiple Phase Flows

A website with full details of the conference program, abstracts and other information can be found at http://fluidos2010.fisica.edu.uy.

We would like to thank all the participants, especially those who contributed with talks, posters and manuscripts, for making FLUDOS2010 such a successful conference. Our thanks also go to our colleagues for their support and encouragement, particularly in the refereeing of papers. We would like to acknowledge additional financial support from *Comisión Sectorial de Investigación Científica (Universidad de la República, Uruguay)*, *Programa de Desarrollo de las Ciencias Básicas (Uruguay)* and the *Centro Latinoamericano de Física (CLAF)*. Our thanks are extended to the local government of Colonia del Sacramento.

The next FLUIDOS conference will be held in November 2013, in Buenos Aires, Argentina. We anticipate enjoying another successful FLUIDOS meeting to be held in one of the main cultural centres of the continent.

Italo Bove, Cecilia Cabeza, Arturo C Martí, and Gustavo Sarasúa
Editors

Wake flow stabilization with DBD plasma actuators for low Re numbers

Juan D'Adamo[‡], **Roberto Sosa**[‡], **Marcos Barceló**[†], **Guillermo Artana**[‡]

[†]Laboratorio de Fluidodinámica, Facultad de Ingeniería. Universidad de Buenos Aires,
[‡]CONICET
CP 1063 Avda. Paseo Colón 850, Buenos Aires, Argentina

E-mail: `jdadamo@fi.uba.ar`

Abstract. We propose to study the stability properties of a wake air flow forced by electrohydrodynamic (EHD) actuators. EHD actuators can add momentum to the flow around an object in regions close to the wall. Hence, we adopt the hypothesis of the forcing actuation modeled as a slip-wall condition.
We study the flow around a cylinder modified by EHD actuators, dielectric barrier discharge (DBD) type, by means of particle image velocimetry (PIV) measures. For the sake of simplicity we observed flows at low Reynolds(Re) numbers ($\sim$ 200) where the flow is mainly two-dimensional. As the forcing frequencies, characteristic of DBD, are much higher than the natural shedding frequency of the flow (kHz vs. Hz), we consider in this work only the forcing actuation as stationary. The actuators are disposed symmetrically near the boundary layer separation point. Thus, flow symmetry is conserved, as it remains weakly non parallel. This aspect allows us to perform stability analysis of the flow and to determine its characteristics which are governed by a strong mean flow correction. We also present results on the global mode evolution under forcing.
The main motivation of this work is to optimize the forcing on the flow, leading to a more effective energy consumption of EHD actuators.

1. Introduction

The simply geometry and the complex behaviour that exhibits the flow around a cylinder represents, for low Reynolds numbers ($Re = U_0 D/\nu < 200$), a prototypical 2D wake flow. The regime is identified with the Bénard von Kármán (BvK) vortex street, when the flow breaks its time continuous invariance at the bifurcation $Re = 48.5$. In this case, the velocity field in the whole flow domain oscillates with the same global frequency and its harmonics, related to the appearance of a sufficiently large region of absolute instability in the near wake. As the oscillation is spatially evolving, its envelope is called the global mode of the instability. Previous studies Zielinska et al. [1] and Wesfreid et al. (1996) [2], characterized the global mode behaviour determining scale laws for its nonlinear evolution as functions of the Reynolds number.
As a self-sustained hydrodynamical oscillator, the flow can be altered by forcing parameters that may be varied smoothly and accurately to give rise different regimes. In this sense, we study this benchmark in order to characterize flow control in wakes by means of electrohydrodynamical actuators (EHD). EHD actuators for flow control have been receiving special attention in the last years, as reviewed by Moreau (2007)[3]. Among all the types of low-energy plasma actuators, one can notice a group of actuators that produces surface discharges. With these devices the goal is

usually to use an electric *wind* produced by the plasma in order to modify the properties of the boundary layer close to the wall. A particular type of discharge is the surface dielectric barrier discharge (DBD). This discharge has been perfected for the first time in air at atmospheric pressure by Masuda(1979) [4] for ionic charging of particles. Roth (1998) et al. used it for airflow applications at the end of the 1990s [5], characterizing the momentum injected for flat plate and airfoil flows. Later, Thomas et al.(2006) [6] applied the DBD for the bluff body flow control. In their results, the authors showed drastic reductions of the flow separation and the associated Karman vortex shedding. On the other hand, some hypothesis on the dynamics of the DBD forcing arose from a computational study validated with PIV experiments in Gonzalez et al.(2009)[7]. From a stability analysis for a bluff body flow, with no vortex shedding, the forcing is introduced numerically as a moving surface boundary condition. In this line, the work by Gronskis (2009)[8] analysed numerical models for DBD forced flows for a cylinder at low Reynolds numbers (< 200).
We intend with this work to optimize the DBD control device by analysing the stability properties of the flow. For this reason, we simplify first by studying the problem at low Reynolds numbers. Global modes evolves as the mean flow is modified by external parameters, Barkley (2006)[9] introduced hypothesis to consider the mean flow as a base flow for 2D stability analysis. In this way, the mean flow represents a marginal stability state and defines the frequency and amplitude of the wake oscillations. The nonlinear saturation of the oscillatory instability is achieved by Reynolds stresses from the mean flow modification. Khor et al. (2008)[10] and Leontini et al. (2010) [11] support these hypothesis from experimental and numerical data for Re up to 600. These ideas have been applied to analyse forced wakes. Previous studies on the stability properties of forced wakes, Thiria and Wesfreid (2007)[12](2009)[13], confirmed that a nonlinear critical behaviour takes place under forcing. Indeed, a bifurcation scenario reappears as the forcing action stabilizes the wake fluctuations, in their case, a rotary oscillation. Figure 1, adapted from[12], shows how the global modes develop under forcing. The scheme is represented by a vertical line, for Re > Rc, that corresponds to forcing from the lower branch up to a critical value, when the wake reaches, or *recovers*, its basic state.
In this context, we pursue the characterization of the DBD control actuator to complete an understanding which will eventually lead to a more effective energy consumption. We organize the work as it follows: we describe the experimental setup to produce the DBD discharge as well as the method to measure the velocity fields of the flow, we present and discuss our results, concerning the global modes evolution, the stability properties of the wake under forcing; at last we resume some conclusions and formulate perspectives for future works.

2. Experimental Setup

The experimental set-up consists of a flow around a circular cylinder at low Reynolds number, $Re \sim 200$.
The velocity field measurements were undertaken with the cylinder placed in a closed loop wind tunnel with a test section of $50 \times 50\text{cm}^2$. The EHD actuator was mounted on the surface of a polymethyl methacrylate cylinder tube with external diameter D =20 mm. The tube wall was 2 mm thick.

The electric circuit is composed of a signal generator, an audio amplifier and an ignition coil to give a discharge at $V_{EHD} = 10kV$ and a frequency of $f_{EHD} = 5.6$ kHz. The signal amplitude is then modulated by bursting the signal as showed by Figure 2. A second frequency arises $f_{Burst} = 1/T_{Burst}$ and the electrical energy delivered by the device is characterized by a duty cycle ($DC = T_1/T_{Burst}$). Special care was taken to assure a stationary input as $f_{Burst} \gg f_{flow}$, being f_{flow} the vortex shedding frequency ($\sim Hz$, as the Strouhal number $St = fD/U_0 \simeq 0.2$). In short, the flow control parameter is the Duty Cycle (DC) input that modulates the "ionic

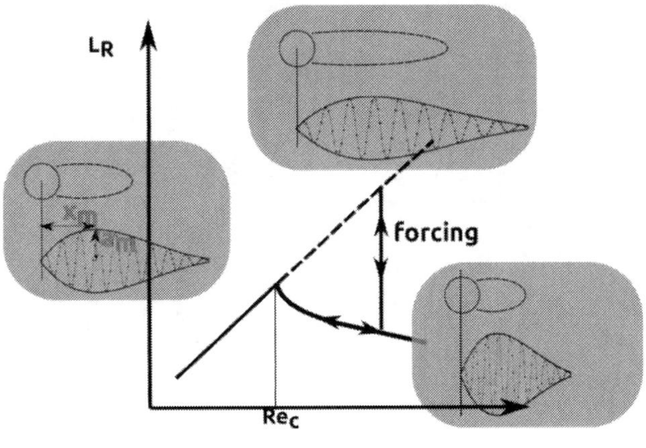

Figure 1. The reverse flow region behind the cylinder evolves nonlinear with Re, as it is modified by the mean flow. A new bifurcation scenario appears under forcing conditions.

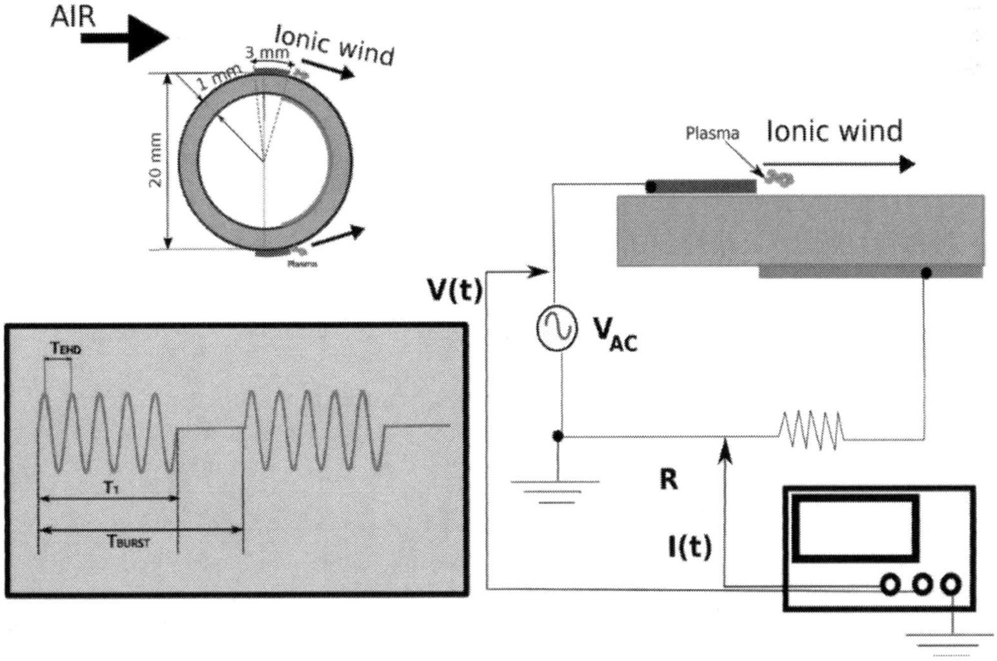

Figure 2. Schematic of the EHD actuator: Electrodes disposed flush-mounted. Electric circuit and input signal detail.

wind" momentum.

Quantitative measurements were performed using 2D Particle Image Velocimetry (PIV) on a

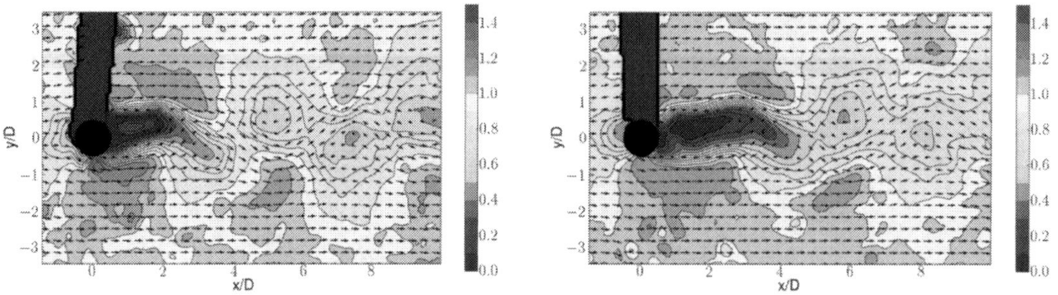

(a) No forcing. (b) Flow under DBD actuation. DC=16

Figure 3. Instantaneous PIV velocity fields. The region of interest contains 140 × 86 vectors, scaled to represent the flow more clearly. Contour levels corresponds to the velocity modulus.

vertical plane placed at mid-span of the cylinder. Image acquisition and PIV calculation were done using a LaVision system, composed of an ImagerPro 1600 × 1200 CCD camera with a 12-bit dynamic range capable of recording double-frame pairs of images at ∼ 8Hz and a two rod Nd:YAG (15mJ) pulsed laser synchronized by a customized PC using LaVision DaVis 7.1 software. Laser sheet width was about 1mm in the test section. The whole 300mm × 200mm imaging region (about 12 × 8D) gives a spatial resolution of 0.06D as showed in 3. All image acquisitions where done using the double-frame mode with a time lapse between the two frames (dt) set to 7ms.

Given the flow natural frequency, its dynamic behaviour is well resolved by the PIV measures for the selected acquisition frequency.

3. Results

3.1. On global modes modification

The modifications introduced by the plasma actuator can be described firstly by the mean flow field, obtained from 400 snapshots which contains al least 30 shedding periods. We observe, in Figure 4 that the near wake region behind the cylinder enlarges under increasing voltage. The streamlines get close further downstream when it is applied the higher EHD actuation, resulting that the mean flow resembles more to the stationary solution, characterized by a long recirculation zone as it in Figure 1 scheme.

Wake flows are characterized by oscillations of a propagating wave which amplitude grows from the origin, the cylinder in our case, reaches a maximum and decays afterwards. The spatial envelope of this coherent oscillation gives the amplitude of the global mode and, for practical purposes, it is reasonable to consider that the flow's first harmonic does, by far, the dominant contribution to the global mode.

Previous works by [1] and [2] studied scaling laws for the global mode in wake flows near the threshold Rc. A typical contour is presented on Figure 5(a) where its maximum amplitude, A_{max} at (X_{max}, Y_{max}) coordinates, is highlighted as it represents an important parameter for scaling . The authors proposed that both A_{max} and X_{max} follow scaling laws of the following form:

$$A_{max} \simeq (Re - Rc)^{1/2}, X_{max} \simeq (Re - Rc)^{-1/2} \tag{1}$$

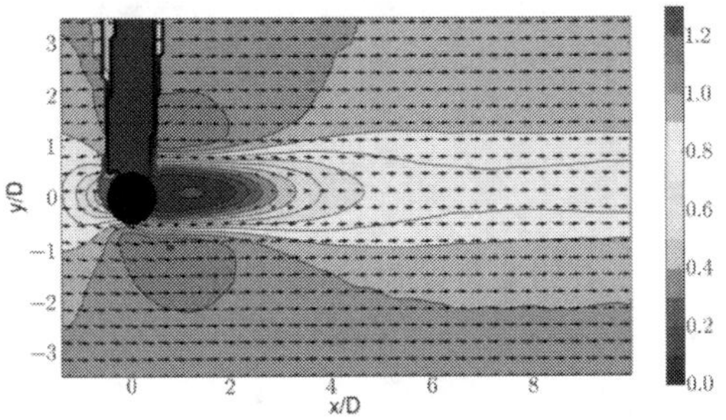

(a) Flow without forcing

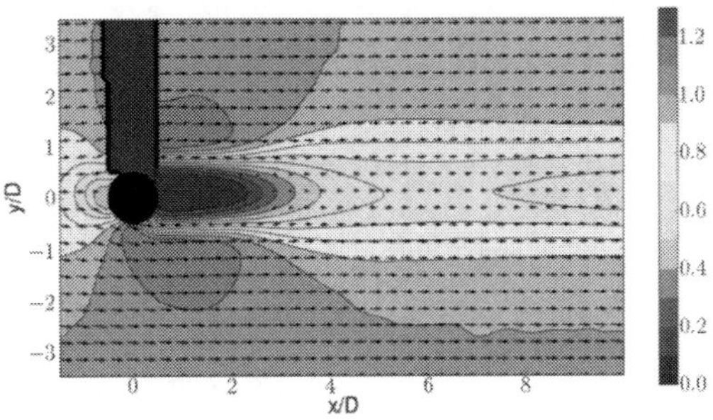

(b) Flow under DBD actuation. DC=10

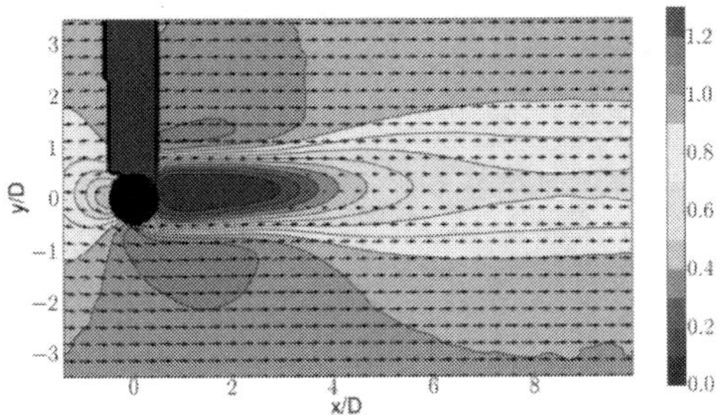

(c) Flow under DBD actuation. DC=16

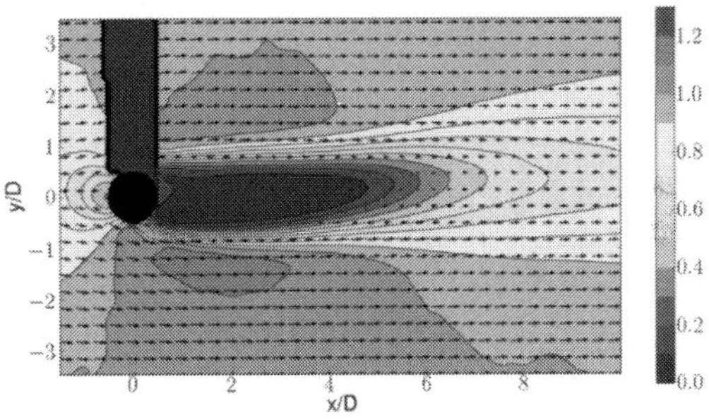

(d) Flow under DBD actuation. DC=22

Figure 4. Mean flow velocity vector field and velocity modulus in contours.

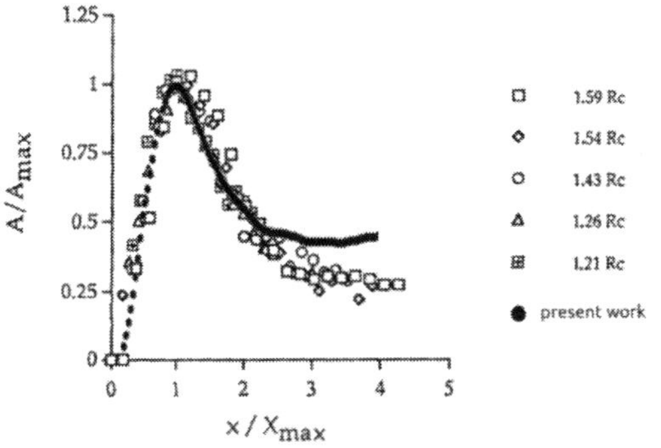

(a) Global modes curves collapse into one curve scaling with A_{max} and X_{max} parameters. Flow around a cylinder without forcing

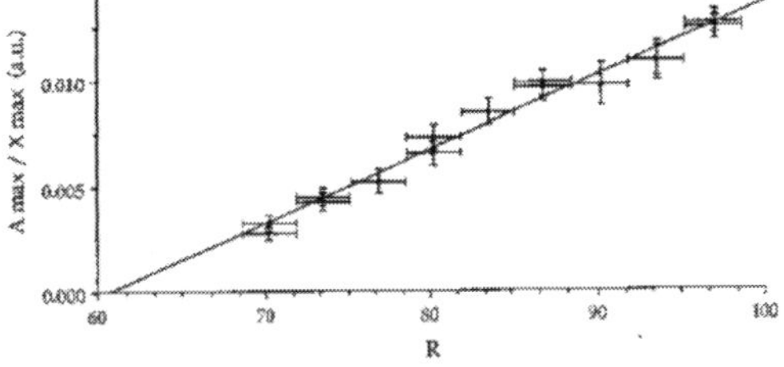

(b) A scaling law emerges for resume the behaviour near the threshold

Figure 5. Scaling laws for Global modes near the threshold, from [2]

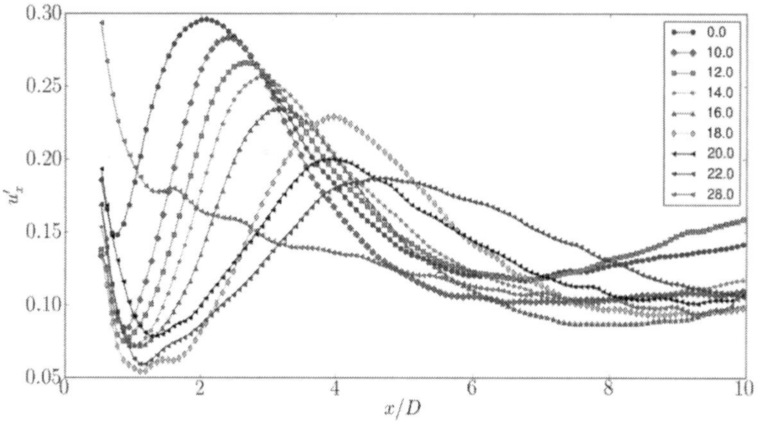

Figure 6. Global mode evolution for increase DBD forcing (DC).

Hence, the curves for different Reynolds numbers collapse into an universal curve. On the other hand, a scaling law, Figure 5(b), that resumes the behaviour as the parameter *Re* approaches the threshold ($Rc \simeq 46$ in cylinders).
In analogy, we determine the global mode evolution for DBD stationary forcing as we increase the duty cycle (DC) parameter. We observe in Figure 6 that increments on DC leads to decrements in the mode amplitude and accordingly its position displaces downstream, proving the flow stabilization. From a certain threshold, the behaviour is quite different as shows the curve corresponding to DC=28.

3.2. Scaling law
From the global mode evolution, we can determine a critical value for the forcing parameter DC. From Figure 7(a), we confirm the hypothesis for the von Karman modes evolution when they are forced toward stabilization. The ratio A_{max}/X_{max} decreases almost linearly from DC=10 up to DC=22 and we can estimate a critical value from the linear approximation ($DC_C \simeq 27.5$). With this value we construct the scaling law for $(DC_C - DC)$ in Figure 7(b).

We remark that further increases of DC parameter produces the flow to behave qualitatively different as we expected from our hypothesis. Figure 8 shows how Von Karman mode has been suppressed and fluctuations remain in the cylinder wall neighbourhood. Indeed, the momentum added from the cylinder wall induces higher velocities and the velocity profiles downstream prevent self-sustained oscillations. The instability is confined to a region very close to the wall, Figure 8(a) shows an instantaneous velocity field for DC=28 and Figure 8(b) the mean flow for the corresponding forcing.

3.3. Stability analysis
As the studied flow is weakly non parallel, we analyse the change on stability properties produced by the forcing on the wake. The analysis of the wake, therefore, is decomposed into several equivalent parallel-flow problems, studied through the inviscid Orr-Sommerfeld, or Rayleigh, equation:

$$(kU(y) - \omega)(\phi'' - k^2\phi) - kU''(y)\phi = 0 \tag{2}$$

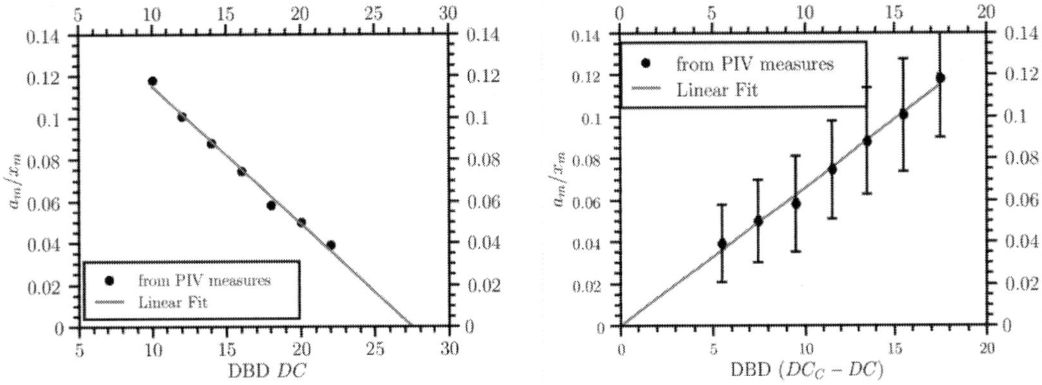

(a) Determination of critical value for DC (b) Scaling law from a_m/x_m evolution.

Figure 7. Scaling law for DC parameter.

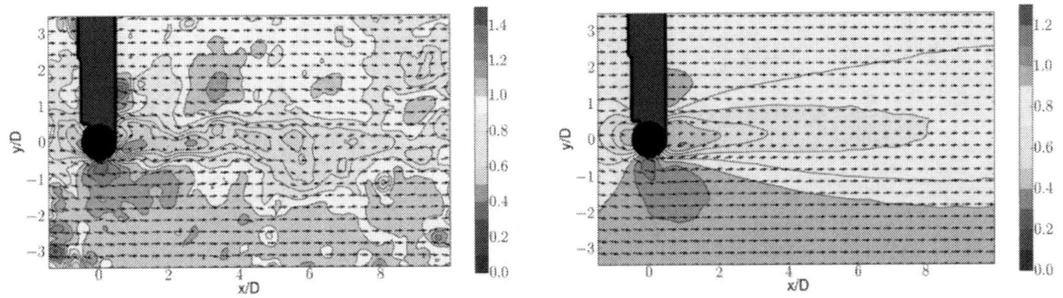

(a) Instantaneous velocity field DC=28. Contours represents velocity modulus. (b) Mean flow under DBD actuation DC=28. Contours represents velocity modulus.

Figure 8. Beyond the bifurcation a qualitatively different dynamics takes place in the flow.

where $U = U(y)$ is the local (for $x = x_1$) velocity profile depending on the transverse coordinate y, k and ω are respectively the complex wavenumber and the complex frequency of the perturbation and ϕ is the associated mode.
An eigenvalue problem arises relating ω and k. The most unstable mode appears for $k = k_0$ where $\frac{\partial \omega}{\partial k} = 0$ and it manifests following the cusp map procedure described i.e. Triantafyllou (1987) [14]. The map in the ω complex plane generated from k complex plane presents a pinch point as showed by Figure 9(c). This determines the mode growth rate ω_i and the real frequency ω_r for the most unstable mode at the x_1 station. The result allows to determine regions whether the flow is locally absolutely unstable (for $\omega_i > 0$) or convectively unstable ($\omega_i < 0$).

The regions are plotted on Figure 10 and we confirm that the global mode evolution is in agreement to a progressive vanish of the absolute instability region.

On the other hand, having one real frequency for each position x do not allow us to determine directly the global frequency of the flow. From Hammond and Redekopp (1997) [15] , the frequency selected is equal to the real part of the absolute frequency ω_0 evaluated at the

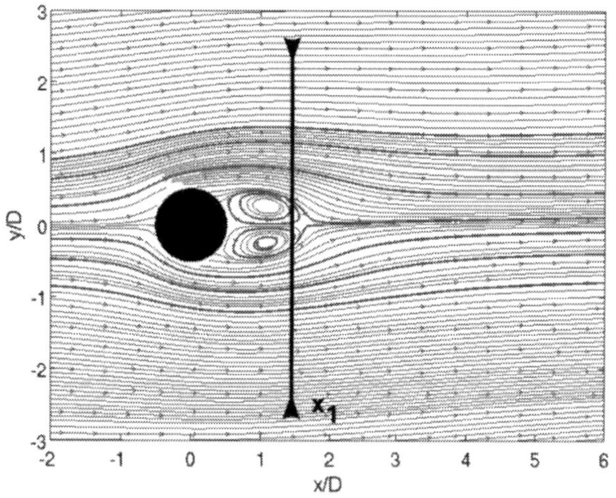

(a) Mean Flow field

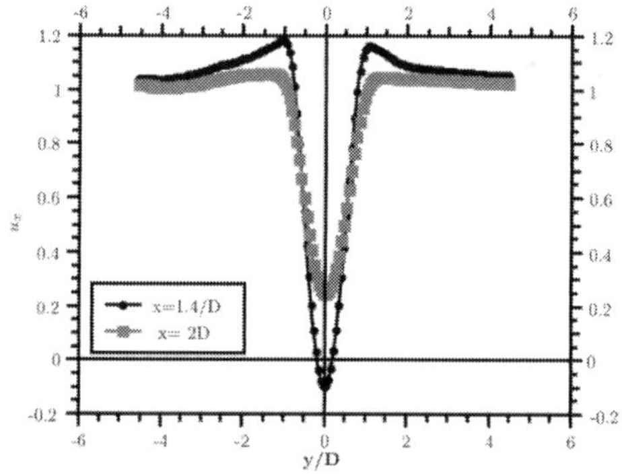

(b) Velocity profiles

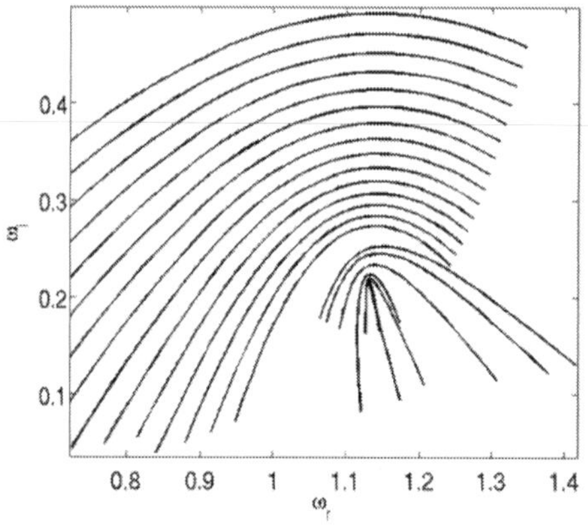

(c) Results of the stability calculation for a local profile, a pinch point in the ω-plane determines the mode complex frequency.

Figure 9. Linear stability analysis from the mean flow.

streamwise position x_s defined as.

$$\left.\frac{\partial \omega}{\partial x}\right|_{x_s} = 0 \tag{3}$$

The saddle point x_s does not lie necessarily on the real axis. Since derivatives of $\omega_0(x)$ are only known along the real x-axis, the location of the saddle point x_s of $\omega_0(x)$ is found through use of the Cauchy–Riemann equations and analytic continuation to complex values of $x = x_r + ix_i$. The expansion is:

$$\left.\begin{aligned} \omega_{0r}(x_s) &= \omega_{0r}(x_r, x_i = 0) - \left.\frac{\partial w_{0i}}{\partial x_r}\right|_{x_i} x_i + O(x_i)^2 \\ \omega_{0i}(x_s) &= \omega_{0i}(x_r, x_i = 0) - \left.\frac{\partial w_{0r}}{\partial x_r}\right|_{x_i} x_i + O(x_i)^2 \end{aligned}\right\} \tag{4}$$

Examining the ω plane for every position in the x-axis, Figure 11(a) shows the evolution of ω_i

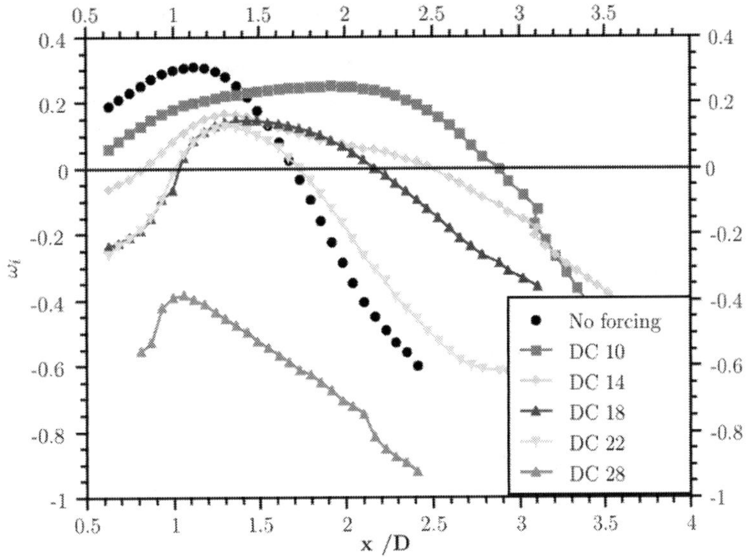

Figure 10. Decreasing size of the absolute instability region

versus ω_r. Expanding (4) produces a curves family (Figure 11(b)). One of these curves presents a pinch point, where the condition (3) is verified.

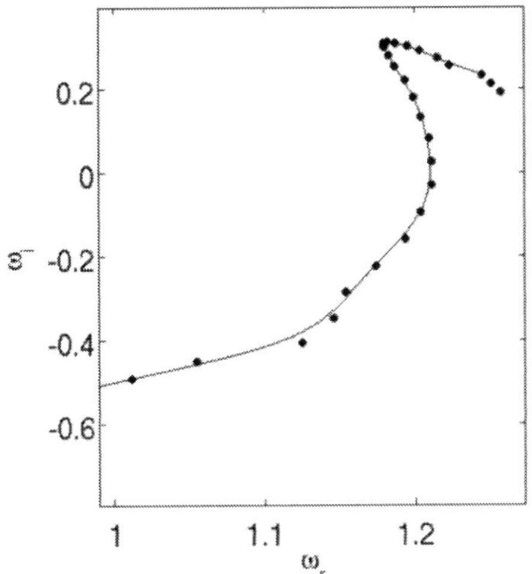

(a) ω_i versus ω_r along x-axis. No forcing case

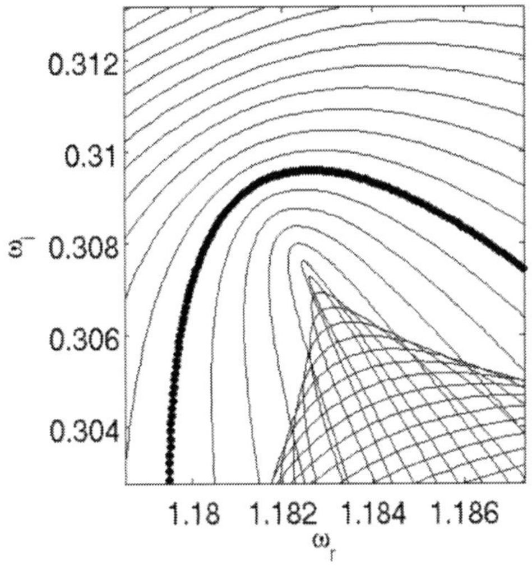

(b) Zoomed Expansion of ω_i versus ω_r following equation (4).

Figure 11. Global frequency selection. [15]

Performing the method for the forced cases, we determine how the global frequency selection is modified. As in other forced wakes [13], the global frequency diminishes as expected for a stabilization. Experimental and numerical errors did not allow us to extract a scaling law for Strouhal, however these results confirm the physical assumptions on stabilization.

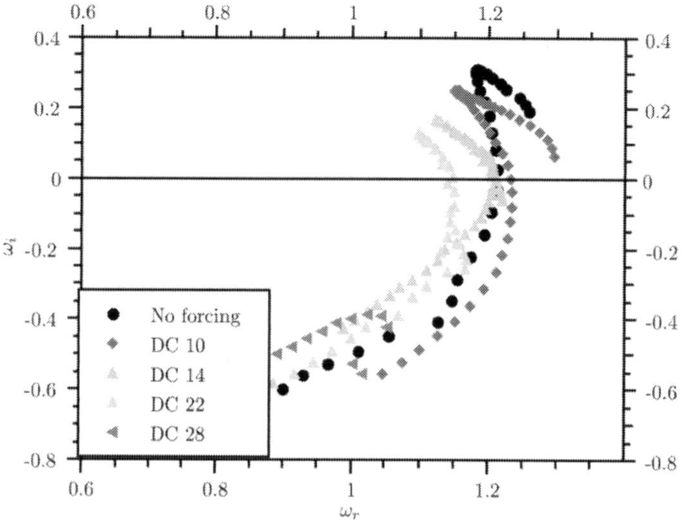

Figure 12. Global frequency selection for forced cases.

4. Conclusions and perspectives

We achieved the analysis of a plasma DBD forced wake flow for low Reynolds numbers. From PIV data of a cylinder flow at $Re \sim 200$, we made use of stability properties of wake flows in order to characterize the forcing. The mean flow modification yields a corresponding evolution of the global modes. In this way, we determine a scaling law in order to optimize the actuator forcing parameter, DC, its duty cycle. On the other hand, the region of absolute instability changes dramatically under forcing as we confirmed by solving the Rayleigh equation. In the same sense, the global frequency (St number) decreases when the forcing approaches the threshold determine by the vanish of the absolute instability region.
Further studies may complete the analysis by increasing Reynolds numbers, and by considering a non-stationary actuation. Indeed, as suggested by [12], the forcing will be more effective energetically when its frequency is around the natural shedding frequency. On the other hand, some recent methods as dynamical mode decomposition (see i.e. Schmid (2010) [16]) presents as a very promising tool for experimental data analysis sharing some properties of linear stability methods.

Acknowledgments

Enlightening discussions with José Eduardo Wesfreid, Ramiro Godoy Diana and Benjamin Thiria are gratefully appreciated.

References

[1] Zielinska B and Wesfreid J 1995 *Physics of Fluids* **7** 1418
[2] Wesfreid J, Goujon-Durand S and Zielinska B 1996 *Journal de Physique Paris II* **6** 1343–1357
[3] Moreau E 2007 *J. Phys. D: Appl. Phys.* **40** 605–36
[4] S Masuda M W 1979 *J. Electrostatics* **6** 57–67

[5] Roth J, Sherman D and Wilkinson S 1998 *AIAA Meeting (Reno, USA, January 1998) paper #98-0328*
[6] Thomas F, Koslov A and Corke T 2006 *AIAA Meeting (San Francisco, USA, June 2006) paper #2006-2845*
[7] Gonzalez L M, Artana G, Gronskis A and D'Adamo J 2009 *Global Flow Instability and Flow Control IV, Crete, Greece, September 2009*
[8] Gronskis A 2009 *PhD Thesis. FI UBA* 1–293
[9] Barkley D 2006 *Europhysics Letters* **75** 750–756
[10] Khor M, Sheridan J, Thompson M and Hourigan K 2008 *J. Fluid Mech* **601** 425–441
[11] Leontini J, Thompson M and Hourigan K 2010 *J. Fluid Mech* **645** 435–446
[12] Thiria B and Wesfreid J 2007 *J. Fluid Mech* **579** 137–161
[13] Thiria B and Wesfreid J E 2009 *J Fluid Struct* **25** 654–665
[14] Triantafyllou G S, Kupfer K and Bers A 1987 *Phys. Rev. Lett.* **59** 1914–1917
[15] Hammond D and Redekopp L 1997 *J. Fluid Mech* **331** 231–260
[16] Schmid P 2010 *J. Fluid Mech.* **656** 5–28

Wall shear stress hot film sensor for use in gases

O D Osorio and N Silin

CONICET / Instituto Balseiro, S C de Bariloche, 8400, Argentina.

Abstract. The purpose of this work is to present the construction and characterization of a wall shear stress hot film sensor for use in gases made with MEMS technology. For this purpose, several associated devices were used, including a constant temperature feedback bridge and a shear stress calibration device that allows the sensor performance evaluation. The sensor design adopted here is simple, economical and is manufactured on a flexible substrate allowing its application to curved surfaces. Stationary and transient wall shear stress tests were carried on by means of the calibration device, determining its performance for different conditions.

1. Introduction

The study of shear stress on the walls that make the boundary of a flow is an area of great interest in Fluid Mechanics and Heat Transfer. However its measurement is complex and is generally limited to experimental models. The measurement of wall shear stresses is relevant when studying the drag of blunt or aerodynamic objects, the pressure drop through conduits, the convective heat transfer of surfaces exposed to flows, etc., but it can also be used to detect the passage of fluid-dynamic structures such as hairpin vortices, vortex shearing behind obstacles or large scale coherent structures [1] [2] [3] [4] [5]. Having a good temporal resolution in this kind of devices is of great interest, one reason for this is that the fluctuations of wall shear are relevant in the understanding of the flow and its effect on walls. In turbulent flows, the fluctuations of shear stress reach values as high as 0.46 of the mean wall shear stress [6]. The second reason is that a slow sensor exposed to a turbulent flow yields average measurement values that do not agree with the mean shear stress. That is, the average introduced by a slow sensor is not satisfactory, requiring therefore a turbulent calibration, i.e. a calibration under the same turbulence intensity that will be found in the case to be measured. To avoid such difficulties, it is necessary to obtain measurements with good temporal resolution and eventually to estimate the average value a posteriori.

A technique for wall shear stress measurement that stands out for its simplicity of implementation and accuracy is the use of thermal sensors. Thermal sensors are heaters whose electrical resistance is a function of temperature, which in turn depends on the convection heat transfer between the sensor and the flow. A trend evident in all the measurement techniques of wall shear stress is that the sensors have their sensing areas ever smaller in order to improve the temporal and spatial resolution. In this process of miniaturization micro-electro-mechanical systems (MEMS) manufacturing technology has played a central role in shear stress sensor technology [7]. MEMS technology is based on the same technology developed for semiconductor device fabrication and is well suited for very small thermal sensors, providing superior performance and lower costs, but also the possibility of manufacturing them on flexible substrates that can be applied on curved surfaces. The sensor that is analyzed in this work was fabricated using such technology.

Despite its advantages, in order to make this technology of practical applicability, the sensor should be insensitive to the thermal characteristics of substrate on which it is installed, and should provide a rapid response to fluctuations of shear stress [8]. In this sense several proposals emerged. Jiang et al [9] achieved a better frequency response of their sensors by manufacturing a vacuum cavity beneath the sensor, thus isolating it from the substrate. On the other hand, Ajagu et al [6] proposed an experimental device consisting of a hot wire sensor located above of a hot film sensor. The hot film acts as thermal shield between the upper sensor and the substrate. Results obtained with this configuration show that this setup can detect very small fluctuations in the measured shear stress. This result served as motivation for pursuing a shear stress hot film sensor of similar performance but of simpler making and constructed on a flexible substrate.

The fabricated sensor was tested for steady state and transient shear stress response by means of a calibration device determining its performance for different conditions.

2. Sensor design and fabrication

The hot-film sensor consists of two flat conductive films: one that will perform as a sensing element, and the other that will serve as a thermal shield between the sensor and the wall. These conductive films were built using a micro fabrication technique. Figure 1 shows a simplified diagram of the sensor exposed to a flow. The flow of air passes over the sensor and parallel to the wall, extracting heat from the sensing element. The strategy used in this work is to keep the sensing element at constant temperature. We do this via a feedback circuit and measure the power dissipated to achieve this, allowing us to measure the shear stress. The shield also operates in constant temperature mode and prevents any variation of wall temperature to affect the sensing element.

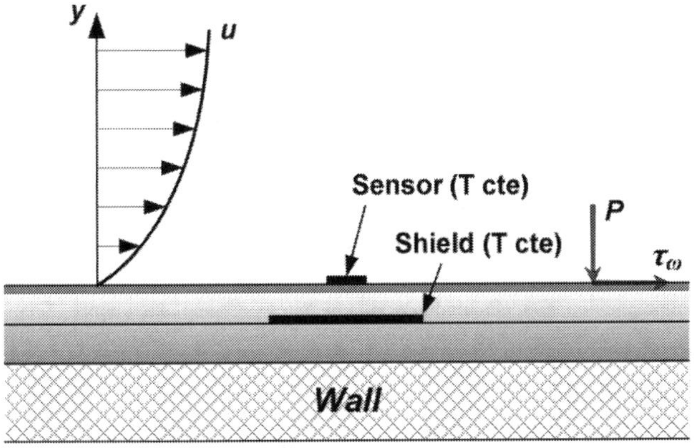

Figure 1. Scheme of the shear stress hot film sensor.

Figure 2. Top view photograph of the sensor and shield.

An important aspect was that the sensing and shield elements were made separately. As substrate we used a 50 μm thick Mylar layer. The sensor and shield patterns were printed on the corresponding substrate using a photolithographic technique called etching. We first applied a thin film of nickel on the substrates, then applied a layer of photoresist and transferred the etching pattern by exposing the photoresist through a pattern mask. As a result an etching mask of photoresist with the desired pattern was created on the Nickel film. Then the exposed areas of the Nickel film were subject to a chemical attack, and finally the photoresist was removed with Acetone. The Nickel layer deposited on the substrate was 100 nm thick. Once the sensor and shield on their respective substrates were obtained they were joined in a single sensor by means of a cyanoacrylate adhesive. A photograph of the final product is shown in figure 2 and a schematic in figure 3. The finished shear stress sensor has a thickness of 140 μm, and is 30 mm in diameter, the sensing element is 1.88 mm long and 50 μm wide. This slender aspect ratio makes the sensor directional, being sensitive for shear stresses in the direction perpendicular to the sensing element length.

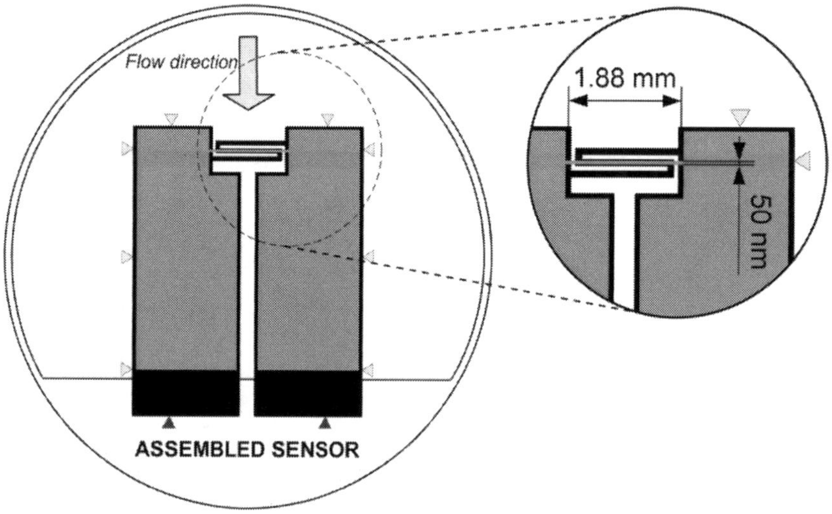

Figure 3. Schematic diagram of the finished shear stress hot film sensor.

Finally in order to stick the sensor to the calibrator wall an 80 μm thick, double sided adhesive tape was placed on the back of the sensor, summing up a total thickness of 220 μm.

3. Experimental arrangement

To evaluate the dynamic performance or the shear stress sensor we need a method to introduce a known perturbation and to measure the sensor output in response. An effective method used for constant temperature hot wire anemometers is to introduce an electrical perturbation to the feedback bridge and observe its settling time. This procedure is commonly called "electronic test" and is a widely accepted evaluation method for hot wire anemometers. Winter [10], Haritonidis [2], Naughton and Sheplak [11], and Löfdahl and Gad-el Hak [12] have presented studies that show that actual time responses of thermal shear stress sensors cannot be predicted by electronic tests alone, and therefore tests based on the variation of flow conditions must be applied. For this purpose we used a relatively simple rotating disk shear stress calibrator. The shear stress calibrator was based on the design presented by Khoo et al [13] [14] [15]. This type of calibrator has also been used previously by Brown and Davey [16] and Chew et al [17] and it consists of two parallel discs, an upper rotating disk and a lower stationary disk, separated by a small distance. One important advantage of this design is that, with a simple modification, it can also be used to test the dynamic performance of the shear stress sensor [13]. The idea behind this device is to generate a known laminar flow for which the wall shear stress can be calculated from the properties of the fluid and simple kinematic measurements.

The device that was developed for the present work is shown in figure 4. Special care was taken to avoid vibrations through careful design and balancing of the rotating components. The rotating disk has a diameter of 160 mm and its axe is supported by two ball bearings, the lower one being conical. A pulley in the upper end of the axe allows coupling the rotating disk through a belt to a DC motor. The lower disk is supported by three regulation screws that allow the adjustment of the gap between the discs. The pulleys where chosen so that the rotation speed of the motor was higher than that of the disc, increasing the system inertia and therefore the steadiness of the rotating speed. The pitch diameter of the motor pulley and the disk pulley were 48 mm and 90 mm respectively. The rotating and the stationary discs were made of transparent acrylic. This makes it possible to observe the sensor once installed. Acrylic is also relatively easy to machine, therefore a high grade finish can be achieved.

The analytical solution for the flow between a stationary and a rotating disk was provided by Stewartson [18] and Mellor et al [19] and it is a power series solution. Following we show the first terms of the series:

$$\frac{v_\theta}{\omega r}=\left[\varepsilon-\left(\frac{\mathrm{Re}_\delta^2}{6300}\right)\left(8\varepsilon+35\varepsilon^4-63\varepsilon^5+20\varepsilon^7\right)+O\left(\mathrm{Re}_\delta^4\right)\right] \tag{1}$$

$$\frac{v_r}{\omega r}=-\left[\left(\frac{\mathrm{Re}_\delta}{60}\right)\left(4\varepsilon-9\varepsilon^2+5\varepsilon^4\right)+O\left(\mathrm{Re}_\delta^3\right)\right] \tag{2}$$

$$\frac{v_z}{\omega r}=2\frac{\delta}{r}\left[\left(\frac{\mathrm{Re}_\delta}{60}\right)\left(2\varepsilon^2-3\varepsilon^3+\varepsilon^5\right)+O\left(\mathrm{Re}_\delta^3\right)\right] \tag{3}$$

where $\mathrm{Re}_\delta \equiv \omega\delta^2/\upsilon$ is the Reynolds number based on the gap between the discs δ, ω is the angular rotating speed, υ is the cinematic viscosity of the fluid, ε ($\equiv z/\delta$) is the non dimensional vertical coordinate taken from the stationary disc (lower disc), r is the radial coordinate, and v_θ, v_r and v_z are the azimuthal velocity, radial velocity and normal velocity to the wall, respectively.

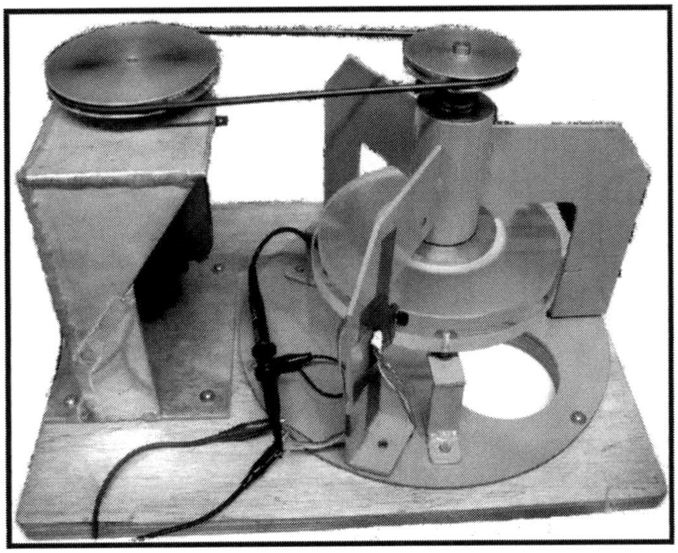

Figure 4. Calibration device.

For $\mathrm{Re}_\delta \leq O(1)$ equations (1), (2), and (3) can be approximated as a Couette flow in the azimuthal direction $\hat{\theta}$. During the experimental work care was taken to keep low values of Re_δ, being smaller than 3.5 for all cases. Under these conditions the radial component of the velocity is negligible and the angle between maximum shear stress and the tangential direction is small (< 10°).Therefore the tangential velocity gradient at the lower wall can be approximated by:

$$\frac{\partial v_\theta}{\partial z}=\frac{\omega . r}{\delta} \tag{4}$$

and the tangential shear stress at the lower disc wall $\tau_{w\theta}$ can be calculated as:

$$\tau_{w\theta}=\mu.\frac{\omega . r}{\delta} \tag{5}$$

where μ is the dynamic viscosity of the fluid.

The shear stress sensor was mounted on the lower disk and oriented with the sensing element in the radial direction. It should be noted that the sensor analyzed here is relatively insensitive to shear stresses parallel to the sensing element, therefore for the sensor mounted as described, the radial component of the shear stress has a negligible effect.

Once the sensor was in place both the sensing element and the shielding element were connected to two constant temperature feedback bridges. This was done by means of thin (50 μm) wires to avoid flow obstruction. The contacts were done with silver conductive paint. Details of the constant temperature feedback bridges construction are presented by Osorio et al [20], these electronic circuits are based on a design by Itsweire [21].

4. Results

As we mentioned before the test using electronic perturbations is a good indicator of the sensor response time in the case of the hot wire anemometers, but that is not the case of wall shear stress sensors [6] [22]. As an alternative to evaluate the dynamic response Khoo et al [13] proposed a test based on the use of a stepped rotating disk, which produces a rapid variation o the wall shear stress. This is the method that will be applied in the present work.

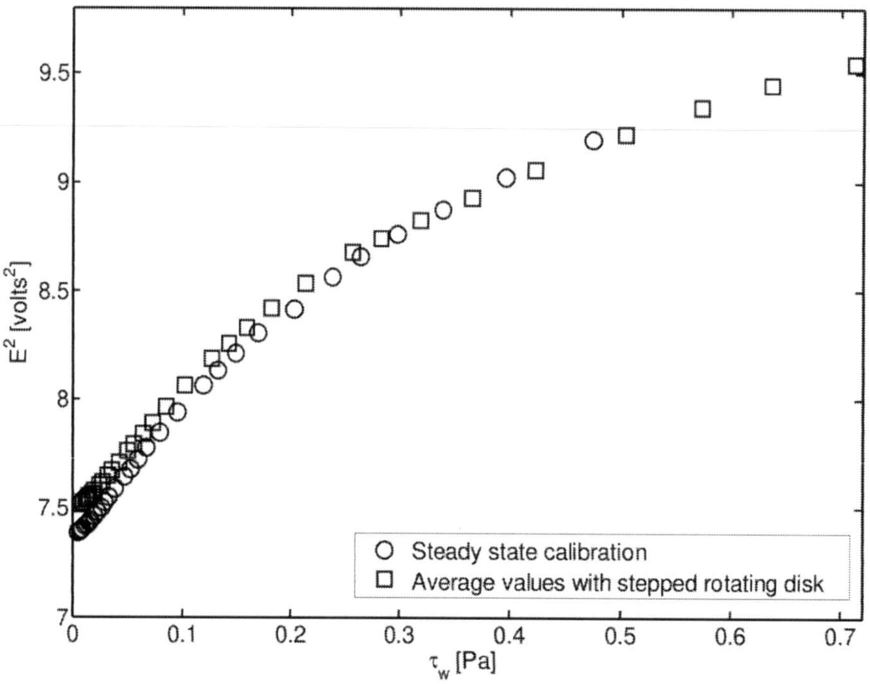

Figure 5. Calibration curve for the wall shear stress sensor for a flat upper disk and for a disk with steps.

First a calibration was performed with a flat rotating disk and for a 380 μm gap between the disk and the sensor. An analysis of the stationary thermal boundary layer brings the following relation between the wall shear stress τ_w, and the bridge output signal E [4] [8]:

$$E^2 = A + B.\tau_w^n \tag{6}$$

where A, B and n are calibration constants. This correlating equation is commonly used for hot wire anemometers and is called King's law.

To perform the calibration the rotating speed was varied so that the imposed shear stress covered the range between 0 and 0.7 Pa. For the present measurements, the overheat ratio was set at 0.06 to avoid any possible alteration of the sensor substrate material. The calibration results using the flat upper disk are shown figure 5. These measurements were fitted by equation (6), obtaining $E^2 = 6.93 + 3.33\tau_w^{0.5}$, where E is expressed in Volts and τ_w in Pascal.

The use of the shield reduces the heat losses to the surface and in some degree it also reduces the heat transfer to the flow, which we have observed as a reduction in the output of the sensor. It has also been observed that the use of the shield sacrifices some sensibility. The shield is intended to isolate the sensing element from the surface temperature, this function was verified by heating the lower disk and recording the sensor output with and without activating the shield. An important change in the sensor output was observed when the shield is not active but the variation is undetectable when the shield is working.

To evaluate the dynamic response the upper disk was modified, incorporating a step of height δ_r, as shown in figure 6.

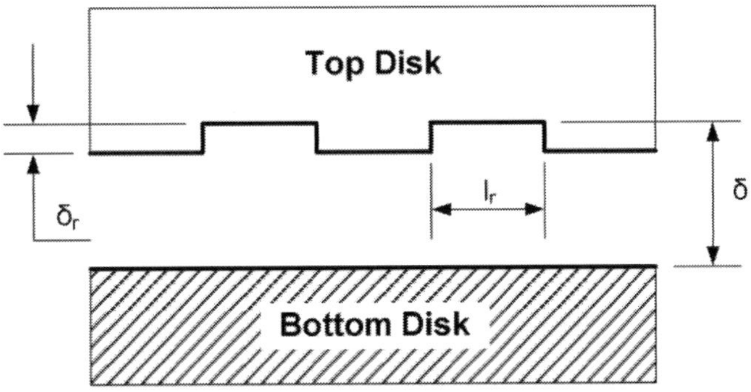

Figure 6. Schematic of an angular section of the modified upper disk.

In this study it was decided to introduce a single step of δ_r = 40 μm so that half of the upper disk offered a gap with the sensor of 340 μm and the other half a gap of 380 μm. The sensor was located at a radius r_m = 48 mm so that the circulation length of the step (l_r) is approximately 150 mm. On the other hand the sensor has a thickness δ_s of 220 μm and the sensing element is located at $l_s \approx$ 21 mm from the discs periphery. From these dimensions the following geometrical ratios can be obtained:

$$l_r/\delta_r \approx 3750 \qquad l_s/\delta_s \approx 95 \tag{7}$$

These ratios show that the azimuthal and radial dimensions are orders of magnitude larger than axial dimensions. This rotating disk with steps was used previously by Isomoto et.al. [23] and Shu et.al. [24]. It was shown that each step involves a sub-region that can be approximated as a fully developed flow [13], though this approximation is only valid relatively far from the step changes. In other words, for the regions far from the discontinuities the flow may be considered as the flow between two walls separated by distances δ or $\delta - \delta_r$. Acording to this the sensor output is expected to resemble a square wave with low and high values corresponding to the shear stresses of the two gap distances. The test proposed by Khoo et al [13] is based on the observation of the sensor response to different disk rotation speeds. If the sensor response shows some kind of attenuation in the square wave amplitude or it is unable to follow the fluctuations it is assumed that the frequency response limit f_D has been reached. This test for our sensor with the shield is shown in Figure 7.

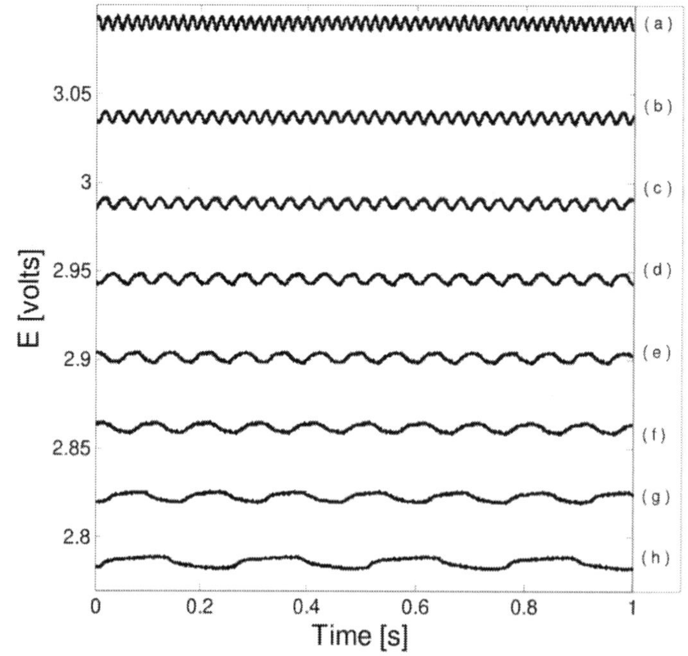

Figure 7. Wall shear stress sensor response to step variations of the shear stress under different fluctuation frequencies: (a) 56 Hz (b) 40 Hz (c) 29 Hz (d) 20 Hz (e) 14 Hz (f) 10 Hz (g) 7 Hz (h) 4 Hz.

In figure 5 we show two calibration curves. One with a flat rotating disc and the second with the modified disc, provided with steps. In the second case we have considered a gap size equal to the average of the two gaps involved. It can be seen that for low rotation speeds the measurements with the modified disc are slightly higher than with the flat disc. For higher velocities both measurements show equal values.

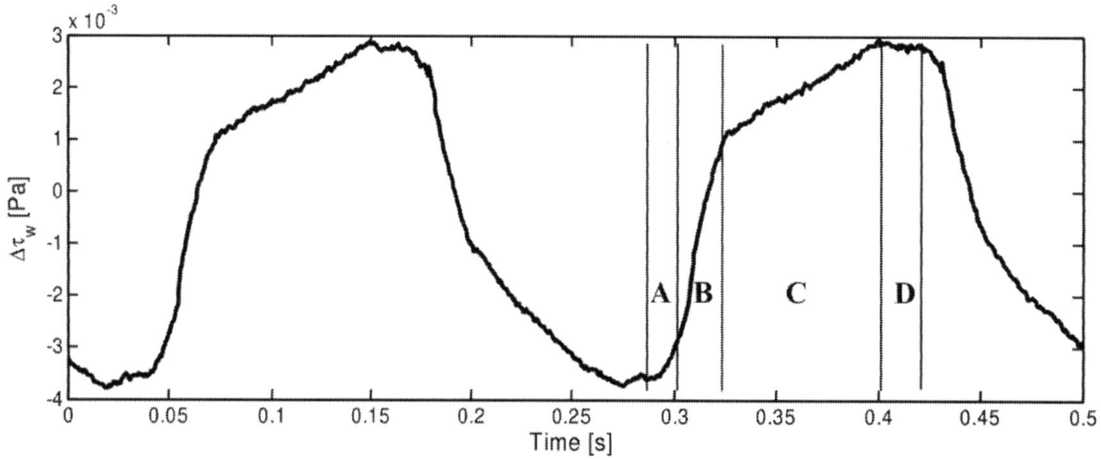

Figure 8. Sensor response due to perturbations imposed by the modified upper disc for a rotation speed of 4 rev/s.

The dynamic response of the system is not simple as it involves various physical phenomena, some associated to the sensor thermal behavior, for example the variation of the temperature field in the neighborhood of the sensing element, and others to the fluid dynamics, for example the flow recirculation in the steps. Figure 8 shows the sensor output in response to the presence of a step in the modified rotating disc.

Three characteristic times can be observed: an initial startup time (region A) possibly associated to fluid dynamic effects, a fast rising time (region B), and a slower response (region C) which ends when the output reaches the steady state D. From figure 8 we can see the behavior cannot be approximated by a first order response.

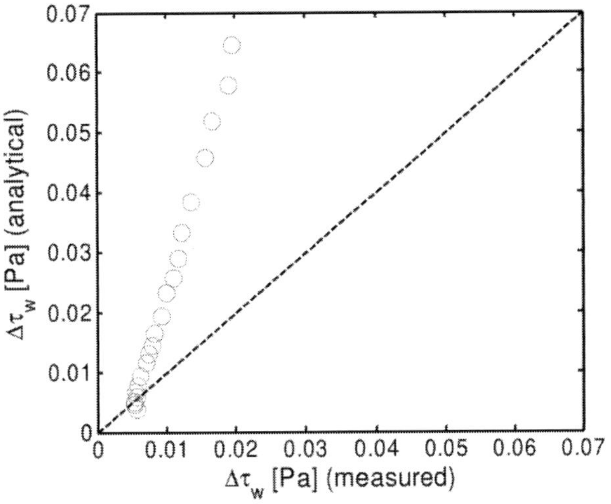

Figure 9. Comparison between the shear stress variation as calculated from the rotating speed and the two gap sizes, and the shear stress variation measured by the sensor.

Figure 9 shows the relation between the shear stress variation as calculated from the rotating speed and the two gap sizes, and the shear stress variation measured by the sensor. For this last value the previous calibration with the flat rotating disk was used. A fast sensor should show a behavior following a straight line of unitary slope till the frequency limit at which the sensor is incapable of following the changes. This last frequency being the sensor dynamic frequency limit, f_D. As can be observed in figure 9 the film sensor developed in this work and incorporating a thermal shield is still not capable of following the shear stress variations, showing some degree of attenuation in the whole range of measurement frequencies.

5. Conclusions

A hot film wall shear stress sensor was built and tested by means of stationary and transient imposed wall shear stress. It was observed that the use of a shield element built in the sensor reduced the sensor element power consumption and eliminated the influence of wall temperature variations by effectively providing active thermal isolation. On the down side it was also noticed that the use of the shield produces some reduction in sensibility. This opens the possibility of performing the shear stress calibration with the sensor placed on a surface different from the surface whose shear stress is to be measured.

The sensor dynamic response was evaluated by means of a modification in the calibrating device which consisted in adding a step of 40 μm height. The test showed a poor dynamic performance of the sensor, equivalent to results reported in previous works using thermal shear stress film sensors. It could be concluded that the losses to the substrate are not the main limitation to the sensor dynamic response and therefore the limitation most probably lies in the characteristics of the convective heat transfer from the surface during a shear stress transient [25]. Based on these results and on the success of shear stress measurements with hot wire anemometers separated from the wall [6], it is expected that better time resolutions may be achieved by detaching the sensing element from the wall by some suitable distance.

6. Acknowledgments

The authors wish to express our recognition to Hernán Pastoriza and Nicolás La Forgia from the Low Temperatures Lab. at the Centro Atómico Bariloche, that fabricated the sensor and helped with various technical aspects.

References

[1] Bellhouse B J and Schultz D L 1966 Determination of Mean and Dynamic Skin Friction, Separation and Transition in Low-Speed Flow with a Thin-Film Heated Element *J. Fluid Mech.* **24** 379–400.

[2] Haritonidis J H 1989 The Measurements of Wall Shear Stress *Advances in Fluid Mechanics Measurements* ed M Gad-El-Hak (New York: Springer-Verlag) pp 229–261.

[3] Bandyopadhyay P R 1994 Development of a Microfabricated Surface for Turbulence Diagnostics and Control *Applied Microfabrication to Fluid Mechanics* (Chicago, IL) pp 67–74.

[4] Goldstein R J 1996 *Fluid Mechanics Measurements* 2nd edition, ed Taylor and Francis, (Washington DC).

[5] Kimura M, Tung S, Lew J, Ho C M, Jiang F and Tai Y C 1999 Measurements of Wall Shear Stress of a Turbulent Boundary Layer Using a Microshear-Stress Imaging Chip *Fluid Dynam. Res.* **24** 329–342.

[6] Ajagu C O, Libby P A and LaRue J C 1982 Modified Gauge for Time-Resolved Skin-Friction Measurements *Rev. Sci. Instrum.* **53** 1920–26.

[7] Ho C M and Tai Y C 1996 Review: MEMS and its Applications for Flow Control *Trans. ASME - J. Fluids Eng.* **118** 437–447.

[8] Alfredsson P H, Johansson A V, Haritonidis J H and Eckelmann H 1988 The Fluctuating wall shear stress and the velocity field in the viscous sublayer *Physics of Fluids* **31** 1026–33.

[9] Jiang F, Tai Y C, Gupta B, Goodman R, Tung S, Huang J B and Ho C M 1996 A surface micromachined shear stress imager *Proc. IEEE Workshop on MEMS* 110–115.

[10] Winter K G 1977 An Outline of the Techniques Available for the Measurement of Skin Friction in Turbulent Boundary Layers *Progr. Aerospace Sci.* **18** 1–57.

[11] Naughton J and Sheplak M 2002 Modern Developments in Shear-Stress Measurement *Prog. Aero. Sci.* **38** 515–570.

[12] Löfdahl L and Gad-el-Hak M 1999 MEMS-Based Pressure and Shear Stress Sensors for Turbulent Flows *Meas. Sci. Technol.* **10** 665–686.

[13] Khoo B C, Chew Y T and Li G L 1995 A New Method by Which to Determine the Dynamic Response of Marginally Elevated Hot-Wire Anemometer Probes for Near-Wall Velocity and Wall Shear Stress Measurements *Meas. Sci. Technol.* **6** 1399–1406.

[14] Khoo B C, Chew Y T, Lim C P and Teo C J 1998 Dynamic Response of a Hot-Wire Anemometer. Part I: A Marginally Elevated Hot-Wire Probe for Near-Wall Velocity Measurements *Meas. Sci. Technol.* **9** 751–763.

[15] Khoo B C, Chew Y T and Lim C P 1998 The Flow Between a Rotating and a Stationary Disc: Application to Near-Wall Hot-Wire Calibration *Meas. Sci. Technol.* **9** 650–658.

[16] Brown G L and Davey R F 1971 The Calibration of Hot Films for Skin Friction Measurement *Rev. Sci. Instrum.* **42** 1729–31.

[17] Chew Y T, Khoo B C and Li G L 1994 A Time-Resolved Hot-Wire Shear Stress Probe for Turbulent Flow: Use of Laminar Flow Calibration *Exp. Fluids* **17** 75–83.

[18] Stewartson K 1953 Flow Between Two Rotating Coaxial Discs *Proc. Cambridge Phil. Soc.* **49** 333–341.

[19] Mellor G L, Chapp P J and Stokes V K 1968 On the Flow Between a Rotating and a Stationary Disk *J. Fluid Mech.* **31** 95–112.

[20] Osorio O D, Silin N and Converti J 2010 Fabrication of Hot-Wire Probes and Electronics for Constant Temperature Anemometers *Latin American Applied Research* **40** 233–239.

[21] Itsweire E C and Helland K N 1983 A High-Performance Low-Cost Constant-Temperature Hot-Wire Anemometer *J. Phys. E: Sci. Instrum.* **16** 549–553.

[22] Sheplak M, Cattafesta L, Nishida T and McGinley C B 2004 MEMS Shear Stress Sensors: Promise and Progress *24th AIAA Aerodynamic Meas. Tech. and Ground Test. Conf. AIAA 2004* (Portland, OR).

[23] Isomoto K and Honami S 1989 The Effect of Inlet Turbulence Intensity on the Reattachment Process Over a Backward-Facing Step *Trans. ASME J. Fluids Eng.* **111** 87–93.

[24] Shu C, Khoo B C and Yeo K S 1994 Numerical Solution of Incompressible Navier–Stokes Equation by Generalized Differential Quadrature *J. Finite Element Anal. Des.* **18** 83–97.

[25] Yoshino T, Suzuki Y, Kasagi N and Kamiunten S 2001 Assessment of the wall shear stress measurement with arrayed micro hot-film sensors in a turbulent channel flow *Proc. 2nd Int. Symp. on Turbulence and Shear Flow Phenomena 2001* (Stockholm) **2**, pp 153–158.

FLUIDOS 2010: XI Meeting on Recent Advances in the Physics of Fluids and their Applications IOP Publishing
Journal of Physics: Conference Series **296** (2011) 012003 doi:10.1088/1742-6596/296/1/012003

Tilt dynamics of an electrostatically actuated microoscillator at a liquid-liquid interface

Alejandro Allievi

Laboratorio de Micro y Nanofluidica, Instituto Tecnológico de Buenos Aires, Av. Eduardo Madero 399, (C1106ACD) Buenos Aires, Argentina

E-mail: aallievi@itba.edu.ar, Alejandro.Allievi@gmail.com

Abstract. We investigate the time-domain tilt response of an electrostatically actuated mechanical microoscillator positioned at a liquid-liquid interface. An analytical model is presented to simulate the microoscillator's rotational motion inside a microchannel completely filled with two immiscible liquids. The model considers two coupled ordinary differential equations; one simulates the mechanical response of the microplate-microbeam assembly making-up the microoscillator and the other provides the behaviour of the electrical charge responsible for the electrostatic moment that tilts the microplate. Results show that remarkable improvements in sampling time and sensitivity can be obtained using a bi-liquid configuration versus its single-liquid counterpart. Therefore, enhanced performance of mechanical microsensors for liquids could be achieved.

Keywords: microfluidics, surface tension, liquid interface, microoscillator, MEMS.

1. Introduction

Mechanical microoscillators are used extensivelly as components in MEMS (Micro Electro Mechanical Systems) that operate with liquids and gases. Their precision, inexpensive fabrication and minute fluid sample requirements are some of the features that generated a considerable volume of research in engineering and sciences. When operating with liquids, it was recognized [1] that resonant frequencies and quality factors Q of microoscillators were very small compared with those in air due to higher viscous effects. As a consequence, rapidly decaying dynamic responses lead to short sampling times and low sensitivity to determine, for instance, a resonant frequency shift due to adsorption of a target substance.

To circumvent these deficiencies, a considerable number of works was carried out. To name just a few contributions, some investigators used feedback forces applied through the gain of a feedback circuit [2, 3] to change the effective damping of the dynamic system. Alternatively, some works used an applied voltage to produce a dynamic change of the spring constant of microcantilevers [4] and of torsional microresonators [5]. In an entirely different approach, liquid was confined to microchannels inside oscillating microbeams operating in air [6], thus avoiding operation in a liquid medium.

In this work, we further our investigation of a simpler and novel concept [7, 8, 9] where surface tension is positively used to improve performance of mechanical microoscillators. In this work, we present an analytical solution to the Lagrangian equations resulting from the application of Hamilton's principle to the proposed electromechanical system. The study of the tilt dynamics

Figure 1. SEM of microoscillator in air ($L = 2\mu m$). Reprinted with permission from [5], Copyright 2000, Amer. Inst. of Physics.

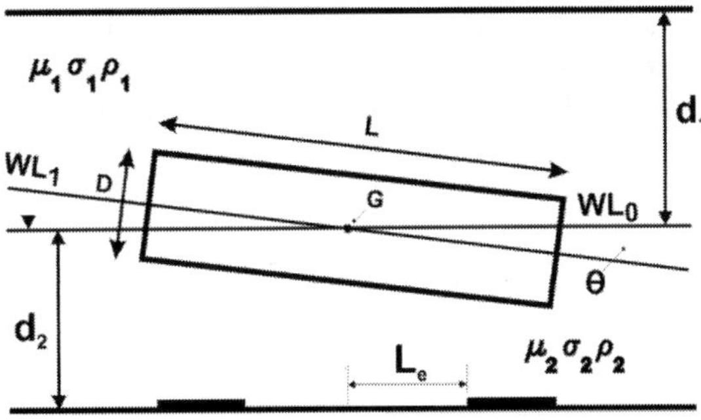

Figure 2. Side view of rotated microplate at liquid-liquid interface showing position of electrodes (microbeam not shown).

of a torsional microbridge demonstrates that improvements by several orders of magnitude can be achieved for extremely small resonators operating at the interface between two inmiscible liquids.

2. Statement of the Problem

Figure 1 shows a Scanning Electron Micrograph (SEM) of a torsional microresonator in air. In our case, Figure 2, this microplate-microbeam assembly is located at the interface of two inmiscible liquids (WL_0). Immediately below the microplate, two electrodes are located at a distance L_e from a torsional axis co-axial with the microbeams' axis that passes through the center of gravity G. The liquid-liquid interface is coplanar with the microplate's horizontal mid-plane. Upper and lower liquids have viscosity, surface tension and density triplets $(\mu_i, \sigma_i, \rho_i)$, with $i = 1, 2$. Application of a voltage V_e between the assembly and one electrode produces an accumulation of charge $q(t)$ in a manner equivalent to that of a parallel moving plate capacitor [10]. The resulting electrostatic moment causes rotation of the microoscillator to a maximum value $\theta(t = t_a) = \theta_a$ (position WL_1 in Figure 2). When V_e is removed, the microoscillator returns to its original initial position at rest. We limit the value of V_e to 1.5V so as to minimize the possibility of lower liquid ionization.

Application of Hamilton's principle to the resulting electromechanical system yields both Lagrangian equations for the computation of $q(t)$ and $\theta(t)$. For the accumulation of charge $q(t)$ the equation reads

$$L_i\ddot{q} + R\dot{q} + \frac{q}{C} = V_e, \tag{1}$$

$$C = \frac{\epsilon\ A}{d_2 - \frac{D}{2} - (\frac{L}{4} + \frac{L_e}{2})\ tan\theta}, \tag{2}$$

where L_i and R are the circuit's inductance and resistance respectively, and ϵ is the permittivity of the lower liquid. L and D are the microplate's side length and thickness respectively, d_2 is the depth of the lower liquid and $A = (L/2 - L_e) \times B$. The underlying assumption in (2) is that the electric field is uniform across the gap between the microplate and the electrode. Field lines are assumed perfectly vertical generating equipotential lines parallel to the microplate/electrode planes. In practice, such parallelism is distorted at the ends of the plates due to non-uniformity

Journal of Physics: Conference Series **296** (2011) 012003 doi:10.1088/1742-6596/296/1/012003

of the electric field. However, such distortion can be readily neglected if the separation gap is small compared to the length $L/2 - L_e$.

For the mechanical microoscillator, the equation that models its tilt response reads:

$$a_1\ddot{\theta} + a_2\dot{\theta} + a_3\theta = \frac{\left(\frac{L}{4} + \frac{L_e}{2}\right) q^2}{2\ \epsilon\ A\ cos^2\theta}. \tag{3}$$

The coefficient a_1 is composed of three terms due to the plate's moment of inertia, the added inertia from surface tension and the fluids' added mass, and can be obtained as [9]:

$$\begin{aligned} a_1 &= \rho^p LBD\frac{(D^2+L^2)}{12} + \\ &+ 2(\sigma_2-\sigma_1)(L+B)\frac{(D^2+L^2)}{12g} + \\ &+ \frac{\pi}{256}(\rho_1+\rho_2)L^4B, \end{aligned} \tag{4}$$

where B is the dimension of the microplate in the direction across the microchannel, ρ^p is the density of the microbeams/microplate material (Pt in this work) and g is the acceleration of gravity.

Contributions to squeezed film damping are from the fluid regions between the microplate and the substrates immediately above and below. They can be computed as [11]:

$$\begin{aligned} a_2 &= \left\{\left[\frac{B[L+1.65(d_1-T_1)]^4\mu_1}{15L(d_1-T_1)^3} + 3.2\mu_1B\sqrt{\frac{L+2.7D}{d_1-T_1}}\right] + \right. \\ &+ \left.\left[\frac{B[L+1.65(d_2-T_2)]^4\mu_2}{15L(d_2-T_2)^3} + 3.2\mu_2B\sqrt{\frac{L+2.7D}{d_2-T_2}}\right]\right\}\left\{\frac{L}{2}\right\}^2, \end{aligned} \tag{5}$$

where T_1, T_2 are the vertical distances from the liquid-liquid interface to the upper and lower surfaces of the microplate respectively.

Finally, the contributions to the restoring term are from buoyancy, torsional moments of the microbeams and surface tension respectively. Their final form is [9]:

$$\begin{aligned} a_3 &= \frac{gLB}{12}\left\{\rho_2[L^2+6T_2(T_2-D)] - \rho_1[L^2+6T_1(T_1-D)]\right\} + \\ &+ 2\frac{\beta l_b t_b^3 G}{b_b} + \\ &+ 2L\left\{\sigma_2\frac{L^2+6T_2(T_2-D)}{12T_2} - \sigma_1\frac{L^2+6T_1(T_1-D)}{12T_1}\right\}\left[\frac{1-\cos\theta(\mathrm{t})}{\cos\theta(\mathrm{t})}\right], \end{aligned} \tag{6}$$

where l_b, t_b are the cross-sectional dimensions of the microbeams, b_b is their length, G is their shear modulus, and β is a coefficient of proportionality that depends on the ratio l_b/t_b[12].

3. Proposed solution

In this section, we present the analytical solutions for equations (1) and (3) when the microplate is electrostatically actuated by a constant voltage V_e. For the oscillators' sizes considered in this work, angles of rotation are sufficiently small to warrant the assumptions $tan\theta \approx 0$ in (1) and $cos^2\theta \approx 1$ in (3). Noting that $(\frac{a_3}{a_1}) - (\frac{a_2}{2a_1})^2 > 0$ must hold for the assembly to oscillate, the solutions of (1) and (3) when the capacitor is charging can be obtained using homogeneous initial conditions:

$$q^c(t=0) = 0,\ \dot{q}^c(t=0) = 0, \tag{7}$$

$$\theta^c(t=0) = 0,\ \dot{\theta}^c(t=0) = 0, \tag{8}$$

as:

$$q^c(t) = c_1^{qc} e^{\lambda_1^{qc} t} + c_2^{qc} e^{\lambda_2^{qc} t} + c_3^{qc}, \tag{9}$$

$$\begin{aligned} \theta^c(t) &= e^{-\frac{a_2}{2a_1}t}\left\{ D_1^{\theta c} sin\left[\sqrt{\frac{a_3}{a_1} - \left(\frac{a_2}{2a_1}\right)^2}\, t\right] + D_2^{\theta c} cos\left[\sqrt{\frac{a_3}{a_1} - \left(\frac{a_2}{2a_1}\right)^2}\, t\right]\right\} + \\ &+ c_0^{\theta c} + c_1^{\theta c} e^{\lambda_1^{\theta c} t} + c_2^{\theta c} e^{\lambda_2^{\theta c} t} + c_3^{\theta c} e^{\lambda_3^{\theta c} t} + c_4^{\theta c} e^{\lambda_4^{\theta c} t} + c_5^{\theta c} e^{\lambda_5^{\theta c} t}. \end{aligned} \tag{10}$$

At certain $t = t_a$, the capacitor reaches a steady state charge q_a. Subsequently, at $t = t_b$ the microplate stops moving at $\theta = \theta_a$. At this point, the voltage V_e is removed and the capacitor discharges. The initial conditions to solve (1) and (3) are then:

$$q^d(t=t_b) = q_a,\ \dot{q}^d(t=t_b) = 0, \tag{11}$$

$$\theta^d(t=t_b) = \theta_a,\ \dot{\theta}^d(t=t_b) = 0, \tag{12}$$

and the solutions become:

$$q^d(t) = c_1^{qd} e^{\lambda_1^{qd} t} + c_2^{qd} e^{\lambda_2^{qd} t}, \tag{13}$$

$$\begin{aligned} \theta^d(t) &= e^{-\frac{a_2}{2a_1}t}\left\{ D_1^{\theta d} sin\left[\sqrt{\frac{a_3}{a_1} - \left(\frac{a_2}{2a_1}\right)^2}\, t\right] + D_2^{\theta d} cos\left[\sqrt{\frac{a_3}{a_1} - \left(\frac{a_2}{2a_1}\right)^2}\, t\right]\right\} + \\ &+ c_1^{\theta d} e^{\lambda_1^{\theta d} t} + c_2^{\theta d} e^{\lambda_2^{\theta d} t} + c_3^{\theta d} e^{\lambda_3^{\theta d} t} \end{aligned} \tag{14}$$

Expressions for all constants in (9), (10), (13) and (14) are provided in the Appendix.

4. Results

In this section, we present results for charge q and tilt θ as a function of time. Given the large number of parameters involved in the problem at hand, we restrict ourselves to varying only the data provided in Tables 1 and 2. Where appropriate, we have taken the quality factor to be $Q = a_3/(2\pi f_0 a_2) = \sqrt{a_3\, a_1}/a_2$, and the undamped natural frequency of the oscillator as $f_0 = [1/2\pi]\sqrt{a_3/a_1}$.

Figure 3 shows the electromechanical response of the assembly at a water-silicon oil interface (W-SO1) and in water only (W-W) for $L = 1\mu m$. The coefficients in (1) are such that cause the charge q to grow exponentially and very rapidly, with full charge condition reached in around

Table 1. Problem data (all lengths in μm), material Pt.

Microbeam			Microplate			Microchannel		Electrical			
l_b	b_b	t_b	L	B	D	d_1	d_2	L_e	$L_i\ [H]$	$R\ [\Omega]$	$V_e[V]$
0.2	$L/2$	0.2	As in Figs.	L	0.2	As in Figs.	d_1	$L/3$	1.88×10^{-14}	5000	1.5

Table 2. Liquid properties.

Case	Liquid	Density $\rho[kg\ m^{-3}]$	Kin. Viscosity $\nu[m^2 sec^{-1}]$	Dyn. Viscosity $\mu[Nm^{-2}sec]$	Surface Tension $\sigma[Nm^{-1}]$
SO1	Silicon Oil 1	0.91×10^3	5.0×10^{-6}	4.55×10^{-3}	19.7×10^{-3}
OO	Olive Oil	0.92×10^3	91.3×10^{-6}	84.0×10^{-3}	32.0×10^{-3}
W	Water	1.0×10^3	1.0×10^{-6}	1.0×10^{-3}	72.8×10^{-3}

8 picosec (shown in the Figure's inset). The resulting impulsive electrostatic moment in (3) causes the tilt response to overshoot to a maximum value before its oscillatory decay to a constant $\theta_a \approx 7 \times 10^{-5}$ for both W-SO1 ($f_0 = 2.744 \times 10^4 Hz, Q = 1.644 \times 10^4$) and W-W ($f_0 = 5.971 \times 10^7 Hz, Q = 20.961$). For W-SO1, the amount of time required for this periodic

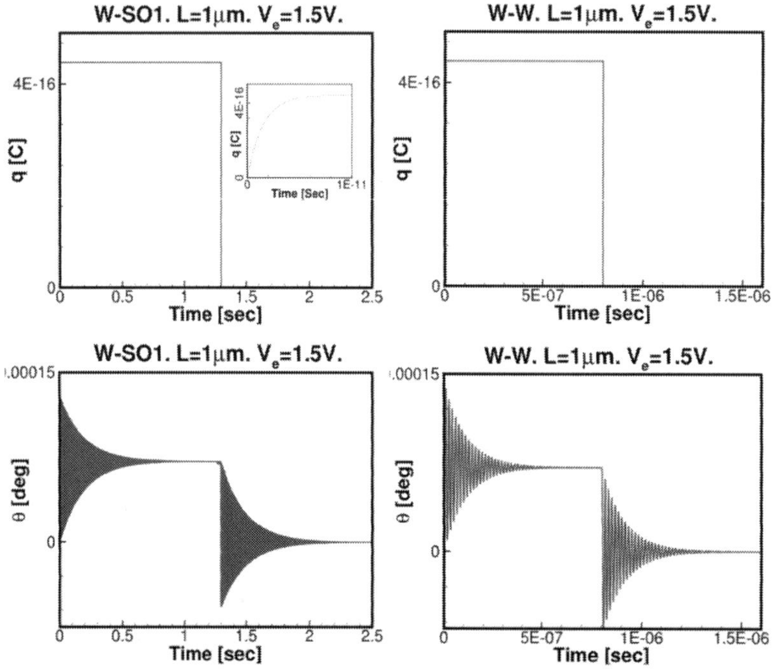

Figure 3. Electromechanical response for water-silicon oil interface (W-SO1) and water only (W-W). $L = 1 \times 10^{-6}m, d_1 = 0.5 \times 10^{-6}m$.

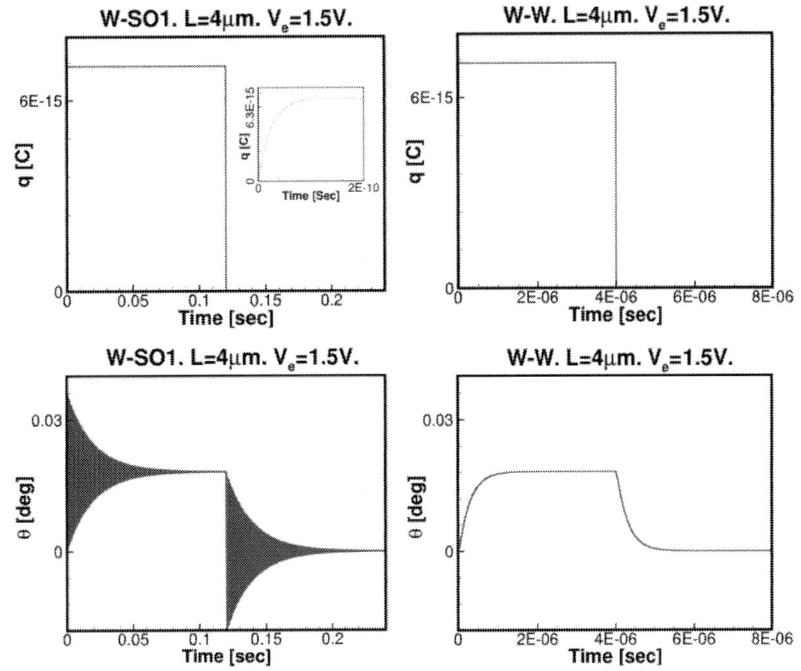

Figure 4. Electromechanical response for water-silicon oil interface (W-SO1) and water only (W-W). $L = 4 \times 10^{-6}m, d_1 = 0.5 \times 10^{-6}m$.

decay to take place is $t_a \approx 1$ sec. However, for W-W is $t_a \approx 7.25 \times 10^{-7}$, a remarkable difference of six orders of magnitude caused by the effect of surface tension. The obvious consequence for a microdevice working with liquid-liquid interfaces is a much greater sampling time as the assembly oscillates, together with a near one thousand-fold increase in the quality factor Q. We note that, when the voltage V_e is removed, the charge q also decays very rapidly, causing the assembly to oscillate back to equilibrium initial conditions $\theta = \dot{\theta} = 0$. This second periodic motion due to the capacitor's discharge should be used as a verification of, for instance, the resonant frequency shift due to adsorption of a target substance.

Figure 4 shows electromechanical response of the assembly at a water-silicon oil interface (W-SO1) ($f_0 = 1.747 \times 10^3 Hz, Q = 1.071 \times 10^2$) and in water only (W-W) for $L = 4\mu m$. For this assembly of larger dimensions, the mechanical response in a single liquid is no longer oscillatory, confirming previous evidence of mechanical microoscillator's poor performance in liquids. In practice, this behaviour would render ineffective a microdevice working on the principle of resonant frequency shift or on dynamic response. On the other hand, a microdevice based on a bi-liquid concept would still serve its intended purpose.

To ease microfabrication, it is sometimes convenient to decrease the depth-width ratio of a microchannel. Figure 5 shows the effect on the response of reduced depth in the microchannel for the microresonator of Figure 3. It can be seen that the consequences of microplate-electrode gap reduction from $4\mu m$ to $1\mu m$ are a significant increase in damping and a strikingly diminished time t_b to reach steady state conditions. For (W-SO1) ($f_0 = 2.744 \times 10^4 Hz, Q = 1.723 \times 10^3$) there is a one order of magnitue reduction in the quality factor. However, the microresonator still fulfills its intended purpose. However, for a single liquid (W-W) the microdevice is no longer useful in a dynamic operative mode.

Figure 6 shows results for a different liquid-liquid combination: water and olive oil ($f_0 =$

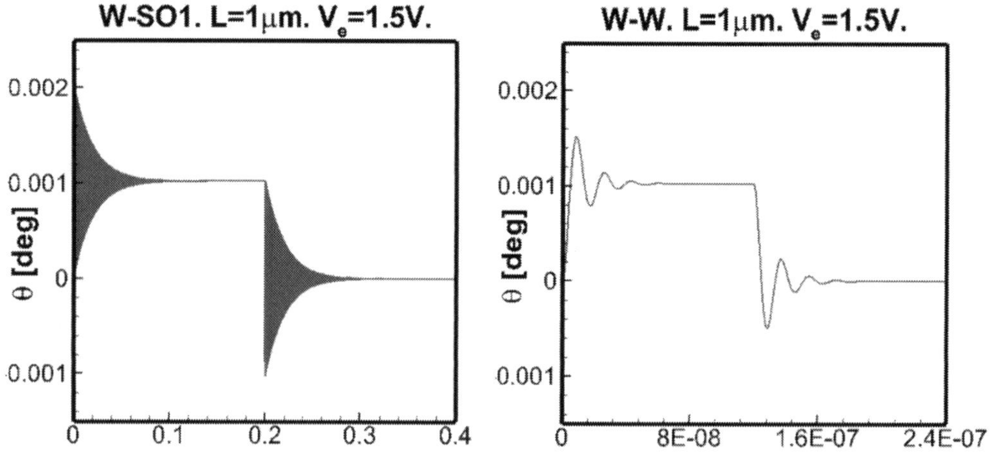

Figure 5. Electromechanical response for water-silicon oil interface and water only (W-W). $L = 1 \times 10^{-6}m, d_1 = 0.2 \times 10^{-6}m$.

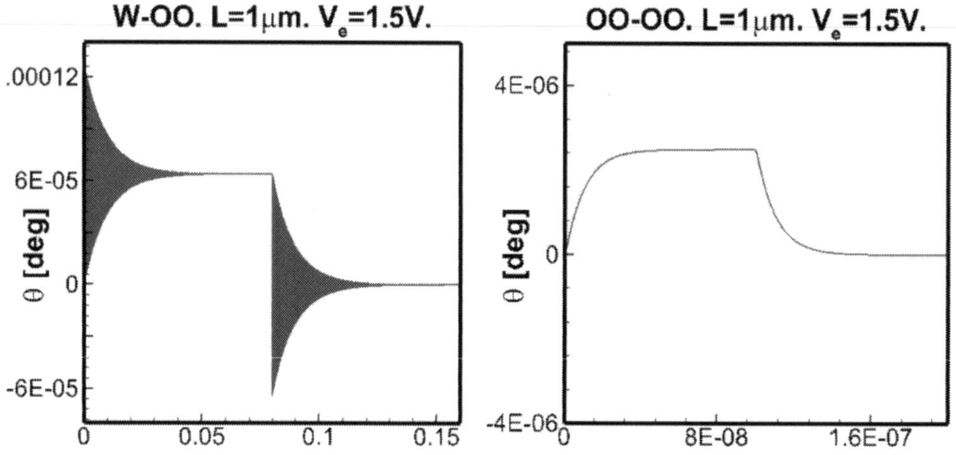

Figure 6. Electromechanical response for water-olive oil interface and olive oil only (OO-OO). $L = 1 \times 10^{-6}m, d_1 = 0.2 \times 10^{-6}m$.

$3.130 \times 10^4 Hz, Q = 940.8$). The later fluid exhibits an over 20-fold increase in viscosity and approximately 50% greater surface tension. These property changes cause much greater viscous effects and a considerably decreased time t_b to reach steady state conditions, as well as an increase in the microoscillator's natural frequency with respect to the case of Figure 3. When the single liquid is olive oil alone (OO-OO) the assembly microplate-microbeam no longer oscillates, again rendering the device totally ineffective.

As a final remark, we note that surface properties, most notably wetting, have considerable effects on the motion and control of liquids inside micrometer scale channels. Each substrate of our microdevice is intended to be made of hydrophilic glass. Despite much controversy in the literature, there are indications that platinum, the material used for the microbeams and the microplate, has a hydrophilic nature with near zero contact angle in the absence of inorganic and organic contaminants [13]. Contamination caused by some of the liquids used in this work might induce hydrophobicity in platinum, leading to an increase of the contact angle. In such situation, the zero-contact angle assumption in the surface tension terms implied in (4) and (6) should be revised. Higher contact angles would result in somewhat reduced surface tension forces, which will affect natural frequencies and quality factors Q accordingly.

5. Conclusions

An analytical model was developed to study time-domain tilt response of an electrostatically actuated torsional microoscillator positioned at a liquid-liquid interface. Two coupled ordinary differential equations were developed to simulate the tilt response of the microoscillator and the electrical charge of the electromechanical system. Viscosity, density and surface tension of both liquids were taken into account. Several orders of magnitude improvements in sampling times and sensitivity were observed with respect to a single liquid configuration. As a result, much improved sensors based on dynamic response or on resonant frequency shifts could be developed using the concept proposed in this work.

Acknowledgments

The author thanks Professor Harold Craighead (Nanobiotechnology Center and Dept. of Applied and Engineering Physics, Cornell University) for the micrograph of Figure 1.

Appendix

Capacitor charging

$$
\begin{aligned}
c_1^{qc} &= -(c_2^{qc} + V_e C),\ c_2^{qc} = \frac{V_e C \lambda_1^{qc}}{\lambda_2^{qc} - \lambda_1^{qc}},\ c_3^{qc} = V_e C \\
\lambda_1^{qc} &= -\frac{R}{2L_i} + \sqrt{\left(\frac{R}{2L_i}\right)^2 - \frac{1}{CL_i}},\ \lambda_2^{qc} = -\frac{R}{2L_i} - \sqrt{\left(\frac{R}{2L_i}\right)^2 - \frac{1}{CL_i}} \\
\lambda_1^{\theta c} &= \lambda_1^{qc},\ \lambda_2^{\theta c} = \lambda_2^{qc},\ \lambda_3^{\theta c} = \lambda_1^{qc} + \lambda_2^{qc},\ \lambda_4^{\theta c} = 2\lambda_1^{qc},\ \lambda_5^{\theta c} = 2\lambda_2^{qc} \\
c_0^{\theta c} &= \frac{B_0^c a_1}{a_3},\ c_i^{\theta c} = \frac{B_i^c}{(\lambda_i^{\theta c})^2 + \frac{a_2}{a_1}\lambda_i^{\theta c} + \frac{a_3}{a_1}} \\
D_1^{\theta c} &= \frac{1}{\sqrt{\frac{a_3}{a_1} - (\frac{a_2}{2a_1})^2}} \left(\frac{a_2 D_2^{\theta c}}{2a_1} - \sum_{i=1}^{5} c_i^{\theta c} \lambda_i^{\theta c} \right) \\
D_2^{\theta c} &= -\left(c_0^{\theta c} + \sum_{i=1}^{5} c_i^{\theta c} \right) \\
B_0^c &= (c_3^{qc})^2 \beta,\ B_1^c = 2c_3^{qc} c_1^{qc} \beta,\ B_2^c = 2c_3^{qc} c_2^{qc} \beta,\ B_3^c = 2c_1^{qc} c_2^{qc} \beta,\ B_4^c = (c_1^{qc})^2 \beta,\ B_5^c = (c_2^{qc})^2 \beta \\
\beta &= \frac{\frac{L}{2} + L_e}{4\epsilon A a_1}
\end{aligned}
$$

Capacitor discharging

$$
\lambda_1^{qd} = = \lambda_1^{qc},\ \lambda_2^{qd} = \lambda_2^{qc}
$$

$$
\begin{aligned}
c_1^{qd} &= -\frac{c_2^{qd}\lambda_2^{qd}}{\lambda_1^{qd}},\ c_2^{qd} = \frac{q_0\lambda_1^{qd}}{\lambda_1^{qd}-\lambda_2^{qd}} \\
\lambda_1^{\theta d} &= 2\lambda_1^{qd},\ \lambda_2^{\theta d} = 2\lambda_2^{qd},\ \lambda_3^{\theta d} = \lambda_1^{qd}+\lambda_2^{qd} \\
c_i^{\theta d} &= \frac{B_i^d}{(\lambda_i^{\theta d})^2 + \frac{a_2}{a_1}\lambda_i^{\theta d} + \frac{a_3}{a_1}} \\
D_1^{\theta d} &= \frac{1}{\sqrt{(\frac{a_3}{a_1}) - (\frac{a_2}{2a_1})^2}}\left(\frac{a_2 D_2^{\theta d}}{2a_1} - \sum_{i=1}^{5} c_i^{\theta d}\lambda_i^{\theta d}\right),\ D_2^{\theta d} = \theta_0 - \sum_{i=1}^{3} c_i^{\theta d} \\
B_1^d &= (c_1^{qd})^2\beta,\ B_2^d = (c_2^{qd})^2\beta,\ B_3^d = 2c_1^{qd}c_2^{qd}\beta
\end{aligned}
$$

References

[1] Inaba S, Akaishi K, Mori T and Hane K 1993 *J. Appl. Phys.* **73** 2654–2658
[2] Anczykowski B, Cleveland J, Krger D, Elings V and Fuchs H 1998 *Appl. Phys. A - Mater.* **66** 885–889
[3] Tamayo J, Humphris A, Malloy A and Miles M 2001 *Ultramicroscopy* **86** 167–173
[4] Rugar D and Grutter P 1991 *Phys. Rev. Lett.* **67** 699702
[5] Carr D W, Evoy S, Sekaric L, Craighead H G and Parpia J M 2000 *Appl. Phys. Lett.* **77** 1545–1547
[6] Burg T and Manalis S 2003 *Appl. Phys. Lett.* **83** 2698–2700
[7] Allievi A 2005 *J. Micromech. Microeng.* **15** 168–175
[8] Allievi A 2005 *J. Micromech. Microeng.* **15** 2326–2338
[9] Allievi A 2008 *J. Fluid. Struct.* **24** 1135–1150
[10] Preumont A 2006 *Mechatronics - Dynamics of Electromechanical and Piezoelectric Systems* (Springer)
[11] Veijola T, Pursula A and Raback P 2005 *J. Micromech. Microeng.* **15** 1624–1636
[12] Timoshenko S 1984 *Strength of Materials* (Krieger)
[13] Gardner J R and Woods R 1977 *J. Electroanal. Chem.* **81** 285–290

Heat transfer model for quenching by submerging

D N Passarella[1], F Varas[1] and E B Martín[2]

[1] Departamento de Matemática Aplicada II, E.T.S. de Ing. de Telecomunicación, Universidad de Vigo, Campus Marcosende, 36310 Vigo, España

[2] Área de Mecánica de Fluidos, E.T.S. de Ing. Industriales, Universidad de Vigo, Campus Marcosende, 36310 Vigo, España

E-mail: diego@dma.uvigo.es; fvaras@uvigo.es; emortega@uvigo.es

Abstract. In quenching by submerging the workpiece is cooled due to vaporization, convective flow and interaction of both mechanisms. The dynamics of these phenomena is very complex and the corresponding heat fluxes are strongly dependent on local flow variables such as velocity of fluid and vapor fraction. This local dependence may produce very different cooling rates along the piece, responsible for inappropriate metallurgical transformations, variability of material properties and residual stresses. In order to obtain an accurate description of cooling during quenching, a mathematical model of heat transfer is presented here. The model is based on the drift-flux mixture-model for multiphase flows, including an equation of conservation of energy for the liquid phase and specific boundary conditions that account for evaporation and presence of vapor phase on the surface of the piece. The model was implemented on *Comsol Multiphysics* software. Generation of appropriate initial and boundary conditions, as well as numerical resolution details, is briefly discussed. To test the model, a simple flow condition was analyzed. The effect of vapor fraction on heat transfer is assessed. The presence of the typical vapor blanket and its collapse can be recovered by the model, and its effect on the cooling rates on different parts of the piece is analyzed. Comparisons between numerical results and data from literature are made.

1. Introduction

Quenching is a technological process that involves cooling of a heated workpiece in order to modify its mechanical properties. Depending on the material of the workpiece, the cooling rate needs to be very fast or can be mild. In particular, for low to medium alloy steels the cooling rates should be as fast as possible.

Focusing on steels, quenching produces metallurgical transformations that are responsible for modifications in mechanical properties (hardness, yield and ultimate stresses and toughness). This feature allow us to tailor the steel properties in a broad range that is not easy to replicate in common engineering materials.

In addition to metallurgical transformations that lead to mechanical modifications, geometrical distortions and generation of residual stresses can be developed during the process. In order to get a successful quenching, the final metallurgical phases and mechanical properties should be the desired ones, and distortions and residual stresses should be kept at a minimum.

As any other process of technological interest, several efforts were made in order to describe it through mathematical models. In this area it is important to remark that quenching of steels can be divided in at least three sub-problems. One major problem is the modeling of metallurgical transformations. This part is well known and the available models are robust and reliable [1].

On another side we have the generation of residual stresses and geometrical distortions. This complex phenomena are also well covered, even though new refined models can be found in specialized literature. Finally, these two sub-problems strongly rely on an accurate description of the temperature evolution inside the piece. Several attempts to put together a thorough description of the quenching process can be found in literature, see refs. [2]-[6]. These and other works take this approach, where each sub-problem is solved using the most adequate model. But, the thermal problem is usually solved taken a very rough description of heat exchange with the fluid through the boundaries conditions. It is important to remark that all the complexity and accurateness of each sub-model is lost if the thermal problem is over-simplified. In this work the analysis of heat transfer in the workpiece-fluid interface is tackled. The microstructure and residual stresses problems will be coupled in further works.

In quenching by immersion the piece transfers heat to the quenching media through vaporization of the fluid, convective flow and interaction of both mechanisms. Generalities about subcooled boiling stages can be found in [7]-[9]. The description of this phenomenon has to take into account the dynamic of a multiphase vapor-liquid flow and specific heat transfer contributions due to vaporization. The problem of multiphase flows with heat exchange is described in general terms in works such as [10]-[12]. Based on these models and including the heat partition model for boiling, a heat transfer model for quenching by submerging is developed here. The model includes the presence of vapor phase and how it affects the heat transfer from the piece to the quenching media.

2. Model

The analysis of multiphase flows presented here was based on the general methodology presented in [10]-[12]. Before the definition of the particular multiphase model it was necessary to analyze coupling parameters between phases, characteristic times and finally characteristic non-dimensional numbers of the specific problem.

Different characteristic times were considered, such as those related to:

- Mass transfer, τ_M, the ratio between the latent heat of condensation of a bubble and the rate of heat transfered from it to the fluid.
- Momentum transfer, τ_V, which is the characteristic response time of a bubble.
- Energy transfer, τ_T, the ratio between the sensible heat of a bubble and the rate of heat transfered from it to the fluid.

Those characteristic exchange times were compared to the characteristic residence time of the flow, τ_r. These comparisons allow us to assess if each quantity is rapidly transfered from one phase to another, therefore reaching equilibrium, or if the dynamic of the transference needs to be described. Under our assumptions of dispersed vapor (say a fraction of liquid, α_l, higher than 0.55) and with a bubble diameter (D_b) ranging from 10^{-4} to 10^{-3} m, all the characteristic times are remarkably smaller than the mean flow residence time.

Based on the characteristic transfer times of each quantity, the corresponding coupling parameters were analyzed according to [10]. The obtained parameters indicated that mass and energy transfered due to phase change is highly coupled between phases, while momentum and energy transfered due to temperature differences have a negligible coupling.

Finally, the non-dimensional analysis showed that the drift number (N_D) and the phase change number (N_{ph}) are in the same order of magnitude, and the Eckert (N_{Ec}) number very small (see appendix for a detailed explanation of the meaning of these characteristic numbers). These results imply that the description of velocities for both phases is needed, and that the interfacial terms in the equation of energy are negligible.

Based on these results, for our particular problem, the main conclusions that were drawn are:

- Liquid and vapor phases are in mechanical equilibrium, therefore the velocities of both phases are strongly correlated.
- The heat delivered from vapor to liquid corresponds mainly to latent heat of condensation. Heat corresponding to cooling of vapor from temperatures higher than saturation temperature are negligible.
- The characteristic time of cooling of vapor phase is very small. Based on this conclusion and the previous one, the vapor temperature can be assumed constant at saturation value.
- Both phases are strongly coupled through exchanges of mass and energy due to phase change.

Based on these conclusions, the most convenient multiphase model selected was the drift-flux mixture-model. This model is easier to solve than the fully detailed Euler-Euler model, and captures all the characteristics of the two-phase flow. Due to the constant vapor temperature assumption, only the conservation of energy of the liquid phase has to be described. The system of partial differential equations (PDE's) that describe this problem is presented in section 2.1. Subindexes m, v and l correspond to mixture, vapor and liquid respectively. The fraction of each phase is represented by α and the $\overline{\overline{\ }}$ and $\widehat{\ }$ symbols correspond to phase average and density-weighted average respectively.

2.1. Conservation equations

The multi-phase problem derived in previous section can be modeled by means of four main conservation equations, which are [12]-[13]:

$$\frac{\partial \rho_m}{\partial t} + \nabla \cdot (\boldsymbol{v}_m \rho_m) = 0 \tag{1}$$

$$\frac{\partial \alpha_v \overline{\overline{\rho_v}}}{\partial t} + \nabla \cdot \left(\alpha_v \overline{\overline{\rho_v}} \boldsymbol{v}_m\right) = \Gamma_v - \nabla \cdot \left(\alpha_v \frac{\overline{\overline{\rho_l}}\,\overline{\overline{\rho_v}}}{\rho_m} \boldsymbol{V}_{vj}\right) \tag{2}$$

$$\frac{\partial \rho_m \boldsymbol{v}_m}{\partial t} + \nabla \cdot (\rho_m \boldsymbol{v}_m \boldsymbol{v}_m) = -\nabla p_m + \nabla \cdot \left(\overline{\overline{\mathbb{T}}} + \mathbb{T}^T\right) + \rho_m \boldsymbol{g} - \nabla \cdot \left(\frac{\alpha_v}{1-\alpha_v} \frac{\overline{\overline{\rho_l}}\,\overline{\overline{\rho_v}}}{\rho_m} \boldsymbol{V}_{vj} \boldsymbol{V}_{vj}\right) \tag{3}$$

$$\frac{\partial \alpha_l \overline{\overline{\rho_l}}\,\widehat{h}_l}{\partial t} + \nabla \cdot \left(\alpha_l \overline{\overline{\rho_l}}\,\widehat{h}_l \widehat{\boldsymbol{v}}_l\right) = -\nabla \cdot \alpha_l \left(\overline{\overline{\boldsymbol{q}_l}} + \boldsymbol{q}_l^T\right) - \Gamma_v \left(h_{sat} - \widehat{h}_l\right) \tag{4}$$

This model needs the description of the drift velocity (V_{vj}) and the mass transfer (Γ_v) between the phases. The selected expression for V_{vj} obtained from [12] is presented in Eq. (5). Is important to note that the drift velocity only acts in the same direction of the gravity.

$$V_{vj} = \sqrt{2}\left(\frac{\sigma g \Delta\rho}{\rho_l^2}\right)^{1/4} (1-\alpha_v)^{1.75} \tag{5}$$

The rest of the closure relationships in order to get a well-posed problem were taken from [12]-[22]. In addition, the $k-\epsilon$ turbulence model for the liquid phase (taking into account the presence of disperse vapor) was used to describe the Reynolds stress tensor $\mathbb{T}^T$. The evolution of the temperature in the solid is solved by a separate heat conduction equation.

2.2. Boundary conditions

The most common ways to estimate the heat flux at the boundaries of the workpiece are:

- by means of a heat transfer coefficient (h) depending on the global temperatures of the piece and cooling fluid, as it is presented in [2] - [5] or,
- calculating the heat flux emitted by the piece (q_w) using the heat flux partition model, which is a function of global values of piece and fluid temperatures and fluid velocity, in a similar way as [17] - [22].

Those approaches are completely insensitive to local variations of fluid temperature and velocity or the presence of vapor in the fluid. In this work we propose to include local dependences in the formulation of the boundary conditions. We take the heat partition model as a base and expand it in the way explained in following paragraphs. It is important to remark that, up to this point, is not our aim to describe early stages such as vapor blanket formation. That specific phenomenon will be included in further studies.

The heat transfered during boiling was described in a similar way than it is presented in [15]-[17], [19]-[22]. It is important to remark that due to the fact that our cooling fluid is oil, the functions presented for water were replaced for expressions based on physical properties of the fluid, rather than fitting constants [21]. The total heat flux extracted from the solid (q_w) result as the addition of three different mechanisms, as it is presented in its original form in Eq. (6), q_{1f} corresponds to the heat flux devoted to heat the fluid, while the others represent boiling.

$$q_w = q_{1f} + q_e + q_q \tag{6}$$

The coupling between the temperature of the solid, the temperature of the liquid and the vapor fraction generated due to boiling is made in the interface solid-mixture. The heat fluxes related to evaporation (q_e, q_q) are responsible for the flux of vapor generated, which in this case is considered injected through the boundary layer on the surface of the solid to the quenching flow. This balance of energy is represented in Eq. (7).

$$q_e + q_q = \alpha_v \rho_v h_{lv} (\mathbf{v}_v \cdot \mathbf{n}) \tag{7}$$

Due to the local characteristic of the formulations used in this work, the heat flux associated to the heating of the liquid (q_{1f}) is obtained by means of a wall-law formulation instead of the global correlations typically used. The wall temperature was modified in order to capture the presence of vapor. The wall-law and the modified temperature are presented in Eqs. (8) and (9).

$$q_{1f} = \frac{\rho_l\, c_{p,l}\, u_w \left(\widetilde{T}_w - T_l\right)}{\frac{Pr^T}{\kappa} ln(y^+) + \beta} \tag{8}$$

$$\widetilde{T}_w = \alpha_l T_w + \alpha_v T_{sat} \tag{9}$$

The relationship presented in Eq. (7) is exact but do not capture the effect of thermal insulation that produces a vapor blanket surrounding the piece. In order to damp the vapor generation in the presence of a considerable vapor fraction on the piece, the following modification on the injected vapor flux is proposed in Eq. (10).

$$\alpha_v \mathbf{v}_v \cdot \mathbf{n} = \frac{(q_e + q_q)}{\rho_v h_{lv}} \alpha_l^m \tag{10}$$

According to this, the vapor injection is stopped if there is no liquid available. Therefore, the heat extraction due to q_q and q_e is also stopped. This physical aspect of the boiling phenomena usually

Table 1. Material properties

	Liquid		Vapor	Solid
ρ $[kg/m^3]$	870		2	8470
μ $[kg/(m.s)]$	9.40×10^{-2}	(293 K)	2.3×10^{-5}	–
	2.35×10^{-2}	(313 K)	–	–
	4.35×10^{-3}	(373 K)	–	–
	1.00×10^{-4}	(490 K)	–	–
k $[W/(m.K)]$	0.14		0.014	14.9
c_p $[J/(kg.K)]$	1800	(290 K)	3000	444
	2300	(420 K)	–	–
σ $[N/m]$	3×10^{-2}		–	–
h_{lv} $[J/kg]$	–		10^6	–
D_b $[m]$	10^{-3}		–	–

is not accounted for in the current models. The final form of the modified heat partition model is presented in Eq. (11), taking into account that q_{1f} is obtained through wall-law formulations. The damping function presented here is a very basic first proposal and further analysis on it will be performed on future work.

$$q_w = q_{1f} + (q_e + q_q)\alpha_l^m \tag{11}$$

3. Case studied

The case of a sphere submerged in an upward fluid flow was used to test the model developed here. The upward flow intends to represent the forced-flow condition of an industrial bath. The flow velocity was set at 1 m/s, based on characteristic flow velocities obtained during the specific analysis of the model.

3.1. Materials and geometry

The fluid was quenching oil, whose properties were obtained from [23], and the material of the sphere was Inconel 600. This alloy was selected in order to make experimental results form [23] and numerical simulations comparable. Summary of oil, its vapor, and Inconel properties is presented in Table 1. The initial temperatures of fluid and piece were 313 and 1150 K respectively.

The domain was described on the $r - z$ plane taking axial symmetry along the z-axis. A sketch of the domain, showing the spherical piece and the surrounding fluid is presented in Fig. 1. The sphere was 30 mm of diameter and the flow domain was a 200 mm long and 200 mm diameter cylinder. The sphere was placed in the centerline, at 60 mm from the bottom. The fluid domain extends in the back of the piece in order to allow an appropriate development of the wake.

3.2. Numerical solution

The PDE system presented in Eqs. (1 - 4) can be assimilated to a regular fluid mechanics problem, with some extra sources, plus two additional conservation equations, one for the vapor phase and another for the liquid phase energy. *Comsol Multiphysics* allows the resolution of

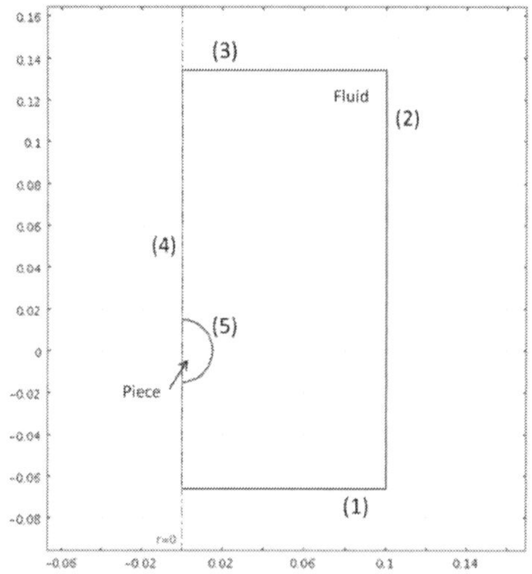

Figure 1. Analyzed domain

multiphase flows, including the mixture-model, and general diffusion-convection heat transfer problems. Based on these features, the corresponding application modes were used to solve the problem presented here. In addition, the cooling of the sphere was solved using an extra heat transfer application mode for the solid temperature. The translation of the model to the Comsol application modes was rather straightforward. For the mixture model, the extra terms due to the drift velocity are already incorporated in the model. The mass transfer and the drift velocity were user-defined according to the development of the model. The only major modification was made to the internal definition of the turbulent viscosity, where the effect of second phases was included according to [12]. For the conservation of energy on liquid phase, the extra terms due to vapor phase were included as user-defined sources.

The resolution of the whole problem was made using the direct solver UMFPACK [24] and the time integration was made along 10 seconds using an implicit scheme of second order. The boundaries of the domain marked in Fig. 1 correspond to: (1) inlet of oil liquid at constant temperature, (2) sliding wall, (3) outlet, (4) symmetry axis, (5) dynamic and thermal wall-laws, and vapor injection due to evaporation. The mesh was generated in *Comsol* using triangular elements. The maximum length of the elements at each boundary was: 1×10^{-3} for (1-4) and 3×10^{-4} for (5). The maximum length in the domain was 3×10^{-3}, with a growth ratio of 1.2. All measurements in m. The whole mesh consisted by 20990 elements, giving place to 222164 degrees of freedom. The transient of 10 seconds was solved on approximately 20 minutes on a desktop PC of one processor with four cores at 2.66 GHz and 4 Gb of RAM.

Numerical diffusion was needed to stabilize the finite element scheme due to the highly convective behavior of the conservation equations of vapor phase and energy of liquid. Isotropic diffusion was used in turbulence, dispersed phase and energy of the liquid; while streamline diffusion (anisotropic in the streamline direction for the dynamics of the flow and SUPG for thermal problems) was used in the equations. Diffusion coefficients were kept at a minimum ($\delta < 0.5$, where δ is a tuneable parameter that controls the amount of artificial diffusion).

The dynamics of the initial boiling events are represented by characteristic times different than the ones studied here. Because of this reason, the model cannot cope with an initial condition of zero vapor and develop the whole boiling process. As first approximation to solve the problem, an initial distribution of vapor phase corresponding to the initial temperature jump was considered. The description of the initial stages of boiling and the development of the vapor

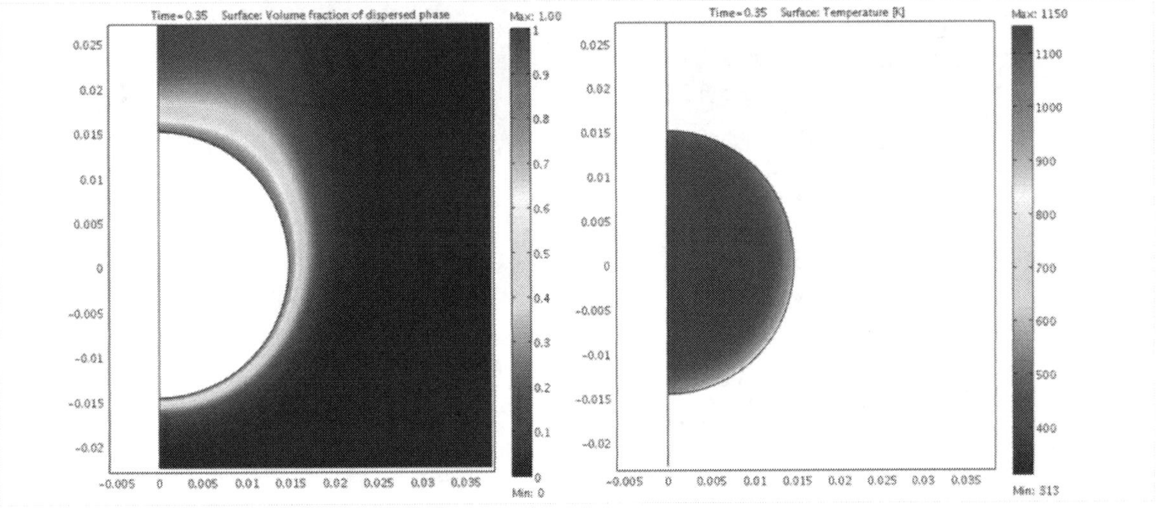

Figure 2. Vapor blanket and piece temperature at t=0.35 seg.

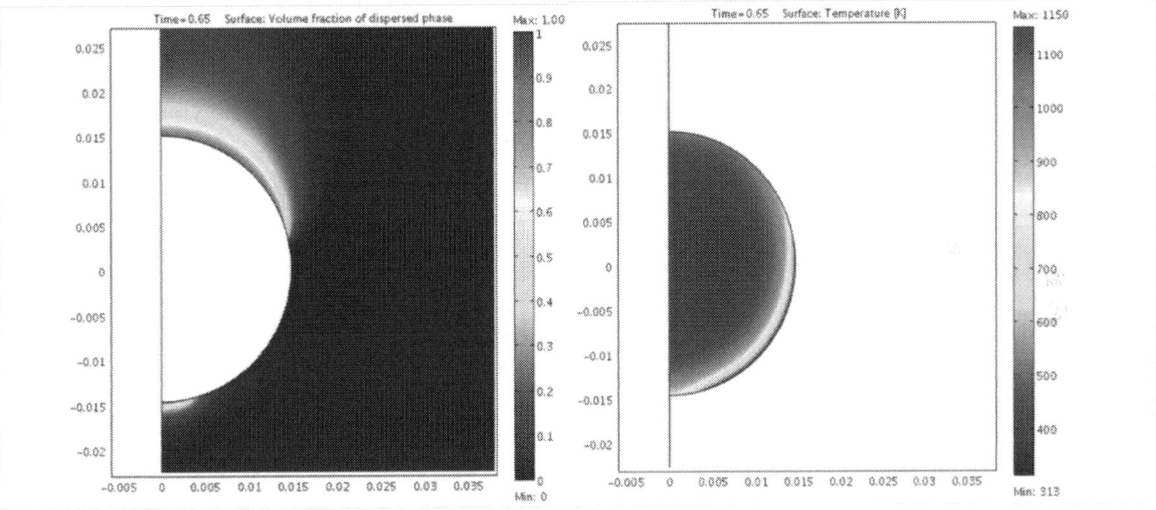

Figure 3. Vapor blanket and piece temperature at t=0.65 seg.

blanket is a much more complicated process that will be studied in further works.

4. Results

4.1. Vapor blanket and temperature fields

The evolution of vapor around the sphere and its temperature field is presented through snapshots at three different times. In Fig. 2 the early stages of the cooling are presented. Most of the surface of the piece is densely covered by vapor. In Fig. 3 the partial collapse of the vapor blanket and its thermal insulation effect is observed. In Fig. 4 the last stages of presence of vapor is depicted. Except for a small region in the wake that is still boiling, the temperature looks radially distributed.

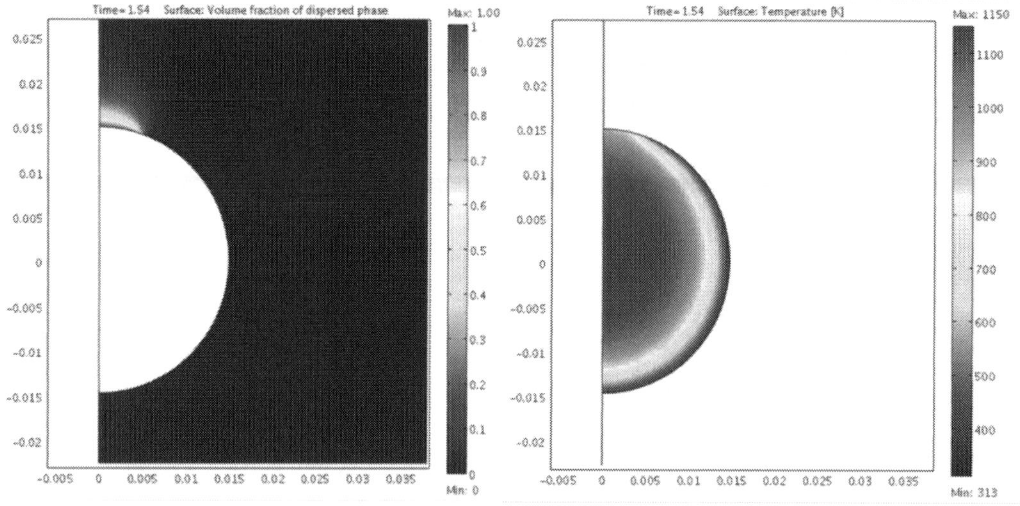

Figure 4. Vapor blanket and piece temperature at t=1.54 seg.

4.2. Cooling curves, heat transfer coefficient and heat fluxes

In order to assess the local effect of the heat transfer model four different points of the sphere were analyzed. They are located: one at the center of the sphere (*center*); and three at 1 mm below the surface, two along the symmetry axis (*upper* and *lower*), and one in the horizontal plane (*lateral*). The evolution of the temperature and the corresponding cooling rate at each point is presented in Fig. 5. There it is observed that lower and lateral zones had similar evolutions, while the upper point is clearly delayed. This difference is even more evident when the cooling velocities curves (T vs. dT/dt) are considered. These results highlight the effect of vapor as thermal insulator and how the heat transfer is locally affected along the surface of the piece. Time and spatial variation of the heat transfer coefficient (h) is depicted in Fig. 6, where the lower part of the circumference is the left side of the axis. This behavior is unlikely to be captured by the typical correlations of h depending on mean parameters of the flow.

The evolution of heat fluxes on the lower, lateral and upper points as a function of the overheating temperature of the wall was also made (see Fig. 7). It is observed that for lower and lateral points, the heat flux recovers the trend of boiling heat transfer curve. Differences in the shape of the curves may be due to the geometry of the piece (sphere vs. plane surfaces/wires) and also the need of fine tunning of the model (damping of heat transfer due to the presence of vapor in the boundary layer of the piece).

4.3. Comparison to experimental data

Results for the center point are compared to data extracted from experimental data presented in [23]. Both geometries are different in shape but with similar volumes, and oil and alloy properties are fairly the same. This allowed us just to compare the order of magnitude of the numerical results. In Fig. 8 the evolution of temperatures and cooling velocities curves are presented. The shapes of the curves are not exactly the same, but the matching of the order of magnitude obtained for both curves is highly satisfactory. Differences at high temperature are attributed to the heat extracted due to latent heat of evaporation during early stages of boiling, that part of the process could still not be captured, as it was stated in the initial conditions discussion.

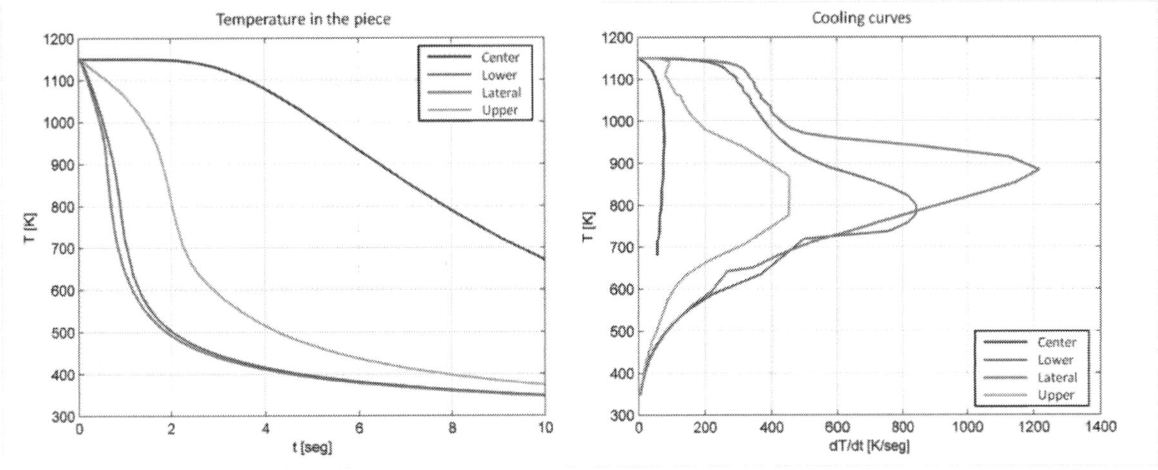

Figure 5. Temperature evolution and cooling rates in different zones of the piece

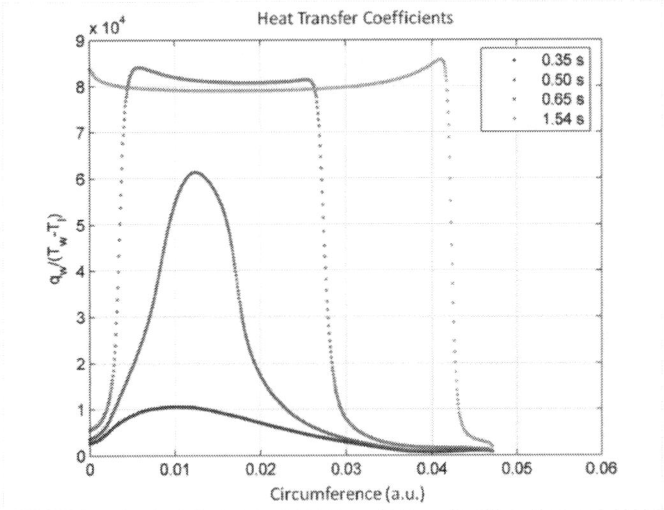

Figure 6. Variation of heat transfer coefficient along the circumference

5. Summary and conclusions

Heat transfer during quenching should be described including boiling phenomena; this should be done through an appropriate multiphase flow model and adequate local description of heat fluxes exchanged with the treated workpiece. The reduction of the specific model from a general frame depends on characteristic values of the problem.

A model that couples vapor dynamics, energy conservation of the liquid phase and heat transferred from the quenched piece was presented. Our model is based on the mixture model for multiphase flows and a heat partition model for heat transfer. Vapor fraction evolution and heat fluxes are obtained as main results.

The effect of thermal insulation of the vapor phase is clearly pointed out. Local analysis shows that cooling velocities are strongly correlated to the presence of vapor, and the resulting heat transfer coefficients are quite complex to be described using regular heat transfer correlations.

The developed model adequately captures the order of magnitude of the physical problem.

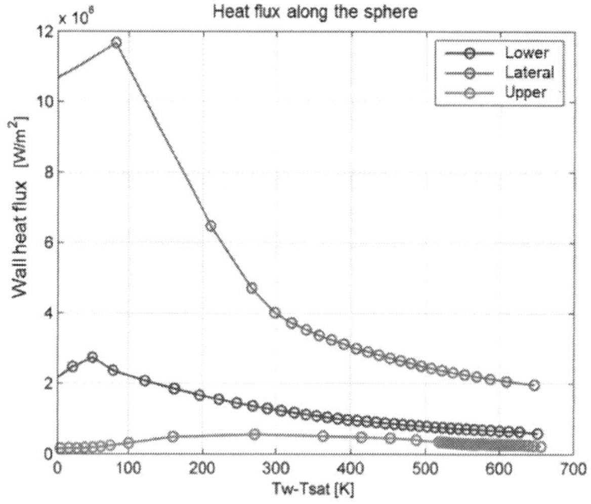

Figure 7. Heat flux vs. overheating temperature at different zones of the piece

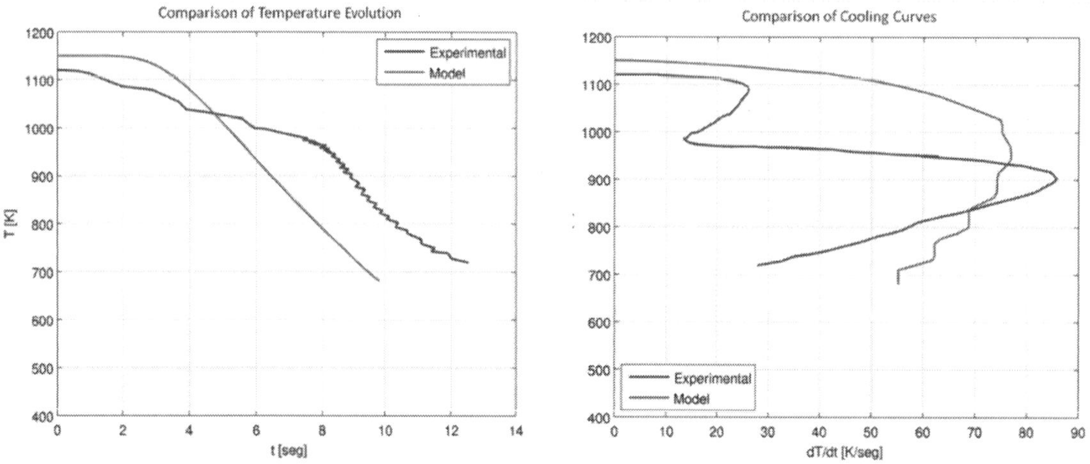

Figure 8. Comparison of numerical results and experimental data

Further analyses on different conditions are needed in order to obtain a quantitative calibration.

Acknowledgments
DNP acknowledges his PhD scholarship from MAEC-AECID (Spain).

Appendix

Definition of non-dimensional numbers used to characterize the multiphase flow.

- Drift Number (N_D): ratio between the characteristic momentum due to the drift velocity vs. the characteristic momentum of the mixture.

$$N_D = \frac{\rho_{l0} V_{vj0}}{\rho_{m0} v_{m0}}$$

- Eckert Number (N_{Ec}): ratio between the characteristic kinetic energy of the mixture vs. the characteristic enthalpy jump between the phases.

$$N_{Ec} = \frac{v_{m0}^2}{\Delta h_{lv0}}$$

- Phase change Number (N_{ph}): ratio between the characteristic momentum due to change of phase vs. the characteristic momentun of the vapor phase in the mixture.

$$N_{Ec} = \frac{\Gamma_{v0} L_0}{\rho_{v0} v_{m0}}$$

References

[1] Porter D A, Easterling K E 1992 *Phase Transformations in Metals and Alloys, 2nd Ed.* (London: Chapman & Hall)
[2] Denis S, Farias D and Simon A 1992 Mathematical model coupling phase transformations and temperature evolutions in steels *Iron and Steel Institute of Japan* **32** 316-25
[3] Song D-L, Gu J-F, Pan J-S and Hu M-J 2004 Numerical simulation of quenching of large sized blocks of 718 steel used for plastic dies *Materials Science and Technology* 1567-72
[4] Kang S-H and Im Y-T 2007 Thermo-elasto-plastic finite element analysis of quenching process of carbon steel *Journal of Materials Processing Technology* **192-3** 381-90
[5] Simsir C and Hakan Gr C 2008 A FEM based framework for simulation of thermal treatments: Application to steel quenching *Computational Materials Science* **44** 588-600
[6] Sugianto A, Narazaki M, Kogawara M and Shirayori A 2009 Failure analysis and prevention of quench crack of eccentric holed disk by experimental study and computer simulation *Engineering Failure Analysis* **16** 70-84
[7] Tensi H M, Stich A and Totten G E 1994 Fundamentals about quenching by submerging *Proceedings on International Heat Treating Conference: Equipment and Processes*
[8] Dhir V K 1998 Boiling heat transfer *Annu. Rev. Fluid Mech.* **30** 365-401
[9] Incropera F P, Dewitt D P, Bergman T L and Lavine A S 2007 *Fundamentals of Heat and Mass Transfer* (John Wiley & Sons, 6th Ed.)
[10] Clift R, Grace J R and Weber M E 1978 *Bubbles, drops and particles* (New York: Dover Publications Inc.)
[11] Crowe C, Sommerfeld M and Tsuji Y 1998 *Multiphase flows with droplets and particles* (Florida: CRC Press)
[12] Ishii M and Hibiki T 2006 *Thermo-fluid dynamics of two-phase flow* (New York: Springer)
[13] M. Manninen, V. Taivassalo, S. Kallio, *On the mixture model for multiphase flow*, VTT Plublications, VTT Energy, Nuclear Energy, Technical Research Center of Finland (VTT) (1996).
[14] Kocamustafaogullari G and Ishii M 1995 Foundation of the interfacial area transport equations and its closure relations *Int. J. Heat Mass Transfer* **38** 481-93
[15] Tu J Y and Yeoh G H 2002 On numerical modelling of low-pressure subcooled boiling flows *Int. J. Heat Mass Transfer* **45** 1197-209
[16] Yeoh G H and Tu J Y 2004 Population balance modelling for bubbly flows with heat and mass transfer *Chemical Engineering Science* **59** 3125-39
[17] Koncar B, Kljenak I and Mavko B 2004 Modelling of local two-phase flow parameters in upward subcooled flow boiling at low pressure *Int. J. Heat Mass Transfer* **47** 1499-513
[18] Krepper E, Koncar B and Egorov Y 2007 CFD modelling of subcooled boiling – Concept, validation and application to the fuel assembly design *Nuclear Engineering and Design* **237** 716-31
[19] Koncar B and Mavko B 2008 Simulation of boiling flow experiments close to CHF with the Neptune_CFD code *Science and Technology of Nuclear Installations* Article ID 732158
[20] Kocar C and Skmen C M 2009 A CFD model of two-phase flow in BWRs with results of sensitivity and uncertainty analysis *Nuclear Engineering and Design* **239** 1839-46
[21] Li X, Wei W, Wang R and Shi Y 2004 Numerical and experimental investigation of heat transfer on heating surface during subcooled boiling flow of liquid nitrogen *Int. J. Heat Mass Transfer* **52** 1510-16
[22] Koncar B and Tiselj I 2010 Influence of near-wall modelling on boiling flow simulation *Nuclear Engineering and Design* **240** 275-83
[23] Houghton Ibérica S.A. datasheet of Houghto-Quench 125 oil (http://www.houghtonintl.com/index.php)
[24] Tim Davis (http://www.cise.ufl.edu/research/sparse/umfpack/)

FLUIDOS 2010: XI Meeting on Recent Advances in the Physics of Fluids and their Applications IOP Publishing
Journal of Physics: Conference Series **296** (2011) 012005 doi:10.1088/1742-6596/296/1/012005

Correlation methods in cutting arcs

L. Prevosto[1] and H. Kelly[1,2,*]

[1]Grupo de Descargas Eléctricas, Departamento Ing. Electromecánica, Universidad Tecnológica Nacional, Regional Venado Tuerto, Laprida 651, Venado Tuerto (2600), Santa Fe, Argentina

[2]Instituto de Física del Plasma (CONICET), Departamento de Física, Facultad de Ciencias Exactas y Naturales (UBA) Ciudad Universitaria Pab. I, (1428) Buenos Aires, Argentina

[*]Member of the CONICET

E-mail: prevosto@waycom.com.ar

Abstract The present work applies similarity theory to the plasma emanating from transferred arc, gas-vortex stabilized plasma cutting torches, to analyze the existing correlation between the arc temperature and the physical parameters of such torches. It has been found that the enthalpy number significantly influence the temperature of the electric arc. The obtained correlation shows an average deviation of 3 % from the temperature data points. Such correlation can be used, for instance, to predict changes in the peak value of the arc temperature at the nozzle exit of a geometrically similar cutting torch due to changes in its operation parameters.

1. Introduction

Plasma cutting is a process of metal cutting at atmospheric pressure by an arc plasma jet, where a transferred arc is generated between a cathode and a work-piece (the metal to be cut) acting as the anode. Small nozzle bore, extremely high enthalpy and operation at relatively low arc current ($\approx 10 \div 200$) are a few of the primary features of these torches [1]. The physics involved in such arcs is very complicated. The conversion of electric energy into heat within small volumes causes high temperatures and steep gradients [2]. Dissociation, ionization, large heat transfer rates (including losses by transparent radiation), fluid turbulence and electromagnetic phenomena are all involved. In addition, wide variations of physical properties, such as density, thermal conductivity, electric conductivity and viscosity have to be taken into account.

These factors make hopeless to get an analytical solution for such arcs. Recently, commercial computational fluid dynamic software has become widely available and then numerical plasma modelling has reached a state advanced enough to be of practical use in the study of the cutting arc processes. However, as such programs were not initially developed for the description of thermal plasma process, they need to be complemented with electromagnetic and plasma properties packages [3]. These packages are not easy to obtain. Also the commercial software results somewhat expensive.

On the other hand, experimental data on cutting arcs are hard to obtain due to the harsh ambient conditions. Most of the available experimental data are related to spectroscopic [1] and probe measurements [4-6] in the external plasma region, giving direct information of only part of the variables involved, mainly the temperature.

In this scenario, the knowledge of a proper correlation method to make advantageous use of the experiments already performed on cutting arcs becomes important from the engineering point of view. The present work applies similarity theory [7,8] to the plasma emanating from transferred arc, gas vortex stabilized plasma cutting torches, to analyze the existing correlation between the plasma temperature (a quantity of primary importance in such devices) and the physical parameters of such torches.

2. Arc dimensionless numbers

2.1. Mathematical model

Since a number of complex, interacting, nonlinear processes are simultaneously occurring in cutting arcs (electromagnetic, gas dynamic as well as thermal effects), the conservation equations were used as the starting point for determining dimensionless numbers.

The most frequently used cutting arc models assume the local thermodynamics equilibrium (LTE) approximation [3], so the plasma is characterized by a single temperature T. The fluid part of the thermal plasma model describing such an arc can be expressed as follows.

Total mass conservation

$$\nabla\cdot(\rho\bar{u})=0. \tag{1}$$

Momentum conservation

$$\nabla\cdot\left(\rho\bar{u}\otimes\bar{u}-\bar{\bar{\tau}}\right)+\nabla p-\bar{J}\times\bar{B}=0. \tag{2}$$

Energy conservation

$$\nabla\cdot\left(\rho\bar{u}\,h+\bar{q}\right)-\bar{J}\cdot\bar{E}-\nabla p\cdot\bar{u}+\dot{Q}_{rad}=0, \tag{3}$$

where ρ represents the total mass density, $\bar{u}$ the fluid velocity, p the pressure, $\bar{\bar{\tau}}$ the stress tensor, $\bar{J}$ the current density, $\bar{B}$ the magnetic field, h the enthalpy, $\bar{q}$ the total heat flux, $\bar{E}$ the electric field and $\dot{Q}_{rad}$ the power lost by transparent radiation.

Two more equations are required to describe the electromagnetic part of the plasma model. The first is the current continuity equation

$$\nabla\cdot\bar{J}=0\,, \tag{4}$$

where

$$\bar{J}=\sigma\bar{E}\,, \tag{5}$$

while the second is one of Maxwell's equations

$$\nabla\times\bar{B}=\tilde{\mu}_0\,\bar{J}, \tag{6}$$

where σ is the electric conductivity and $\tilde{\mu}_0$ the free space magnetic permeability.

The total heat flux in equation (3) describes the heat transported by conduction and the enthalpy transport by mass diffusion, and is defined as [3]

$$\bar{q}\equiv-\kappa_e\nabla T+\bar{J}_e h_e, \tag{7}$$

where κ_e is the effective thermal conductivity and $\bar{J}_e$ is the electron mass diffusion that can be approximated by

$$\bar{J}_e\approx-\frac{m}{e}\bar{J}\,, \tag{8}$$

where e is the elementary electric charge and m is the electron mass. Equation (8) neglects the charge transported by ions. In equation (7) $h_e=5k_BT/(2m)$ represents the specific electron enthalpy (k_B is the Boltzmann's constant).

The effective viscosity is

$$\mu_e=\mu_l+\mu_t, \tag{9}$$

and the effective thermal conductivity is

$$\kappa_e=\kappa+\frac{\mu_t C_p}{P_r}, \tag{10}$$

where C_p, μ_l and κ are the constant-pressure plasma specific heat, viscosity and thermal conductivity, respectively. The turbulent viscosity μ_t and the Prandtl number P_r require extra relationships (which are commonly referred to as the turbulence model) to calculate the turbulent enhanced viscosity and thermal conductivity. Relationships expressing the dependence of gas composition, thermodynamic properties, transport coefficients and radiation power losses on temperature and pressure; must be also added in order to make the set closed. To make the problem definite it is necessary also to specify boundary conditions for all of such variables.

2.2. Dimensionless numbers

In what follows, the set of equations are expressed in dimensionless form by making use of reference values (denoted by the subscript "zero") for the above plasma quantities, and with L as a scale length:

$$\nabla^* \cdot \left(\rho^* \bar{u}^*\right) = 0, \tag{11}$$

$$\nabla^* \cdot \left(\rho^* \bar{u}^* \otimes \bar{u}^*\right) - \frac{\tau_0}{\rho_0 u_0^2} \nabla^* \cdot \bar{\bar{\tau}}^* + \frac{p_0}{\rho_0 u_0^2} \nabla^* p^* - \frac{J_0 B_0 L}{\rho_0 u_0^2} \bar{J}^* \times \bar{B}^* = 0, \tag{12}$$

$$\frac{\rho_0 h_0 u_0}{L J_0 E_0} \nabla^* \cdot \left(\rho^* \bar{u}^* h^*\right) + \frac{q_0}{L J_0 E_0} \nabla \cdot \bar{q}^* - \bar{J}^* \cdot \bar{E}^* - \frac{p_0 u_0}{L J_0 E_0} \nabla^* p^* \cdot \bar{u}^* + \frac{\dot{Q}_{rad0}}{J_0 E_0} \dot{Q}_{rad}^* = 0, \tag{13}$$

$$\nabla^* \cdot \bar{J}^* = 0, \tag{14}$$

$$\bar{J}^* = \frac{\sigma_0 E_0}{J_0} \bar{E}^*, \tag{15}$$

$$\nabla^* \times \bar{B}^* = \frac{\mu_0 J_0 L}{B_0} \bar{J}^*. \tag{16}$$

The superscript (*) indicate dimensionless numbers, obtained by dividing dimensional variables by the appropriate scale values ($\rho^* \equiv \rho/\rho_0$, $\bar{u}^* \equiv \bar{u}/u_0$ and so on). The equation (12) gives three dimensionless numbers,

$$\Pi_{N_E} \equiv \frac{p_0}{\rho_0 u_0^2}, \qquad \Pi_{N_R} \equiv \frac{\tau_0}{\rho_0 u_0^2}, \qquad \Pi_{em} \equiv \frac{J_0 B_0 L}{\rho_0 u_0^2}.$$

The first and the second are the well-known Euler and Reynolds numbers. The third number shows the relative weight between the electromagnetic and inertial forces. The equation (13) gives four dimensionless numbers,

$$\Pi_h \equiv \frac{\rho_0 u_0 h_0}{J_0 E_0 L}, \qquad \Pi_{wf} \equiv \frac{p_0 u_0}{J_0 E_0 L}, \qquad \Pi_q \equiv \frac{q_0}{J_0 E_0 L}, \qquad \Pi_{rad} \equiv \frac{Q_0}{J_0 E_0}.$$

These numbers reflect the conversion of electric energy into other forms. The enthalpy number Π_h takes into account energy which is used for heating the plasma flow, Π_{wf} times $\left(\Pi_{N_E}\right)^{-1}$ is based on energy used for acceleration, and Π_q and Π_{rad} involve energy losses by conduction and mass diffusion; and energy losses by electromagnetic radiation, respectively. The Ohm's and Ampere laws give the numbers,

$$\Pi_E \equiv \frac{\sigma_0 E_0}{J_0}, \quad \Pi_B \equiv \frac{\mu_0 J_0 L}{B_0}.$$

These numbers reflect the relation between self and external magnetic fields (if any), and the process of charged particle transfer in an electric field.

In addition to these numbers, from the boundary conditions various parametric numbers can be obtained. The boundary condition for the arc current (I)

$$I = \int_S \bar{J} \cdot d\bar{S}, \tag{17}$$

(S represents the area of the arc cross section) in dimensionless form becomes

$$I = J_0 L^2 \int_{S^*} \bar{J}^* \cdot d\bar{S}^* , \qquad (18)$$

and hence gives the current number Π_I

$$\Pi_I \equiv \frac{J_0 L^2}{I} .$$

Also, the numbers

$$\Pi_d \equiv \frac{d}{L}, \quad \Pi_{ip} \equiv \frac{P_{rel}}{p_0}$$

account for the nozzle geometry (diameter (d)) and the influence of the inlet gas pressure (P_{rel}) on the arc quantities.

The set of dimensionless equations (11-16) gives nine independent dimensionless numbers in agreement with the Buckingham π-theorem, since the number of quantities is thirteen and the number of dimensions four. Besides, the arc boundary conditions introduce two new parametric numbers.

Most of these numbers contain physical properties, such as density and electric conductivity, which vary widely because of the large temperature range, and certain reference values have therefore to be taken. In this work, the scale values of the plasma physical properties were determined for a mean temperature T_0 which corresponds to an electron concentration of 1 % in the plasma at the ambient pressure [8]. These plasma scales (listed in Table 1), were taken from the calculated values in [9].

Table 1. Scale values of the plasma physical properties.

Physical variables	Oxygen	Nitrogen	Air
Temperature (T_0) ($\times 10^3$ K)	9.2	9.3	9.3
Electrical conductivity (σ_0) (Ω^{-1} m^{-1})	2.23×10^3	1.89×10^3	1.96×10^3
Enthalpy (h_0) (J kg^{-1})	2.86×10^7	4.97×10^7	4.52×10^7
Density (ρ_0) (kg m^{-3})	2.1×10^{-2}	1.83×10^{-2}	1.88×10^{-2}

3. Data on cutting arcs

The experimental data on cutting arcs temperature correspond to the external plasma region (open regions where the pressure has relatively small variations around the atmospheric value). Peak values of the arc temperature close to the nozzle exit for currents in the wide range 12 ÷ 200 A, and for three different working gases (air, nitrogen and oxygen); were taken from the literature.

4. Generalized arc temperature correlation

Not all of the obtained dimensionless numbers are of equal importance. Inside the torch the pressure forces are almost balanced by inertial forces (i.e., other processes as viscosity and electromagnetic acceleration are considerably less relevant) and so only the Euler number becomes important in the momentum equation.

On the other hand, since the electrical power delivered in the arc is almost used for heating the plasma flow (i.e., the conversion of electric energy into kinetic energy, while the energy losses by thermal conduction, mass diffusion and radiation are less important); only the enthalpy number becomes relevant in the energy equation. Also from the electromagnetic equations only the electric field number is important. Hence, the three relevant dimensionless numbers are

$$\Pi_{N_E} \equiv \frac{p_0}{\rho_0\, u_0^{\,2}}, \quad \Pi_h \equiv \frac{\rho_0\, u_0\, h_0}{J_0\, E_0\, L}, \quad \Pi_E \equiv \frac{\sigma_0\, E_0}{J_0}.$$

These numbers can be combined with the parametric numbers to eliminate the unknowns. The following modified form of the enthalpy number can be used as a generalized argument to construct the generalized plasma temperature correlation for the considered type of arc

$$\Pi_h^{(1)} \equiv \Pi_h\, \Pi_E\, \left(\Pi_{N_E}\right)^{0.5} \left(\Pi_I\right)^2 \left(\Pi_d\right)^3 \left(\Pi_{ip}\right)^{0.5} = \rho_0^{\;0.5}\, P_{rel}^{\;0.5}\, h_0\, \sigma_0\, d^3\, I^{-2}. \quad (19)$$

The generalized function of the peak value of the arc temperature at the nozzle exit may be taken to be the simple ratio

$$\Pi_T \equiv \frac{\widehat{T}_{exit}}{T_0}, \quad (20)$$

and hence the generalized arc temperature correlation may be expressed as

$$\Pi_T = f\left(\Pi_h, \Pi_E,\, \Pi_I,\, \Pi_{N_E}, \Pi_d, \Pi_{ip}\right) = f\left(\Pi_h^{(1)}\right). \quad (21)$$

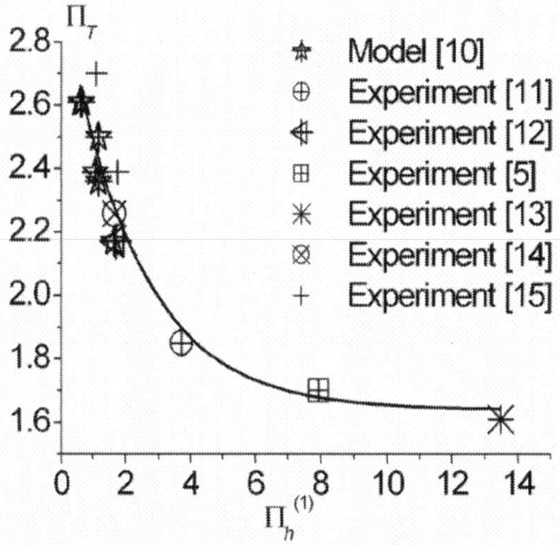

Figure 1. Correlated arc temperature for cutting torches ranging from 12 to 200 A.

The generalized function of the arc temperature for most of the temperature data reported in the literature is shown in Fig. 1. This figure shows that the generalized function Π_T can be accurately described by the relationship (in SI units: p_{rel} in Pa; d in m; I in A; T in K)

$$\widehat{T}_{exit} = T_0 \left(1.6 + 1.3 \exp\left\{-0.44 \left(\rho_0^{\;1/2}\, P_{rel}^{\;1/2}\, h_0\, \sigma_0\, d^3\, I^{-2}\right)\right\}\right). \quad (22)$$

The values of the arc temperature predicted using such correlation show a deviation not larger than 10 % with respect to the experimental data (with an average deviation of 3 %); in spite of the fact that such data include different torches ranging from 12 to 200 A.

Note that (22) shows that the centre-line value of the arc temperature strongly depends on both, arc current and nozzle diameter; but slightly on the type of gas and its inlet pressure. The factor $\rho_0^{\;1/2}\, h_0\, \sigma_0$ is a number equal to $\approx 9.24 \times 10^9$ for oxygen, 1.27×10^{10} for nitrogen, and 1.21×10^{10} kg$^{1/2}$ s^{-2} m$^{-1/2}$ Ω, for air.

5. Conclusions

The theory of dynamic similarity has been used to construct a generalized arc temperature correlation of a vortex-stabilized cutting plasma torch. The enthalpy number obtained from the energy equation has been found to significantly influence the temperature of the electric arc.

The obtained correlation shows an average deviation of 3 % from the temperature data points. Such correlation can be used, for instance, to predict changes in the peak value of the arc temperature at the nozzle exit of a geometrically similar cutting torch due to changes in its operation parameters (electric current, nozzle diameter, working gas specie or/and the inlet gas pressure); at least within the given parameters range. The relation shows that the centre-line value of the arc temperature strongly depends on both, arc current and nozzle diameter; but slightly on the type of gas and its inlet pressure.

ACKNOWLEDGEMENTS

This work was supported by grants from the Universidad de Buenos Aires (PID X108), CONICET (PIP 5378) and Universidad Tecnológica Nacional (PID Z 012).

REFERENCES

[1] Nemchinsky V A and Severance W S 2006 *J. Phys. D: Appl. Phys.* **39**, R423
[2] Prevosto L, Kelly H and Mancinelli B 2009 *J. Appl. Phys.* **105**, 013309
[3] Gleizes A, González J J and Freton P 2005 *J. Phys. D: Appl. Phys.* **38**, R153
[4] Prevosto L, Kelly H and Mancinelli B 2008 *IEEE Trans. Plasma Sci.* **36**, 263
[5] Prevosto L, Kelly H and Minotti F O 2008 *IEEE Trans. Plasma Sci.* **36**, 271
[6] Prevosto L, Kelly H and Mancinelli B 2009 *IEEE Trans. Plasma Sci.* **37**, 1092
[7] Yas'ko O I 1969 *Brit. J. Appl. Phys. (J. Phys. D)* **2**, 733
[8] Paingankar A M, Das A K, Shirodkar V S, Sreekumar K P and Venkatramani N 1999 *Plasma Sources Sci. Technol.* **8**, 100
[9] Boulos M, Fauchais P, Pfender E 1994 *Thermal Plasmas, Fundamentals and Applications Vol. 1* (Plenum Press)
[10] Colombo V, Concetti A, Ghedini E, Dallavalle S and Vancini M 2008 *IEEE Trans. Plasma Sci.* **36**, 389
[11] Freton P, Gonzalez J J, Gleizes A, Camy Peyret F, Caillibotte G and Delzenne M 2002 *J. Phys. D: Appl. Phys.* **35**, 115
[12] Girard L, Teulet Ph, Razafinimanana M, Gleizes, Camy-Peyret, Baillot E and Richard F 2006 *J. Phys. D: Appl. Phys.* **39**, 1543
[13] Sakuragi S 1995 *Eng. Sci. Reports*, Kyushu University **16**, 383
[14] Pardo Sanjurjo C, González-Aguilar J, Rodríguez Yunta A and Calderón M A G 1999 *J. Phys. D: Appl. Phys.* **32**, 2181
[15] Peters J, Heberlein J and Lindsay J 2007 *J. Phys. D: Appl. Phys.* **40**, 3960

The Magnetosphere Mixing Layer: Observations, MHD Stability, and Large Eddy Simulations

F T Gratton[1,2,5], L E Bilbao[1-3], G Gnavi[1-3] and C J Farrugia[4]

[1] Instituto de Física del Plasma, Consejo Nacional de Investigaciones Científicas y Técnicas y Universidad de Buenos Aires, Buenos Aires, Argentina
[2]- Departamento de Física, Facultad de Ciencias Fisicomatemáticas e Ingeniería, Pontificia Universidad Católica Argentina, Buenos Aires, Argentina
[3] Departamento de Física, Facultad de Ciencias Exactas y Naturales, Universidad de Buenos Aires, Buenos Aires, Argentina.
[4] Space Science Center, University of New Hampshire, Durham, NH, 03824, USA

[5]E-mail: fgratton@arnet.com.ar, fgratton@infip.org

Abstract. The work is about the evolution of large vortices in the boundary layer of the terrestrial magnetopause studied by computational physics, in support of interpretive analysis of spacecraft data from an event observed on December 7, 2000. The distinctive features of the configuration examined are i) the very small magnetic shear across the boundary, and ii) the dynamics concerns mainly the vorticity. The magnetic field is only fluted during the process, but it determines the direction of the vortex axis. Considering new elements from computer simulations and observations, an extension of the research of a recent publication in this journal is reported. A magnetohydrodynamic code for large eddy simulations is used to examine the influence of supersonic, and superAlfvénic speeds, on the dynamics of the boundary layer. Correlations of data are studied to identify signatures of the Kelvin-Helmholtz instability, and the presence of whirling plasma. The aim is to improve the understanding of factors that govern the tailward growth of the low latitude boundary layer, and the potentiality of plasma entry into the magnetosphere, during periods of northward interplanetary magnetic field.

1. Introduction

1.1 Preamble

A short outline of this work is as follows. The stage is the terrestrial magnetopause (MP). We focus the attention on the boundary layer (BL) at the low latitude magnetopause flanks. The paper deals mainly with computational magnetohydrodynamics (MHD) of large vortices. These often form at the BL, and appear more frequently when the interplanetary magnetic field (IMF) shows a significant north component. In addition, we discuss a matter of considerable interest, in view of the scarcity of *in situ* observations. Can records of a single spacecraft that crosses the BL during the transit of quasi-periodic perturbations, pass a judgment of vortex detection? Under what conditions a forecast of the presence of vortices is possible. The paper is a continuation of an investigation recently published in this journal [1] along two directions: *i*) the effect of increasing the magnetosheath flow up to supersonic, and superAlfvénic speeds on vortex simulations, and *ii*) the correlations of experimental data that may disclose the existence of a vortical BL.

1.2 The general context

As appropriate to an introduction, we outline the topic that presently is a cause of concern in magnetospheric physics. This work considers special elements of the larger problem. A cold, dense, magnetized plasma flows around the magnetosphere, gathering speed as it progresses tailward. A tutorial description of the system properties, the global configuration of magnetic fields, and flows can be found, for instance, in [2]. Increasingly, attention has been directed to the presence of magnetosheath plasma deep inside the Earth's plasma sheet when the IMF stays northward for long spells. This plasma is relatively (with respect to ordinary values in the same regions) cold, dense, and stagnant (for example, $T \cong 1$keV, $n \sim 1\text{cm}^{-3}$). See, among other references, [3, 4, 5, 6]. The phenomenon is known as the cold dense plasma sheath (CDPS).

The difficulty is to explain how did this CDPS get there. The presence in the Earth's geomagnetic tail of plasma with properties similar to that of the adjoining magnetosheath makes the flanks a likely entry locale. Indeed, flank entry has often been advocated to explain episodes of CDPS (e.g., [7]). The data records are also well correlated with periods of northward IMF. Hence, the problem is the mechanism of mass entry from solar wind into magnetosphere during northward IMF periods [4, 7].

Two main processes have been suggested. The first one (*a*) is reconnection occurring simultaneously poleward of both cusps. This process adds closed flux tubes filled with magnetosheath plasma to the magnetosphere, into which they slowly sink, as they are convected tailward. See, for instance, [8, 9, 10, 11] and the references quoted in these papers.

The second (*b*) is the velocity shear instability (Kelvin-Helmholtz, KH) in the non-linear stage, when rolled-up vortices that entrain magnetosheath material and broaden the boundary layer are formed. See, [1, 12, 13, 14, 15, 16, 17, 18, 19, 20, 21, 22, 23, 24] and the literature quoted therein. Inside large vortices, the close mingling of magnetosheath with magnetosphere plasma favors the enhancement of particle diffusion at the microscopic scale. Teresawa, Fujimoto, and coworkers (see [4], [7]) pioneered the study of the CDPS formation and the influence of the KH instability on this phenomenon.

Under particular conditions, stretched magnetic field lines may even give rise to small-scale tearing inside a big eddy, and hence to a local de-freezing of plasma and field with consequences on plasma diffusion [12, 13]. However, note that one must distinguish this secondary effect, a consequence of the primary event (the build-up of a KH vortex), from the concept of large-scale reconnection of $\boldsymbol{B}$ field lines poleward of the cusps.

In this paper we consider a different KH scenario, based on a plasma and magnetic field configuration recorded by the Cluster mission during an event of December 7, 2001. The distinctive features are: i) a very small magnetic shear across the boundary, and ii) a dynamics that concerns mainly the vorticity. The process only flutes the magnetic field, although the latter determines the direction of the vortex axis [1, 26, 27]. In another aspect, our simulations differ from other investigations: we use a localized MP model with two free boundaries. We work with two periodicities: along the flow and normal to it (in the sense of the geomagnetic field). The perturbations are free, not anchored, also in the direction normal to the flow.

The two mentioned transfer processes, i.e., (*a*) reconnection poleward of the cusps, and (*b*) KH in its non-linear phase, have been generally considered separately. Several research teams have tried to establish consistency of a small set of observations with models, and simulation ideas of one or other of these two entry mechanisms. These attempts have not yet given a satisfactory elucidation of the issue. It has been shown that one mechanism is more or less sufficient to account for some broad features of the observations in a given case, but it has not been established the necessity by showing the inadequacy of the "*rival*" mechanism in the same situation.

An intrinsic difficulty of the matter of discussion is the dearth of "*in situ*" records. The boundary layer is not observable from afar with other methods, as it happens with several astrophysical systems. For the greatest part one spacecraft alone, if at all, is crossing the MP at interesting times. Moreover, during a spacecraft crossing of the BL it would be desirable to have the support of another in the neighbourhood, to know the state of the magnetosheath at the same time, but this is a rare occurrence.

Hence, the interest to have criteria to identify vortices with data from one spacecraft alone. On the other hand, if available at the proper time and place, the use of multiple spacecraft missions like Cluster and Themis may offer further analysis tools. Thus, for example, some of them might be able, given a favorable satellite constellation, to monitor the magnetosphere and magnetosheath regions adjacent to the vortices. In addition, possibilities of speeds determined by triangulation methods may be feasible.

1.3 The aims of this work

Many aspects of the complex issue just outlined remain obscure, and need further investigations. Considering process (*b*) primary attention should be given to the following factors. The strength and orientation of the magnetosheath's magnetic field, both magnetic and velocity shears across the MP, the flow speed (supersonic, superAlfvénic?), the density gradients across the MP, the scale lengths of the magnetic and velocity transitions (which may be different (e.g., [25, 29]), and the influence of supersonic flows where compressibility, a stabilizing agent, comes into play. When is a flank region KH-stable under supersonic flows? How does the broadening of the flank BL produced by the KH instability depend on (i) the local orientation of ***B*** and ***V***; (ii) the physical conditions at the boundary; and (iii) its distance from the subsolar point? Some of these questions have been addressed by members of our team, and by coworkers in recent studies [1, 23, 24, 26, 27, 28, 29, 30].

On the other hand, the KH instability is associated with other telltale signatures, beside the information provided by "*in situ*" observations. Such are, for example, the geomagnetic micro-

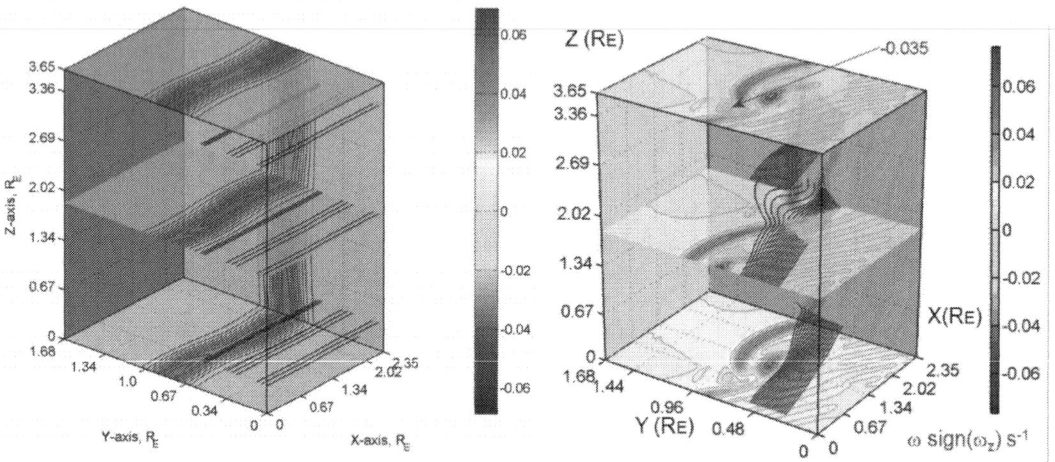

Figure 1a. Left panel. Initial perturbation of the velocity field for the computational run C42. Runs' numbers are explained at the end of Section 2, and their properties are listed in Table 1. Perturbed vorticity contours (color code units s^{-1}) and a set of (black) streamlines on the magnetosheath side. **Figure 1b.** Right panel. LES results for C42 at t=132 s, vorticity contours and a set of (blue) streamlines.

pulsations in the Pc5 range (2-10 mHz). These are quasi-periodic perturbations in the few R_E (Earth radius) wavelength range at the MP flank and BL when the IMF points strongly north. A meaningful example of this phenomenon, with a precise correlation of the frequencies of power spectrum peaks of GOES 8 (Geostationary Operational Environmental Satellites), and of ground magnetometers of the CANOPUS X meridional chain, with the characteristic frequency of the activity observed "*in situ*" by Cluster 3 is reported in [26, 27]. The presence of such signatures constitutes a severe constraint on any interpretation that does not somehow involve the KH instability mechanism.

To recapitulate, (*i*) the rolling-up of the vorticity sheet in the boundary layer must be investigated with compressible, MHD codes. Data analysis and interpretation must be assisted by theoretical and

Journal of Physics: Conference Series **296** (2011) 012006 doi:10.1088/1742-6596/296/1/012006

computational work. Besides, a concurrent theme is (*ii*) to strengthen the knowledge of KH signatures, by making apparent the kind of data correlations that help to diagnose the presence of whirling plasma from *in situ* records of one spacecraft alone. Thus, the paper is oriented toward points (*i*) and (*ii*).

The content of the paper is as follows. After the Introduction, Section 2 summarizes information about December 7, 2000, our KH reference event, and the computational code. Characteristic features of that event are incorporated in the models of the numerical studies. Section 3 presents the main computational results, and their dependence on the sonic, and Alfvén Mach numbers of the magnetosheath flow. Subsection 3.4 focuses on the evolution of strata of different temperature, and density in the BL, and paves the way to Section 4, which deals with the study of correlations of December 7, 2000 data, and their interpretation. Discussion and conclusions are in Section 5.

2. Notes on December 7, 2000, and models for that event

2.1 The reference event

The event that suggests the boundary layer models, the particular disposition of velocity and magnetic fields, and the density and temperature stratifications employed in this paper, is a crossing of the MP recorded by Cluster on December 7, 2000. The phenomenology of this event and the associated theoretical and computational models are given in detail in [1, 26, 27, 29]. Only a minimal description is given here to allow the reader to proceed, skipping a search of references if he/she so prefers.

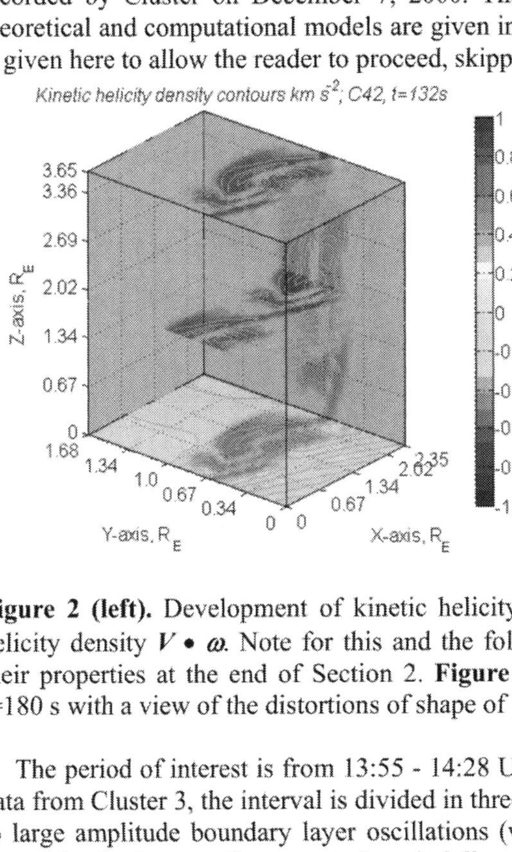

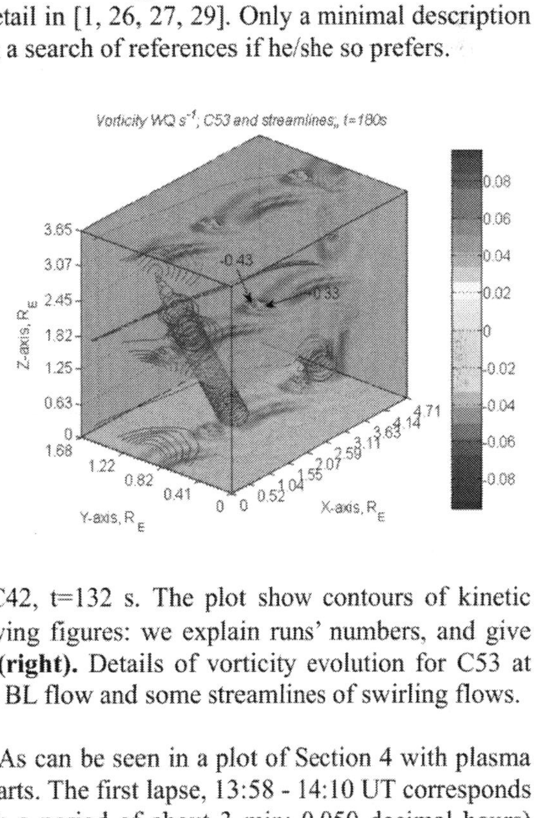

Figure 2 (left). Development of kinetic helicity, C42, t=132 s. The plot show contours of kinetic helicity density $\boldsymbol{V} \bullet \boldsymbol{\omega}$. Note for this and the following figures: we explain runs' numbers, and give their properties at the end of Section 2. **Figure 3 (right).** Details of vorticity evolution for C53 at t=180 s with a view of the distortions of shape of the BL flow and some streamlines of swirling flows.

The period of interest is from 13:55 - 14:28 UT. As can be seen in a plot of Section 4 with plasma data from Cluster 3, the interval is divided in three parts. The first lapse, 13:58 - 14:10 UT corresponds to large amplitude boundary layer oscillations (with a period of about 3 min; 0.050 decimal hours) caused by the impact of a strong solar wind discontinuity onto the MP. The large excursions of the MP are used to obtain an estimate of the speed of the tailward flow of the adjacent magnetosheath, and to gauge the intensity and the change of direction of the magnetic field across the boundary layer, from the inner edge to the magnetosheath [26, 27]. The survey reveals a MP configuration with a very small magnetic shear, and an external flow still subsonic and subAlfvénic.

Monitoring the IMF clock angle with ACE, the permanence of a small magnetic shear angle during the rest of the event can be reasonably assumed. The remaining lapses are then characterized by a very small magnetic shear across the BL. We cannot ensure the outer speed estimate, except as a tentative evaluation, because of the absence of another spacecraft at the same magnetosheath flank. However, ACE does not register substantial changes of the solar wind speed for the remaining intervals after the first period, which encourages our assumption that the speed estimate holds during the interval of interest. The distinctive aspect of the second and third parts, lasting from 14:10 - 14:28 are quasi-periodic oscillations with a shorter period, 79 s (0.0219 decimal hours). These are strong perturbations of all vector and scalar fields, with intensities decreasing during the third part of the event, from 14:20 to 14:28. After this, a change of the IMF clock angle attested by ACE, determines magnetic conditions unfavorable to the KH instability, and in fact, the former boundary layer activity observed by Cluster 3 extinguishes after 14:28 UT.

2.2 LES and computational code

Eddies generated by the KH mechanism have been extensively studied in fluid dynamics, both with experiments and with numerical simulations. In space physics, the knowledge available on the MHD version of the KH instability has also achieved maturity. In 1984, Miura [31] initiated the computational studies of KH in the MP context. Since then, the steady progress in this area can be appraised in a set of reviews (for instance [32, 33, 34, 14]), and in recent research papers (among others: [1, 12, 13, 15, 16, 17, 18, 19, 20, 22, 23, 24]).

We base the large eddy simulations (LES) on non-dissipative, compressible, MHD equations in 3D and time. The computational program, developed for complex magneto-fluid dynamics problems of plasma physics [37, 38, 39], was adapted for the present task. The in-house code starts from the integral conservation form of the MHD equations, and employs a *finite volume, Lagrangian-Eulerian* technique. The set of equations and some computational features can be found in [1] (this same journal), and more technical details are in [35, 36].

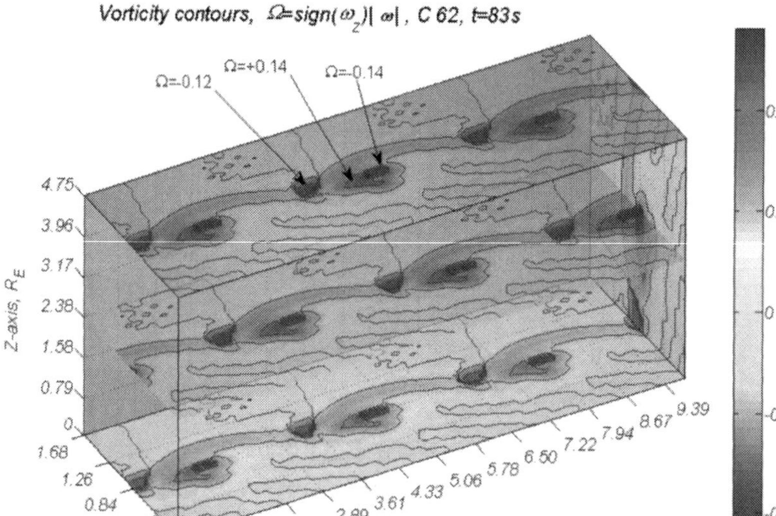

Figure 4. Vorticity contours for C62 at t=83 s. The magnetosheath speed is twice that corresponding to C42, and C53, and the computing box accepts three wavelengths.

2.3 Initial and boundary conditions

Due to the minimal magnetic shear, the MHD theory of the Kelvin-Helmholtz instability predicts significant growth rates for flute modes that have a most favorable angle between the ***k*** – vector and the direction of the bulk flow at φ=32° [26, 29]. Modes with deviations of $\boldsymbol{k}$ by 10 – 15 degrees from this orientation do not grow. The precise growth rate values depend on the averaged steady state model of fluid and flow employed. We assume hyperbolic tangent functions to connect continuously the vector and scalar fields (magnetic and velocity fields, density and temperature) of the outer magnetosheath limit with those of the inner magnetosphere edge. The same configuration is used as the initial state for the LES, with the addition as initial starter of one perturbation mode only to the velocity field, with amplitudes ranging about 5% the magnetosheath speed. During the simulation, the scalar and vector fields are fixed to constant values as boundary conditions at the outer and the inner limit of the BL.

The computational results are presented in 3D boxes, with the *X*-axis pointing tailward, aligned with the magnetosheath flow, the *Y*-axis is normal to the local part of the MP surface, and points inward across the BL. The *Z*-axis completes a right-hand Cartesian triad (in this case, roughly a north direction). Thus, relative to the BL flow direction, *X* is streamwise, *Y* is transverse (across the flow, pointing into the magnetosphere), and *Z* is spanwise. Note that for LES we chose the positive *X*-axis in a direction contrary to the GSM *x*-axis orientation used in Section 4. Distances along the three axes are in units of the Earth radius (R_E)

The terrestrial MP is perhaps the most important case of *mixing layer* of space physics. In fluid dynamics, a mixing layer is a region between flows of different velocity, often with different density.

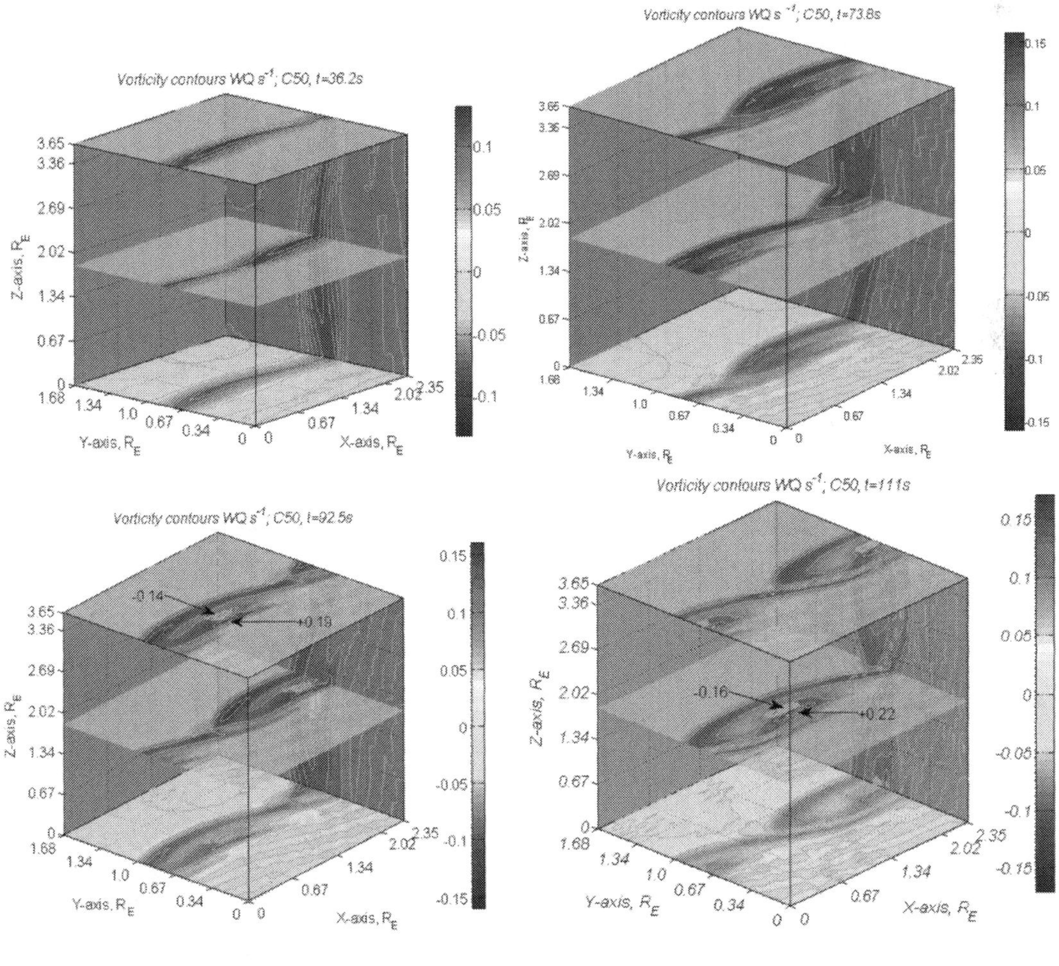

Figure 5. Vorticity rollover for C50, t=36.2, 73.8 (upper panels from left to right), t=92.5, and t=111 (bottom panels from left to right).

In computational fluid dynamics, a *temporal* mixing layer has periodic conditions in the main flow direction (on the box sides' normal to the *X*-axis). The LES of this paper are all of the temporal kind. A difference of our LES with respect to other simulation studies is that we assume periodicity conditions also along the Z-axis, so that perturbations travel freely in a direction that points in the sense of the geomagnetic field.

Unstable temporal mixing layers develop into large vortical patterns, which are called *coherent structures* because they remain for long periods of time, and can be recognized at positions far downstream with approximately similar shapes (see, for instance, [40, 41, 42]).

We compute the non-linear evolution of the KH instability with local BL models that portrait features observed during the Cluster crossing of December 7, 2000 [26, 27]. To investigate the influence of supersonic and superAlfvénic flow regimes on the boundary layer, we then extend the LES to higher magnetosheath speeds, preserving the same configuration of magnetic field, temperature and density boundary values. We also use the LES results to interpret correlations of plasma data measured during the December 7, 2000 event, which reveal the presence of vortical structures that mix magnetosheath and magnetosphere matter.

The computer runs are identified by a number, like C42 that stand for case 42; the number originates from the book-keeping of the physical parameters of particular simulation runs in a large size directory of numerical experiments. The results of only a few computer runs are presented here, and their physical properties are given in Table 1 (at section 3).

3. Computational results: LES dependence on M and M_A

3.1 Vorticity evolution

The initial configuration for the simulations is shown in Figure1a, left panel. The vorticity $\boldsymbol{\omega}$=curl($\boldsymbol{V}$) is represented by the intensity $\omega=|\boldsymbol{\omega}|$, to which the sign of the *Z* component is associated with the formula $\Omega \equiv \omega \times \mathrm{sign}\ (\omega_z)\ \mathrm{s}^{-1}$. This is because ω_z is the main component that determines the orientation of the vorticity vector. The figure shows vorticity contours Ω associated with the initial perturbation, and a set of streamlines starting at the magnetosheath side of the BL. The computer run is C42, t=0, prepared with one perturbation wavelength that fits into the box, and with the estimated magnetosheath speed for the December 7, 2000 event. The wavevector of the perturbation is set at 32° with respect to the *X*-axis. Only one wavelength fits into the computing box (lengths are in Earth radii, R_E); the *Y*-axis scale length is twice that of the *X*, *Z*-axis to facilitate the view of the MP evolution. The

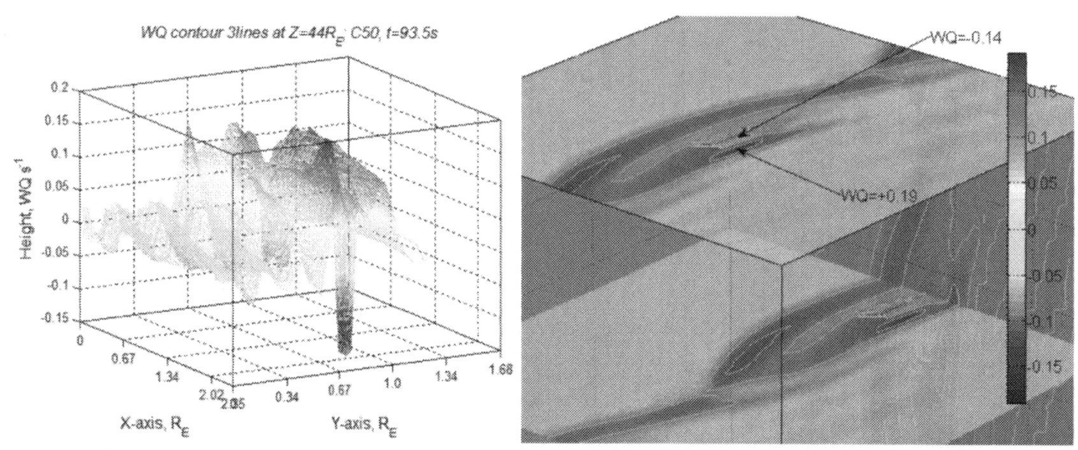

Figure 6. Details of vorticity distribution for C50, t=92.5s. Left panel: vorticity surface on a plane slice *Z*=44 R_E; the height is $\Omega = \omega \times \mathrm{sign}(\omega_z)\ \mathrm{s}^{-1}$, and the colors range from red for large positive Ω, to blue for negative Ω. Right panel: a close-up view of Ω vorticity contours, with a color scale.

sonic and Alfvén Mach numbers are M=0.40 and M_A=0.68 respectively, evaluated on the magnetosheath side. The scenario is moderately subsonic and subAlfvénic. Compressibility effects are present but are not strong, as we shall see. The magnetic field, although not visibly affected in shape, because of the quasi-flute arrangement of the fastest growing mode, is nevertheless a dominant factor and influences the topology of the non linear evolution, fixing the orientation of the vortex structures.

In Figure1 b, we show with Ω vorticity contours the evolution of the perturbed vorticity layer after t=132 s, (roughly ¾ of a turnover time for the vortex). Note the concentration of positive vorticity in a vortex core (with intensity amplified by the 3D stretching), and the strained vorticity layer. Two negative vorticity patches have also developed, and are visible at positions of large strain of the shear layer, one closely associated with the positive vortex core. A set of (blue) streamlines, launched from positions near the external side of the boundary layer, are also represented. We observe that the flow is still laminar, but is becoming increasingly 3D at this stage.

The next Figure 2 shows the development of kinetic helicity, for C42, t=132 s, contours of kinetic helicity density $\boldsymbol{V} \bullet \boldsymbol{\omega}$ are displayed. Figure 2 illustrates the progression from the simple, parallel shear flow of Figure 1a, where $\boldsymbol{V} \bullet \boldsymbol{\omega}$ is initially zero, to a complex stage with a large vortex core and deformations of great magnitude of the vorticity layer. Kinetic helicities of both sign arise from the initial velocity field perturbation.

Table 1 summarizes some elements of the computer runs. The case number is in the first column, the second indicates the wavelengths of the initial mode fitted in the computing box. The third column

Table 1. Properties of the simulation cases.

Case	wavelengths	V_1 (km/s)	M	M_A	×
C42	λ	139	0.40	0.68	1
C53	2λ	139	0.40	0.68	1
C62	3λ	278	0.80	1.36	2
C50	λ	420	1.21	2.05	3
C57	2λ	556	1.60	2.72	4

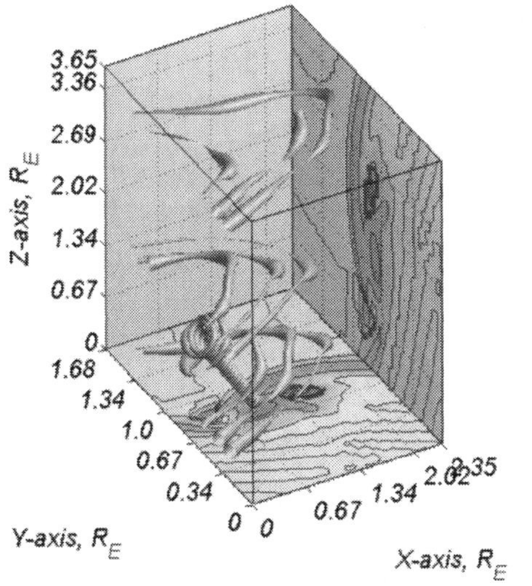

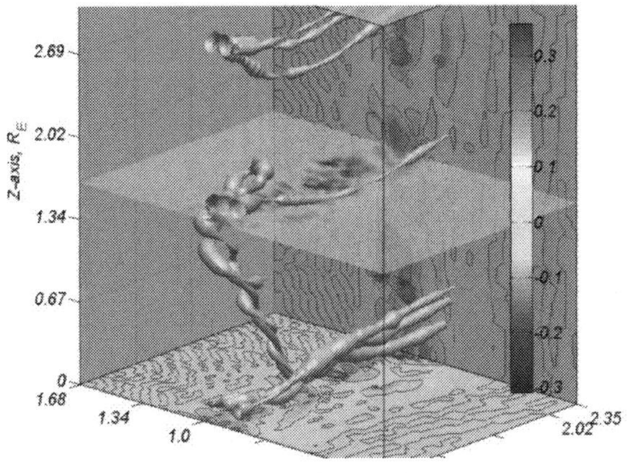

Figure 7 left. Vorticity contours on Z=0, and X=2.35 R_E with stream-tubes, C42, t=132s. The tube section is proportional to div($\boldsymbol{V}$). **Figure 8 right.** Close-up view of vorticity contours and stream-tubes, C50, t=300s; same format as Figure 7

gives V_l the magnetosheath speed in km/s, and the last column indicates the increased speed factor with respect to the reference case C42. The fourth and fifth columns show the sonic, and Alfvén Mach numbers, respectively.

The Mach numbers, sonic M and Alfvén M_A, of the runs characterize the external magnetosheath flow. Since at the other end, the inner edge of the boundary layer, the magnetosphere velocity is approximately zero, the physical state of motion inside the BL is governed by smaller values of the Mach parameters. This is due not only to the decreasing average velocity in the boundary layer, but also to the increase of temperature, and the diminution of density, as we approach the magnetosphere side. Hence, the sonic Mach decreases in the BL by the action of two causes, the reduction of the average flow velocity, and the increment of the speed of sound. Similarly, the Alfvén Mach number decreases by the diminution of velocity and by the growth of the Alfvén speed due to the smaller density near the magnetosphere side (if the magnetic field intensity remains comparable to the magnetosheath value, as in this case). Therefore, only in the runs with the highest speeds, like the cases C50 and C57, the large outer Mach values begin to exert influence on the physics of the BL. The next vorticity plot, Figure 3, is for C53 at t=180 s (roughly one turn over time). The computer box is prepared to host two wavelengths for the initial perturbation. The plot displays vorticity contours, and shows some streamlines that prove the formation of swirling flows. It shows details of the vorticity evolution, with a close-up view of the distorsions of the boundary layer flow. Note the division of the vortex layer, the warping of the original shape, and the rise of counter-rotating centers.

In Figure 4 , that shows vorticity contours for C62 at t=83 s, the magnetosheath speed is 278 km/s, two times larger than the previous cases C42, and C53. The magnetosheath flow is still subsonic M=0.80, albeit not by a wide margin, and is superAlfvénic M_A=1.36. The computing box elongated over the X-axis contains three wavelengths of the initial velocity perturbation. The plot shows the intensification of a row of positive vortices, and their pairing with negative vorticity centers, which are places strongly bent out of shape by the rolling over of the vortex sheet.

In Figures 5 and 6, we examine the process of coupling of vorticity centers of opposite signs at even higher speed. In Figure 5 the rollover of vorticity is for the case C50, in which the magnetosheath speed is 420 km/s (increased three times). It is illustrated by four panels at the times (in seconds) t=36.2, 73.8, 92.5, and t=111 from left to right above, and then again from left to right at the bottom. The external flow is now supersonic, and superAlfvénic, M=1.210, M_A=2.053. The formation of an intense vortex core, and the warping of the initial vorticity sheet are accelerated in this run, and the main features appear earlier. The pairing of large positive and negative vorticity concentrations in small regions is also present, with intensified values. The vortex core is always aligned with the magnetic field direction, as we shall see in other plots.

A closer view on vorticity, for the same case C50 at t=92.5s, is shown in Figure 6. In the left hand side the vorticity distribution on a plane slice, at the position Z=44 R_E of the computing box, is given. The height is $\Omega = \omega \times \mathrm{sign}(\omega_z)$ s^{-1}, and the surface is shaded with colors, red for large positive Ω, down to blue for negative Ω. The positive peak (vorticity amplified by 3D stretching into a strong vortex core) has a close companion, a deep pit of negative vorticity, with a depth in absolute value close to the peak height. The right hand side of Figure 6 shows a close-up of vorticity contours, $\Omega = \omega \times \mathrm{sign}(\omega_z)$ s^{-1}, with a color scale. The close pair of vorticity kernels have Ω=0.19 and Ω=-0.14 s^{-1}.

3.2 Compressibility

We turn now the attention to the effects of compressibility, and the influence of the magnetic field on the mixing layer evolution. Consider again the reference configuration, C42 with 139 km/s as the magnetosheath speed. Figure 7 shows vorticity contours on two sides of the computing box (the bottom Z=0, and the backside X=2.35 R_E) and a set of stream-tubes for t=132s. The tubes are similar to streamlines but with a finite section proportional to div(V). The stream-tubes are launched at three different heights Z from the plane X=0 (the view is from a reference system moving tailward at half the magnetosheath speed). The tube section is normalized with the initial value at X=0, and the growth or diminution depends on the local value of div(V). Therefore, the tube section reveals the relative rate of change of density with time, at each place along the tube. Alternatively, stated with a formula, div(V)=- ($1/\rho$) ($d\rho/dt$), where the time derivative is a Lagrangian quantity, i.e., is computed moving along with the flow. Stream-tubes depict graphically the presence of compressibility effects. The figure shows that the boundary layer is, in fact, subject to compressive activity, evidenced by the significant changes of tube sections. Thus even at a subsonic magnetosheath speed, M=0.40 in this case, the rolling-over dynamics, and 3-D stretching, produce important compressive action in the boundary layer.

However, when we examine the stream-tube captured by the vortex, so to speak, which exhibits a swirling pattern, we note that the tube section remains constant as it coils around the vortex axis. We have examined for subsonic magnetosheath flows several other cases, and different times, omitted for brevity. The pattern of compressibility effects is qualitatively similar to that of Figure 7. Thus, for subsonic flows, we can state that during the rollover compressibility effects in the BL are significant, but do not appear to affect the vortex core.

To contrast, Figure 8 gives results of C50 at t=300s to emphasize differences. The speed of the outer flow is increased three times, 420 km/s, and the magnetosheath is supersonic, and superAlfvénic M=1.21, M_A=2.05. With the same format as Figure 7, the plot shows a close-up of stream-tubes. We observe inside the BL, as before, significant Lagrangian changes of density in action. However, now we also note tube section changes of the stream-tube that coils around a vortex. Thus, at the magnetosheath speed level of C50, compressibility has influence also on the vortex kernel dynamics.

3.3 Magnetic field

The magnetic field lines remain straight in all the cases examined. This is due to the small magnetic shear of the field configuration studied here. The fastest growing mode does not bend the magnetic field lines, the perturbations are flute-like, and maintain a flute pattern in the non linear phase. As time goes by, a slight swaying of the magnetic lines can be observed, similar to a wheat field undulating under a light breeze. Nevertheless, the presence of magnetic field is important as it determines the direction of the vortex axis. It also inhibits the development of streamwise vorticity, which is a common feature of ordinary fluid mixing layers. These are the so-called *hairpin* vortices (also known as longitudinal vortices) of the fluid dynamic literature, e.g., [41].

We examine streamwise vorticity with stream-ribbons, as shown in the example of Figure 9. A

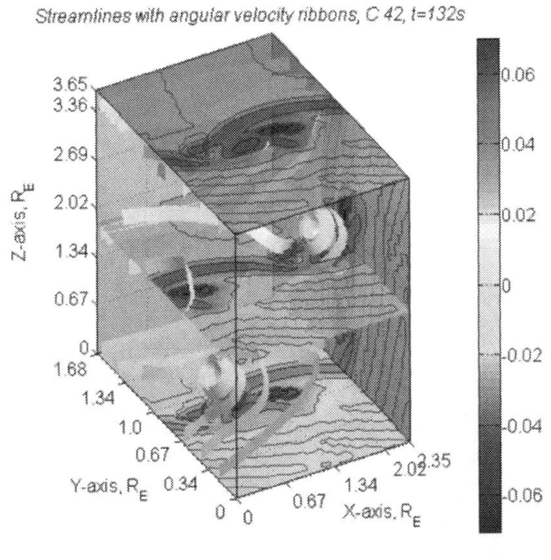

Figure 9. Vorticity contours, and stream-ribbons with twist proportional to |curl(**v**)| for C42, t=132 s.

stream-ribbon is defined like a streamline, with the line replaced by a ribbon of a given width. When there is vorticity along a streamline, the ribbon appears twisted in agreement with the projection of the angular velocity on the line direction. Figure 9 is for C42, t=132 s, it shows vorticity contours, and stream-ribbons where the twist is proportional to |curl(**v**)|. However, the ribbons do not exhibit any worth noting degree of twist, including the ribbon that coils-up around the vortex kernel. We have examined several plots of this kind, always with negative results, reaching the conclusion that the formation of streamwise vortices is absent in our LES. We conjecture that the presence of magnetic field is a hindrance for a process that occurs in ordinary fluids.

Magnetic pressure does not play any noticeable function in the equilibrium of the vortex core in the subsonic cases considered, C42, C53, and C62. However, in supersonic runs, like C57, M=1.60 (fourfold velocity increase) the kernel of largest vorticity becomes also a region of higher magnetic strength. The effect is illustrated in Figure 10 which shows B contours (absolute value of the magnetic field, with a color scale) and a set of magnetic field lines (yellow) for two times. Left panel, t=49s, right panel t=99s. An increase of magnetic field strength in correspondence with vortex loci appears (see also Figure 11), a feature absent in LES at lower speeds.

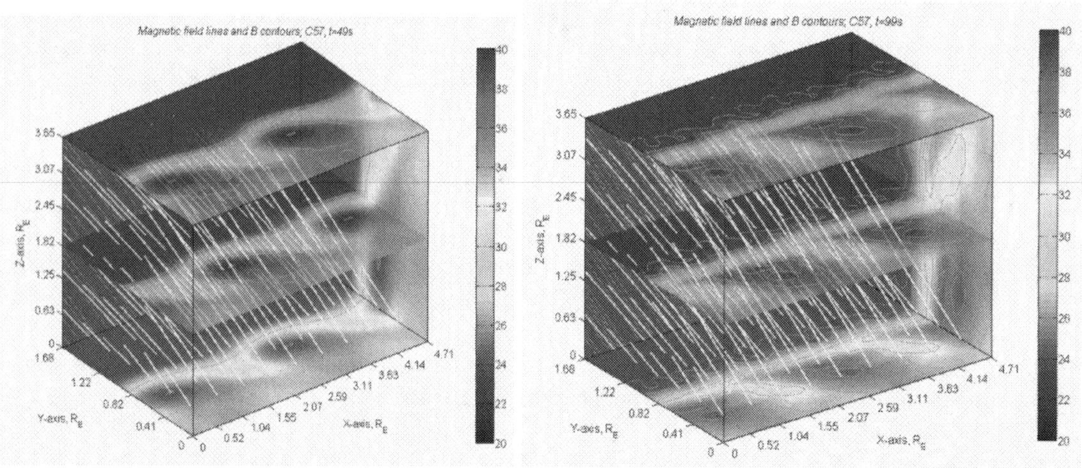

Figure 10. C57, B contours (absolute value of $\boldsymbol{B}$) with a color scale), and magnetic field lines (yellow). Left panel, t=49s; right panel, t=99s.

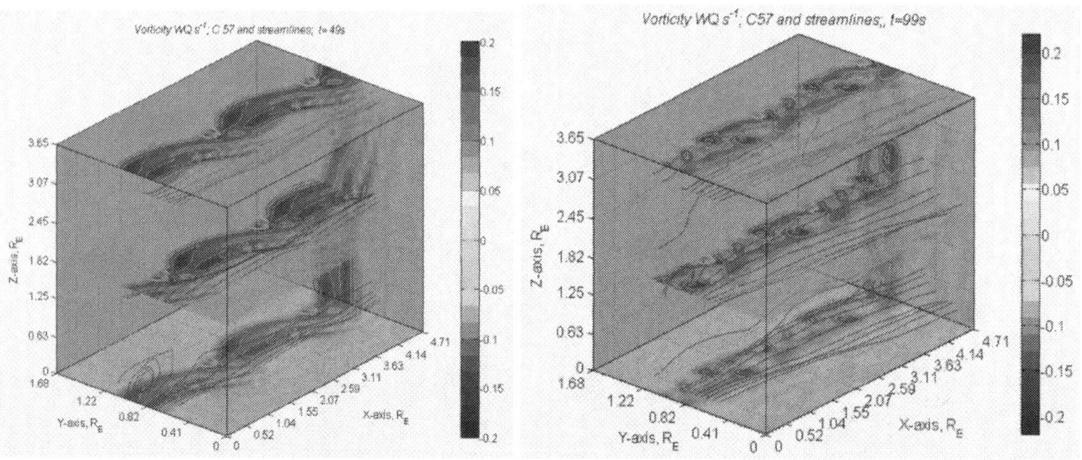

Figure 11. C57 vorticity and streamlines; the left panel is for t=49s, the right panel for t=99s.

Finally, we show, for the same times of Figure 10, the pattern of vorticity contours and some streamlines for C57, the case of greatest speed analyzed. In Figure 11 the left panel is for t=49s, it illustrates the splitting of the vorticity sheet, and the presence of counter-rotating centers. The right panel of Figure 11 is for t=99s, where the pairing of split vortex cores, and the coupling of opposite rotation centers, can be observed. The LES reveals a doubling of positive vortex kernels in the box. A spacecraft in the boundary layer, therefore, should encounter vortices with a twofold increased frequency with respect to the original perturbation.

3.4 Evolution of density and temperature

The vorticity rollover of high speed C50 is examined in Figures 5, and 6. We now consider the mixing of tenuous (closer to magnetosphere values) with dense (nearer to magnetosheath levels) plasma for C50, portrayed in Fig 12 with isodensity surfaces. The time of the run is t=92.5s. Constant (proton) density surfaces for N= 6 (left panel), and N= 11 p cm^{-3} (right panel) are shown. The (blue) velocity cones on the isodensity surfaces indicate the direction of the local velocity field, the size of the cones is proportional to the local speed. The constant density surfaces permit a visualization of vortex tubes, whose inclination is associated with the magnetic field direction. The perspective is rotated around Z, with the Y-axis on the right hand side, for easier visibility of the constant N surfaces.

Details of successive alternations of cold-dense and hot-tenuous portions of matter in the boundary layer can be appreciated in the following two figures. The computer run is C62, with V_1=278km/s, M=0.80 M_A=1.36 (a twofold increase of speed with respect to C42). We plot equal density contours on planar slices at a fixed Z height. Fig 13 shows isodensity contours at two times t=83 s in the upper panel, and t=105 s in bottom right panel. The roll-over is evident, and it is also plain that the density pattern passing over a stationary spacecraft, as the plasma flows downstream, should leave a record of alternate density peaks and valleys, with peculiar shapes that depend on the observation point in the BL, and the evolution time of the vortical structures.

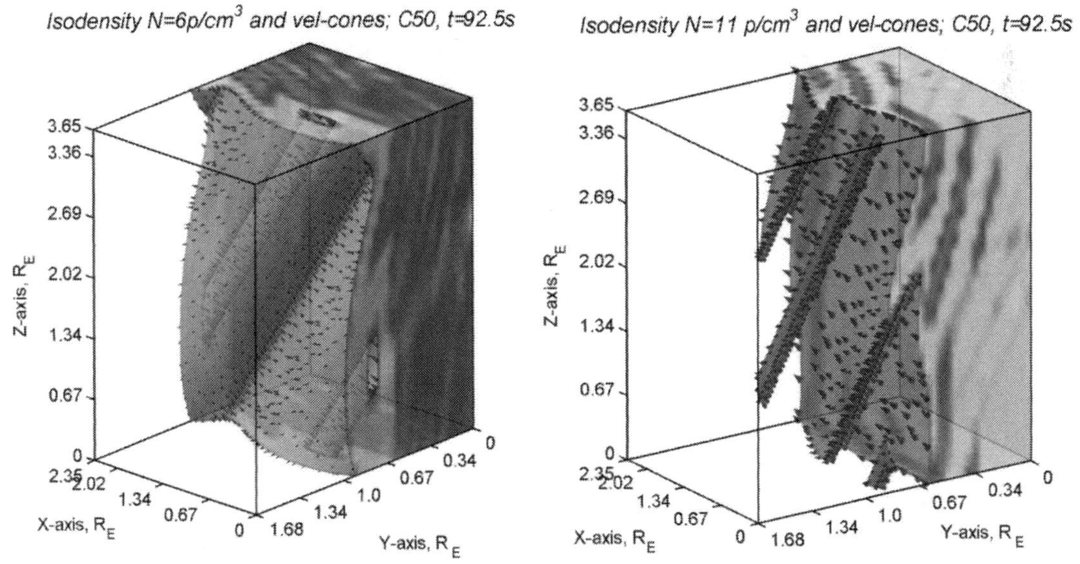

Figure 12. C50, t=92.5s. Constant density surfaces for N= 6, left panel, and N= 11 p cm^{-3}, right panel.

In analogous manner Figure 14 depicts equal temperature contours at a fixed Z plane, at the times 60 s on the left panel, and 105s on the right panel. The alternation of high and low temperatures correlates well with the series of tenuous and dense plasma parts, so that peaks and valleys are always associated in pairs of hot-tenuous and cold-dense sectors. The temperature of C62, at t=105 s, is very large in certain small parts that coincide with positive vortex kernels, so that the color scale of the plot is set with $ln(T)$. The ratio of the hot spot to the magnetosphere temperature is in the 7 – 8 range.

Therefore, ideal MHD predicts the build-up of vortex cores that are also hot spots. At the moment, it is not clear if the highest temperature value is realistic, or if it exceeds the bounds of our numerical resolution. It remains to be confirmed by future, more elaborated dissipative MHD computations the actual extent of the temperature rise predicted by theory in vortex kernels. In any case, our LES indicate that the trend is there, and it poses a challenge to the ingenuity of the experimentalists to find out whether a significant temperature increment, concentrated in small places of the boundary layer, can be detected by spacecraft instruments.

4. Scatter plots of December 7, 2000 data, and their significance

We show plasma data recorded by Cluster 3 on December 7, 2000 in Figure 15. The plot gives the GSM velocity components versus time, UT, in decimal hours (V_x blue, V_y red, and V_z black lines, km/s). Two other lines are for particle density $N \times 10$ cm^{-3}, green, and temperature T/20 eV, magenta. The event consists of three periods of intense boundary layer activity. The time lapses are associated with changes of the IMF clock angle [26, 27]. The first interval (13.975 - 14.167) shows large amplitude oscillations caused by the arrival of a strong interplanetary perturbation. The following (14.167 - 14.467) two time lapses are interpreted as due to strong (14.167-14.333), and moderate Kelvin-Helmholtz activity (14.333 - 14.467), later quenched by an unfavorable clock angle change of

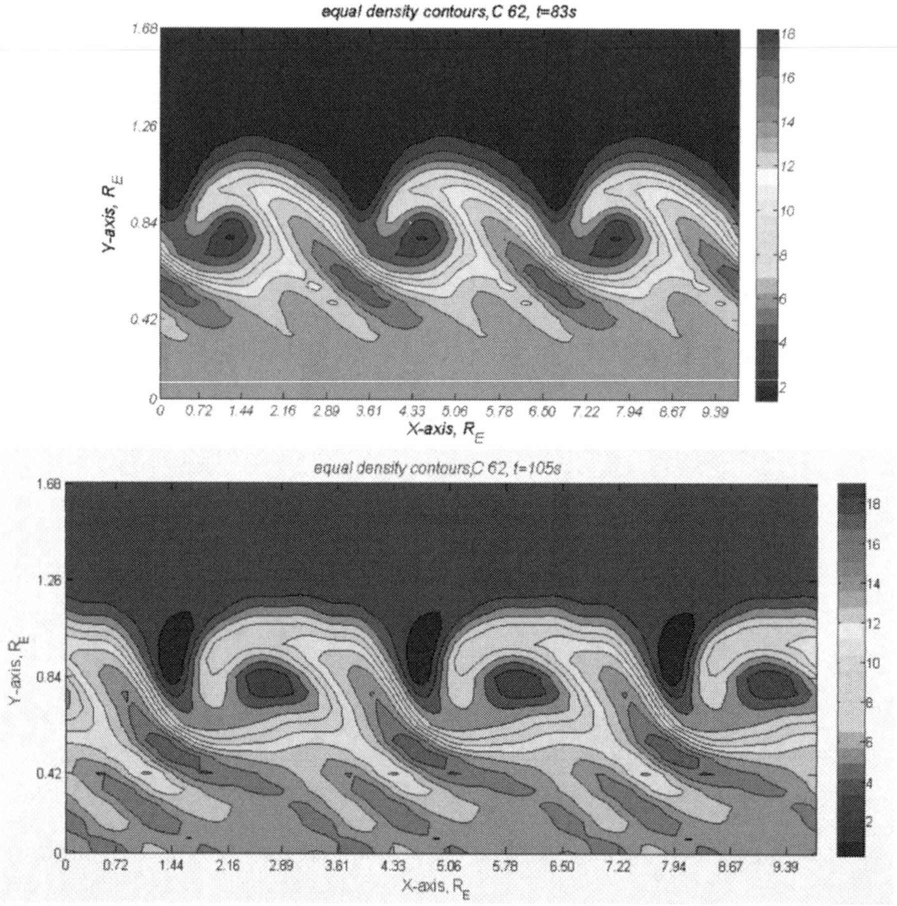

Figure 13. C62, isodensity contours at constant Z; t=83s upper panel, and t=105s bottom panel.

the IMF at the end of the event. The velocity field components (Figure 15) are in the GSM system (Geocentric Solar Magnetospheric [2]), which has its x-axis from Earth to the Sun (with the positive sense contrary to the generic magnetosheath flow direction). The y-axis is perpendicular to the terrestrial magnetic dipole, pointing to dusk, and z is orthogonal to x, y as in right-hand Cartesian coordinates, so that the z-axis is in the same sense as the north magnetic pole.

As can be seen, during the KH interval (14:10-14:28 UT) a principal part of the motion takes place in the x, y GSM plane. We therefore associate the X, Y, plane of the LES with this plane. The average of measured velocity over the period is $\boldsymbol{V}_{av}$ =(-176.7, 85.1, 28.8) km/s. Note that, V_{yav}, the average of V_y data is not zero, because of the flaring of the magnetopause (the increasing span of the magnetosphere in the anti-sunward direction) at the position of Cluster 3, which estimated from the V_{yav}/V_{xav} ratio is about 26° (where V_{xav} is the corresponding V_x average). Thus, the actual y-axis is not normal to the MP, while the simulations assume that Y is normal to a MP model. In the following we work mainly with the x, y velocity components relative to the average, $U_x=V_x-V_{xav}$, $U_y=V_y-V_{yav}$. Hence, we consider the data from a tailward moving frame, with the y-flaring effect subtracted. We analyze scatter plots of data to find vortex signatures, i.e., to recognize patterns that herald the presence of whirling matter in the BL. The analysis is on a similar line as [21].

Figure 16 is a scatter plot where the x velocity component U_x, and density N of each data define a point in the U_x-N plane, marked by a circlet. The radius of the small circle is proportional to the

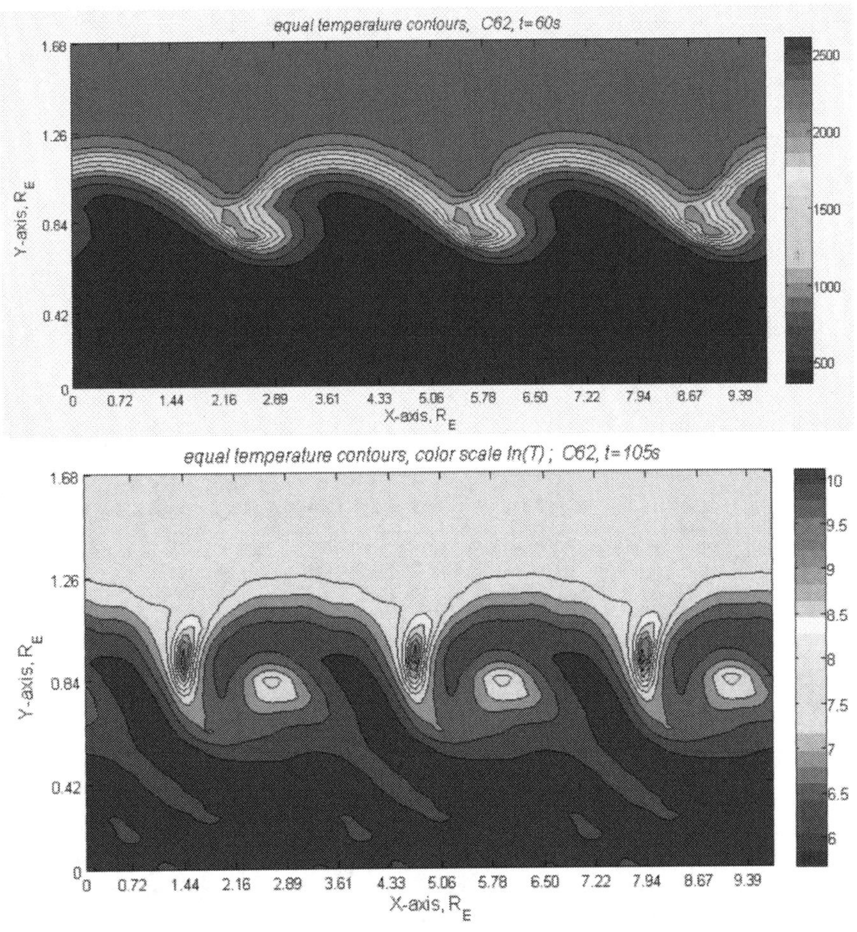

Figure 14. C62, equal temperature contours at constant Z; t=60 s upper panel, and t=105s bottom panel with a color scale set with $ln(T)$.

temperature of the data (larger size corresponds to higher temperature). In addition, to help visibility, we set the color of the marker proportional to $ln(T)$ (red for hot, and blue for cold matter; the logarithm scale is used because of the large range of temperatures).

A linear correlation exists, the majority of low-density plasma moves tailward, and the greater part of high density matter moves sunward. Some tenuous-hot data have very large anti-sunward speeds, and conversely some dense-cold plasma is moving very fast toward the sun.

A scatter plot of temperature, T, versus density, N, is shown in Figure 17. Here the size of the circlets is proportional to $| U_x |$, the absolute value of U_x, i.e., the tailward speed above the average. The color indicates the sign of U_x, red is for sunward and blue for tailward motion. The plot complements the correlation of Figure 16, showing at a glance that a large majority of dense plasma is in motion toward the sun, while on the other hand the greater part of tenuous matter moves anti-sunward. The size shows that there are fast members in both populations. There is a trend from higher to lower temperatures as we shift from lower to higher densities.

Figure 18 is a scatter plot of December 7, 2000, Cluster 3 plasma data. It shows U_y versus U_x (GSM velocity components with average subtracted) where the circlets size is proportional to N^2 (to enhance the tenuous - dense difference), and the color is proportional to $ln(T)$ (red for high temperatures). The time interval is the period of KH activity, 14.167 - 14.467 UT decimal hours (14:20-14:28 UT), when the IMF turns north. The motion is relative to the average of the observed plasma velocity.

The plot shows a linear correlation between $U_x=V_x-V_{xav}$ and $U_y=V_y-V_{yav}$, such that high speed, sunward motions, are predominantly inward oriented (and for the greater part associated with cold-dense plasma). Conversely, fast tailward motions are mainly outward directed (and correspond predominantly to hot-tenuous matter).

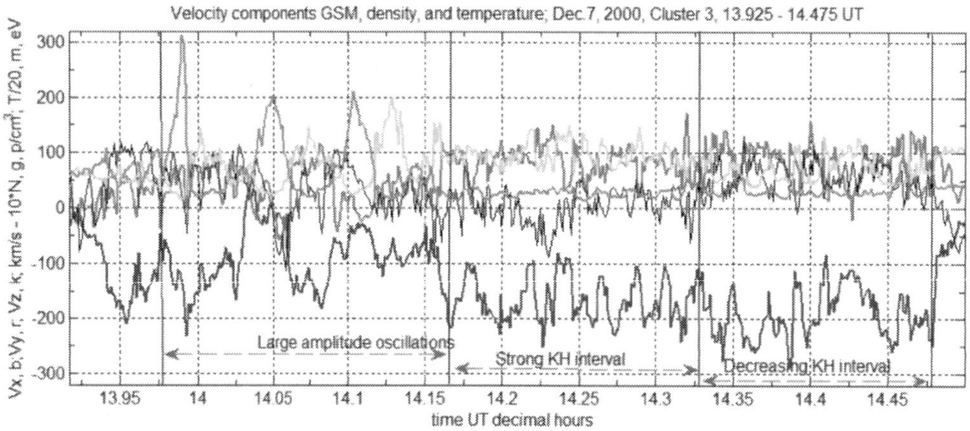

Figure 15. Data recorded by Cluster 3 on December 7, 2000. GSM velocity components versus time, UT in decimal hours (V_x blue, V_y red, and V_z black lines, km/s). The two other lines are for particle density N (scale 10 cm^{-3}), green, and temperature $T/20$ eV, magenta.

The data average of the velocity is the bulk speed of the plasma that passes over the position of the stationary spacecraft (the orbital speed of the device is insignificant compared with that of the plasma). The average is over the KH interval. We have shown with the scatter plots of the BL records, that high temperature, and low-density plasma patches, characteristic of inner layers (close to the magnetosphere) are moving tailward with velocities much higher than the average speed, with increments above average of 50-55 %. On the other hand, cold and high-density plasma, typical of outer strata (close to the magnetosheath), are observed to move sunward, with speeds about 50% higher than the interval average.

Of course, there is also a large set of data values, with intermediate densities and temperatures, which are moving with velocities close to the bulk plasma speed. However, it is striking and we emphasize the fact, that large velocity differences with respect to the average plasma motion, both in sunward and in tailward directions, are observed during the time lapse associated with the KH instability (due to the negligible magnetic shear). Moreover, we call the attention on the evidence that cold-dense plasma patches are moving sunward, against the magnetosheath flow, with speeds substantially larger than the recorded average, while fractions of tenuous-hot matter move tailward, also with velocities well in excess of the observed average.

Furthermore, because of the U_x-U_y linear correlation, the tailward accelerated hot-tenuous matter moves also predominantly outward (toward the magnetosheath). Conversely, the high speed cold-dense plasma moving sunward also moves mostly inward (toward the magnetosphere). Thus, the scatter plots strongly suggest a composition of plasma motions, i.e., a rotational motion plus a bulk

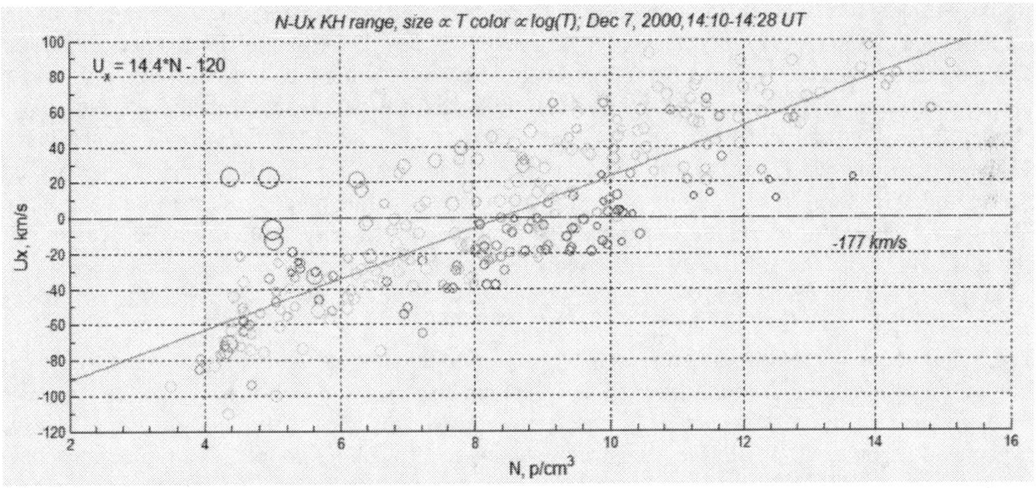

Figure 16. Scatter plot of December 7, 2000 U_x, versus N (data of the KH period). The size of the circlets is proportional to T, the color proportional to $ln(T)$ (see text).

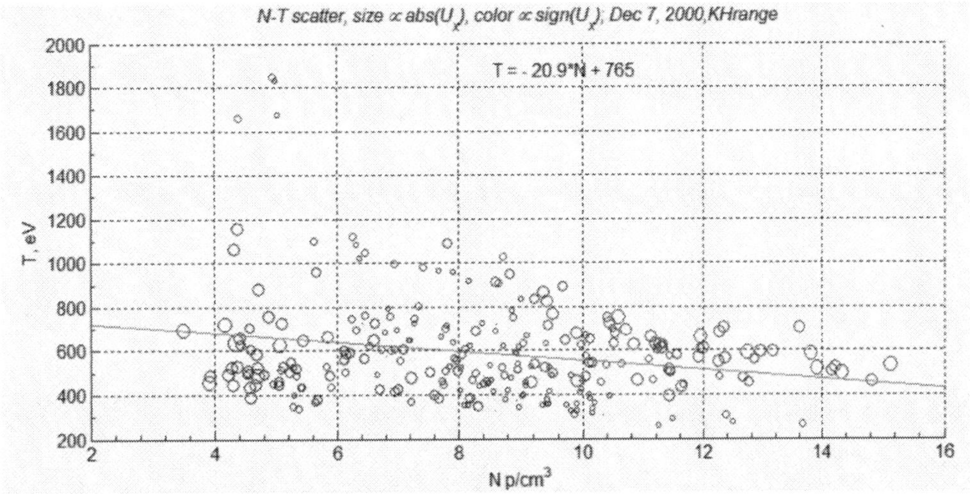

Figure 17. Scatter plot of T versus N. The size of the circlets is proportional to | U_x |, the color indicates the sign of U_x (see text).

tailward translation, together with a rolling-over of plasma strata of different density and temperature. In other words, these are the trends qualitatively described by the LES, in subsection 3.4 for example.

Hence, we consider the set of correlations as embodying a criterion that allows, from records of one spacecraft alone, to decide the presence of vortical motions in the BL(on this see also [21]), which therefore becomes an effective mixing layer of inner and outer plasmas.

5. Discussion, summary, and conclusions

It is generally accepted that on time scales much longer than the ion cyclotron gyro-period it is possible to capture the salient macroscopic aspects of the plasma with a theory much more manageable than the kinetic theory of plasmas, which is certainly more fundamental but entails a prohibitive complexity when applied to large-scale phenomena of the interplanetary space.

The theory is MHD, in which the plasma is portrayed as a conducting fluid. In view of the scarce or negligible efficacy of classical Coulomb collisions in interplanetary plasmas, at first sight the use of a fluid model raises some perplexity. However, in space plasmas the relaxation processes are governed by the ubiquitous fluctuations of the electromagnetic field, which produce anomalous transport properties. Particles collide with the fluctuating fields rather than with themselves. The transport problem falls in the realm of plasma turbulence theory, which is the subject of many analytic and numerical investigations. However, while some kinetic theories have attained a highly sophisticated level, the understanding of what happens at the plasma microscale is still modest to be of direct help for large-scale observations. These, on the other hand, are inherently limited by the availability of spacecrafts, and instruments. In practice, the scenarios are so many and so various, that the analysis of many large-scale phenomena, and even some of the meso-scale, are done perforce with macroscopic models.

The MHD equations used in this research are substantially simpler than those of statistical physics. In the decade of 1960, E. N. Parker argued and explained the plausibility of MHD for space physics. Since then, judiciously employed MHD, has been a powerful help in the study of the solar wind dynamics and its interaction with the terrestrial magnetosphere [2]. Of course, due to the non linearity of MHD, mathematical obstacles remain, and due to the manifestations of the subjacent fluctuations, also physical difficulties persist. Nevertheless, in spite of its limitations, the usefulness of MHD for data analysis and interpretation is undeniable. In space physics MHD has not been exhausted, nor superseded.

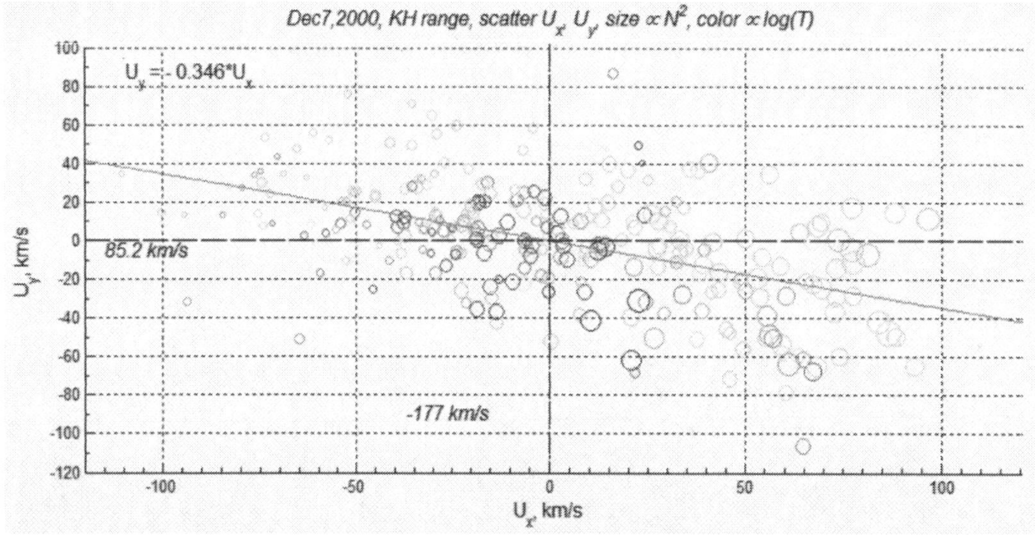

Figure 18. Scatter plot of December 7, 2000 data (KH period) U_y versus U_x. The circlets size is proportional to N^2, and the color is proportional to $ln(T)$.

Moreover, in contrast with what is current in ordinary fluid dynamics, we should recall that the validation of computational and theoretical models in space physics is necessarily indirect and limited. It is often a consequence of the consistent power of the models to explain a variety of different phenomena. Their compatibility with other trusted models or in few special cases their confirmation by a more advanced theory.

The winding-up of the vorticity layer produces large eddies. The 3D vortex stretching intensifies further these structures, as we have seen (also in [1]). The Reynolds number of the large-scale dynamics is given by $R_e = lv/v_{ef}$, where l denotes the vortex diameter, v indicates the rotation speed of the eddy, and v_{ef} is the effective diffusion coefficient for angular momentum transfer. When the rollover time of the vortex $\tau=l/v$ is much smaller than the diffusion time, $t_d=l^2/v_{ef}$, the Reynolds number is large, $R_e = t_d/\tau >> 1$, a common case in space plasmas. Thus, a range of large-scale lengths can be studied with ideal MHD (ignoring dissipative effects). Dissipative processes become important at much smaller lengths, in the dissipation range where the energy of the eddies finally decays.

Under this condition, the vorticity flux is an invariant of non-stratified fluids, and vortex amplification by 3D stretching then operates with full efficiency. However, although the large-scale dynamics of the eddies can be studied during a few turnover times neglecting dissipation, an advise of caution is in order. In MHD mixing layers the vorticity flux may change with time by causes other than dissipation. The Lorentz force, $\boldsymbol{F}_L=\boldsymbol{j}\times\boldsymbol{B}/c$, is in general not ***curl*** free, and $-\nabla\rho\times\boldsymbol{F}_L/\rho^2$, the associated Lorentz baroclinic term, is not zero either. Moreover, the MP plasma is stratified in a way that the entropy across the BL is not uniform. Thus, in the BL even the ordinary baroclinic term $-\nabla\rho \times \nabla p/\rho^2$ [40] is active. All these causes may affect the invariance of the vorticity flux [1], but it is not clear yet the time scale for the action of these effects in the MP context. In any case, the stretching of vorticity lines occurs in our study, as confirmed by the LES results,

The temporal mixing layer approach is useful and practical, but it has shortcomings. After some periodical returns of the developing vortex into the computing box, the plasma of the actual boundary layer has traveled a distance of many R_E downstream, where the physical conditions in the magnetosheath, and the magnetosphere may have changed, but not so in the LES boundary conditions. Moreover, in the simulations the external magnetosheath flow is assumed laminar, while the solar wind is instead in a fluctuating state, or it is often turbulent. Therefore, the LES presented here are aimed to reveal trends, to foresee global features of the system, but cannot be intended as faithful portraits of observed events.

The paper presents new computational physics, data analysis, and interpretation, associated with a general patterns of fields, and flow, suggested by records of December 7, 2000. This event is taken as representative of a set of low latitude, near flank MPs, with very small magnetic shear under northward pointing IMFs. The near absence of shear of the magnetic field and the flute-like variation of the magnetic lines, during the rollover of the vorticity layer, are characteristic features of the configuration studied. It is the main trait that distinguishes this work from other studies that examine different configurations [12, 13, 14, 15, 16, 17, 18, 19, 20, 21]. The process merely flutes the magnetic field, although the latter determines the direction of the vortex axis [1, 26, 27] and leaves its stamp on the eddies. In another aspect our simulations differ from other investigations: we use a localized MP model with two free boundaries. We have two periodicities: one along the flow, the X-axis, and the other normal to it, the Z-axis that extends in the sense of the terrestrial magnetic field. The dynamics of the perturbations is free also along the Z-axis. The evolution is not restrained by "*ad hoc*" conditions on Z-boundary planes.

The research is a follow-up of [1, 26, 27, 29] that produced original results studying the consequences of increasing the flow speed of the basic configuration .We present new computational plots, and also an analysis of data correlations. The following is a summary of the facts and results obtained, in which we examined the influence of increasing M and M_A in all the points from (a) to (h).

a. Deformation of the vorticity sheet and vortex division; stretching, amplification, and frequency doubling.

b. Generation of swirling flows (initially absent).

c. Development of counter-rotating centers, with associations of positive – negative vorticity kernels.
d. Presence of compressibility effects in the boundary layer, even for subsonic flows. In particular, the vortex core is affected in supersonic but not in subsonic flows.
e. Magnetic "flutes" characterize the configuration; the vortex axis is along the magnetic field; LES show only moderate swaying of the magnetic lines.
f. Evidence of magnetic strength enhancement around vortex kernels in supersonic flows, which is negligible at lower speeds.
g. Mixing of magnetospheric and magnetosheath plasma; alternation of high-low density and low-high temperature strata.
h. Formation of plasma hot spots, i.e., high temperature centers of low density.

In addition, guided by results (f-g) as basic interpretive concepts, we examined scatter plots of velocity, density, and temperature of the reference event data to find signatures of the presence of vortices. Hence, we tested:

i. A criterion to decide the presence of whirling plasma in the BL using records of one spacecraft alone.

To conclude, a brief account of the themes studied in this work is as follows.

1. Large eddy simulations in magnetohydrodinamics set in a counterpoint dialog with the available experimental evidence
2. Understand via computational MHD the physics of stratified and velocity sheared BLs controlled by magnetic fields with negligible change of direction across the MP.
3. Computer simulations of MHD mixing layers, oriented to find trends and to derive guidelines that may help the interpretive effort of MP observations.
4. Analysis of BL data directed to prove the presence of vortical motions with plasma mixing in particular events.

Acknowledgements
Work supported by CONICET grant 11220090100608 PIP 2010 - 2012 ("*La Magnetopausa en Campo Magnético Interplanetario Norte: Análisis de Datos, Teoría y Simulación de Grandes Vórtices*"). Work at University of New Hampshire was supported by NASA grant NNX08AD11G.

References

[1] Gratton F T, Bilbao L E, Farrugia C J and Gnavi G 2009 *JOP Conf. Series* **166** 012023 doi:10.1088/1742-6596/166/1/012023
[2] Russel C T and Kivelson M G eds 1995 *Introduction to Space Physics* Cambridge Un. Press, New York
[3] Fairfield D H, Lepping R P, Hones E W, Bame S J and Asbridge J R 1981 *J. Geophys. Res.* **86** 1396
[4] Terasawa T *et al.* 1997 *Geophys. Res. Lett.* **24** 935
Fujimoto M and Terasawa T 1994, *J. Geophys. Res.* **99** 8601
[5] Fujimoto M, Mukai T and Kokubun S 2002 *Adv. Space Res.* **30** 2279
[6] Øieroset M et al. 2005 *Geophys. Res. Lett.* **32**, L12S07, doi:10.1029/2004GL021523
[7] Fujimoto M *et al.* 1998 *J. Geophys. Res.* **103** 4391
[8] Cowley S W H 1983 in *High Latitude Space Plasma Physics*, ed. Hudqvist B and Hagfors T pp. 225 Plenum New York
[9] Song P and Russell C T 1992 *J. Geophys. Res.* **97** 1411
[10] Raeder J, Walker R J and Ashour-Abdalla M 1995 *Geophys. Res. Lett.* **22** 349
Raeder J et al. 1997 *Geophys. Res. Lett.* **24** 951
[11] Sandholt P E, Farrugia C J, Cowley S W H *et al.* 1999 *Geophys. Res. Lett.* **26** 2833
[12] Otto A and Fairfield D H 2000 *J. Geophys. Res.* **105** 21175
[13] Fairfield D H et al. 2000 *J. Geophys. Res.* **105** 21159
[14] Farrugia C J, Gratton F T and Torbert R B 2001 *Space Sci. Rev.* **95** 443

[15] Nykyri K and Otto A 2001 *Geophys. Res. Lett.* **28** 3565
[16] Otto A and K Nykyri K 2003 *Geophys. Monogr.* **133** 53 eds. Newell P T and Onsager T AGU Washington D.C.
[17] Nakamura T K M, Fujimoto M and Otto A 2006 *Geophys. Res. Lett.* **33** L14106 doi:10.1029/2006GL026318
[18] Nakamura T K M, Fujimoto M and Otto A 2008 *J. Geophys. Res.* **113** A09204 doi:10.1029/2007JA012803
[19] Smets R, Delcourt D, Chanteur G and Moore T E 2002 Ann. Geophysicae **20** 757
[20] Hasegawa H, Fujimoto M, Phan T et al. 2004 *Nature* **430**, 755
[21] Hasegawa H, Fujimoto M, Takagi K et al 2006 *J. Geophys. Res.* **111** A09203 doi:10.1029/2006JA011728
Takagi, K., C. Hashimoto, H. Hasegawa, M. Fujimoto, and R. TanDokoro 2006, *J. Geophys. Res., 111*, A08202, doi:10.1029/2006JA011631.
[22] Hasegawa H, Retinò A, Vaivads A et al 2009 *J. Geophys. Res.* **114** A12207 doi: 10.1029/2009JA014042
[23] Gratton F T, Farrugia C J, Bilbao L E, Gnavi G and Lund E 2006 *Am. Inst. of Physics* **C. P. 875** Plasma and Fusion Science pp.300 - 303 ed. Herrera Vázquez J J E Washington DC
[24] Gratton F T, Bender L, Farrugia C J and Gnavi G, 2004 *J. Geophys. Res.*, **109**, A04211 doi:10.1029/2003JA010146
Gratton F T, Gnavi G, Farrugia C J, and Bender L, 2004 *Brazilian J. of Physics*, **34**, 1804
[25] Le G and Russell C T 1994 *Geophys. Res. Lett.* **21** 2451
[26] Farrugia C J, Gratton F T, Lund E J *et al* 2008 *J. Geophys. Res.* **113** A03208 doi:10.1029/2007JA012800
[27] Farrugia C J and Gratton F T 2009 *J. Atmos. Solar-Terr. Phys.* doi:10.1016/j.jastp.2009.10.008
[28] Foullon C, Farrugia C J, Fazakerley A N *et al* 2008 *J. Geophys. Res.***113** A11203 doi:10.1029/2008JA013175.
Foullon C, Farrugia C J, Fazakerley A N *et al* 2010 *J. Geophys. Res.***115** A09203 doi:10.1029/2009JA015189
[29] Gnavi G, Gratton F T, Farrugia C J and Bilbao L E 2009 *J. of Physics* **C.S. 166**, 012022 doi:10.1088/1742-6596/166/1/012023
[30] Farrugia C J, Erkaev N, Torbert R et al. 2010 J. Geophys. Res. **115** A08227 doi:10.1029/2009JA015128
[31] Miura A 1984 *J. Geophys. Res.* **89**, 801
[32] Belmont G and Chanteur G 1989 *Physica Scriptae*, **124** 124
[33] Kivelson M G and Chen S H 1995 in *Physics of the Magnetopause*, Geophys. Monogr. Ser. **90** 257 eds. Song et al AGU Washington, D. C.
[34] Miura A 1995 in *Physics of the Magnetopause*, Geophys. Monogr. Ser. **90**, 285 eds. Song, et al. AGU, Washington DC
[35] Bilbao L E 2006 *Am. Inst. of Physics* **C. P. 875** *Plasma and Fusion Science* pp.467 - 472 ed. Herrera Vázquez J J E Washington DC.
[36] Bilbao L E 2009 *J. of Physics*, **C.S. 166** 012020 doi:10.1088/1742-6596/166/1/012020
[37] Bilbao L E 1990 *J. Computational Physics*, **91**, 361
[38] Bilbao L E and Linhart G 1996 *Plasma Physics Reports* (Fizika Plazmy) **22**, 457
[39] Linhart G and Bilbao L E 2002, *IEEE Trans. Plasma Science*, **30** pp. 460-467
[40] Tritton D J 1988 *Physical Fluid Dynamics*, Clarendon Press, Oxford
[41] Lesieur M 1990 *Turbulence in Fluids*, Kluwer Academic Pub., Dordrecht
[42] Lesieur M and Métais O 1996 *Annual Rev. Fluid Mechanics*, **28** 45.

FLUIDOS 2010: XI Meeting on Recent Advances in the Physics of Fluids and their Applications IOP Publishing
Journal of Physics: Conference Series **296** (2011) 012007 doi:10.1088/1742-6596/296/1/012007

Aspects of the influence of an oscillating mini-flap upon the near wake of an airfoil NACA 4412

J. S. Delnero[1,2]; J. Marañón Di Leo[1,2]; J. Colman[1]; M. García Sainz[1,3]; F. Muñoz[1,3]; N.Hérouard[1,4]; M.E.Camocardi[1,2]

[1]LaCLyFA, Departamento Aeronáutica, Facultad de Ingeniería, Universidad Nacional de La Plata, calle 116 entre 47 y 48, CP 1900, La Plata, Argentina

[2]Conicet, Avda. Rivadavia 1917, CP C1033AAJ, Cdad. de Buenos Aires, Argentina.

[3]Comisión de Investigaciones Científicas de la Pcia. de Bs. As., 526 y 10, CP 1899, La Plata.

[4]Ecole Nationale Supérièure de Mécanique et d´Aérothecnique, Poitiers, France.

E-mail: delnero@ing.unlp.edu.ar

Abstract. A NACA 4412 airfoil was tested, in a boundary layer wind tunnel, with the aim to study the effect of a Gurney mini-flap, as an active and passive flow control device submitted to a turbulent flow field. The main objective was the experimental determination of flow pattern characteristics downstream the airfoil in the near wake. The untwisted wing model used for the experiments had 80cm wingspan and 50cm chord, with airfoil NACA 4412. The mini-flap was located on the lower surface at a distance, from the trailing edge, of 8%c (c airfoil chord). The Reynolds number, based upon the wing chord and the mean free stream velocity was 326,000 and 489,000. The turbulence intensity was 1.8%. The model was located into the wind tunnel between two panels, in order to assure a close approximation to two-dimensional flow over the model. As an active control device a rotating mini-flaps, geared by an electro-mechanical system (which rotate to a 30°) was constructed. The wake pattern and pressure values near the trailing edge were measured. The results obtained, for this mechanism, show us that the oscillating mini-flap change the wake flow pattern, alleviating the near wake turbulence and enhancing the vortex pair near the trailing edge at the mini-flap level and below that level, magnifying the effect described first by Liebeck [1]. That effect grows with the oscillating frequency. Additionally, the wake alleviation probably affects also the far wake. All of these facts suggest us to continue with the experiments, trying to measure the pressure distribution around the airfoil in all the cases, obtaining the lift and drag characteristics.

1. Introduction

The trailing edge camber and its geometry together with its influence upon Kutta condition could be used to the lift, pressure drag and wake management. Researchers analyzed various passive and active flow control mechanisms in order to enhance lift coefficient, in particular, during landing approach and also take off. One of those mechanisms is the mini-flap Gurney, which is a small width plate located on the lower surface of the airfoil, as close as possible to the trailing edge, along wingspan. The goal is to achieve a drift of the rear stagnation point (Kutta condition) with the subsequent circulation enhance and, so, achieving a lift enhance. One of the pioneer researchers was Liebeck [1], whom worked with a symmetrical Newman airfoil with a 1.25%c height Gurney flap, being c the airfoil chord. He founded an important lift increase, associated to circulation enhance due to the downwash increment in the near wake. Further, other researchers performed several studies [2,14], which corroborated that, as other trailing edge passive flow control devices, Gurney flaps (see Figure 2) enhanced the lift, the maximum lift coefficient and the C_l versus angle of attack slope, with little drag and stall angle changes. Moreover, they found that the best height of such devices was the local boundary layer thickness on the lower surface near the trailing edge. Also they found that the near wake was a von Kármàn type wake [11]. There weren´t observed coherent structures in the wake of an airfoil without Gurney flap.

Vortex shedding increment suction on the upper surface near the trailing edge and, at the same time, decelerate the flow on the lower surface with the subsequent pressure increment on that side, near the trailing edge. These pressure differences near the trailing edge enhance the overall circulation around the airfoil and, hence, the lift. A Gurney flap on the lower surface near the trailing edge will delay the stall, promoting grows of the maximum lift coefficient. For that reason many designers considered the use of those mini-flaps Gurney type combined with other flow control devices in order to achieve less complex high lift devices than the classical ones, without performance loss during takeoff and landing. If we observe the wake, in particular, the near wake, we could conclude that a circulation grow is associated with a downwash increment. This requires more asymmetry of the near wake.

From our point of view, it´s not realistic the conception of a classical physical model of the wake flow, as a symmetrical one like a von Kármàn street, because due its symmetry there will not be a net downwash associated with a lift enhance. Moreover, in the near wake region of an airfoil with a min-flap on the lower surface, under lift conditions, the intensity of vortex shed from the upper surface will be different than the one shed from the lower surface. This asymmetry will be responsible of the net extra downwash and, hence, of the lift enhance. Experiments performed by the authors in a previous work [15], showed that in the near wake region, where the vortex system begins its formation, there are an important frequency peak in the instantaneous velocities spectra, generating the so called *wake instability*. Nevertheless, the major part of the works referring the study of flaps Gurney, as passive flow control devices, was made in conditions of laminar or almost laminar free stream flow.

Another authors [10, 15] proposed the use of micro tabs capable to move, to make active flow control upon wind turbines rotor blades. Another author [16], suggested the use of active mini-flaps Gurney type to diminish the wake development and intensity. Tang et al [17] analyzed the aerodynamic behavior of a NACA 0012 with an oscillating Gurney flap, reaching maximum lift and pitching-moment coefficients enhancing, which increase with the oscillating frequency. But all of the cited authors and, in general, the research in such fields are made under a laminar or almost laminar free stream.

So, we were motivated to analyze the use of mini-flaps as active flow control devices, in order to modulate shear layers which generate a circulation grow and, subsequently, a lift enhancement, in the field of turbulent low Reynolds aerodynamics that is, for Reynolds numbers below 10^6, based the number upon the mean free upstream velocity and the airfoil chord, being turbulent the free stream. This is because an important number of aerodynamics problems occur in the low atmospheric boundary layer, which is essentially turbulent under windy conditions (airplanes during takeoff and landings; small size unmanned aerial vehicles; wind turbines; etc).

In this first work, the authors proceed to study in the boundary layer wind tunnel, the near wake of a wing model with a NACA 4412 airfoil, located horizontally inside the test section, between parallel plates in order to have 2D flow over it. The model had a mini-flap of 1.5%c (being "c" the model chord), located along wingspan at 8%c from the trailing edge on the lower surface, capable to oscillate around its axis with variable frequency. The model, also, have 35 pressure taps around its mid-span, one on the leading edge and the other per pairs at the same streamwise position along chord, over the upper and lower surfaces respectively.

Under such circumstances we carried out the experiments with the model inside the wind tunnel test section, looking to qualify and quantify the near wake characteristics and, simultaneously, taking pressure data.

We look forward to finding a relation between the near wake characteristics and the flow circulation which - in fact - will be influenced by the wake.

2. Experimental procedure

The experiments were carried out in the closed circuit boundary layer wind tunnel at the Boundary Layer & Environmental Fluid Dynamics Laboratory, Aeronautical Department, Engineering Faculty, National University of La Plata (Argentina). Test section is 1.4m width, 1m height and 7.5m length. The untwisted wing model, with a NACA 4412 airfoil, had 80cm wingspan, 50cm chord ("c") and a 1.6%c (8 mm) height mini-flap located at 8%c from the trailing edge at the lower surface, along wingspan. The mini-flap was capable to oscillate around its axis, being possible to vary the oscillating frequency.

The wing model was mounted between two double lateral plates, which trailing edges were capable to manually adjust along each axis, in order to have a favorable (or almost zero) pressure gradient along the test section, with the thinnest possible boundary layer (see Figure 1). The aim is to have not only 2D flow over the model, but also little difference between the geometric and the aerodynamic chords.

Figure 1. Front view of the wing model at the test section. We could see the anemometer´s probe to check the mean free stream velocity and the wall panels at each side of the model.

In the experiments we measure the instantaneous velocity in the near wake region, in three "x" positions, 1h, 2h and 5h (Positions 1, 2 and 3 respectively, being h the mini-flap height) and, at each "x" position, the data were measured in 20 vertical points, which upper limit was 1.5%c and lower limit 2%c, separated each 0.04%c. Turbulence intensity was 1.8%. Experiments were performed for two Reynolds numbers, 326000 and 489000, based upon the mean free stream velocity (at 1.5m ahead the model at its height) and the model chord, corresponding to values of 10m/s and 15m/s respectively.

Experiments were carried out in three steps, for each Reynolds number: 1st step with the clean model; 2nd step with the mini-flap deployed but fixed (as a passive flow control device); 3rd step with the mini-flap oscillating at three frequencies each time (22Hz, 38Hz and 44Hz). In all steps, for four values of the angle of attack: -3^0, 0^0, 5^0 and 11^0 (this last near stall angle). We measured instantaneous velocities in the points cited above, by means of a constant temperature hot wire anemometer Dantec Streamline, with X-wire sensor probes. The acquisition frequency was 4000 Hz, filtered at 1000Hz, taken 8192 samples per channel in each measuring point. The mini-flaps frequencies were measured with a laser tachometer. The wing model had two pressure taps at the same "x" position (0.88c), one on the upper surface and the other on the lower surface. The connection between the pressure taps and the Pressure System (NetScanner, equipped with piezoelectric sensors) was by tubes of 1.8mm inside diameter, each of the same length. Such pressure taps were located upstream of the perturbation zone (mini-flap position), with the aim to obtain instantaneous pressure data not perturbed, directly, by the mini-flap position itself.

Figure 2 shows a schema of the wake measuring positions, along "x" axis and the corresponding vertical points, indicating in this Figure only the 0 and -10 points the separation between vertical adjacent points is 0.25H (2 mm).

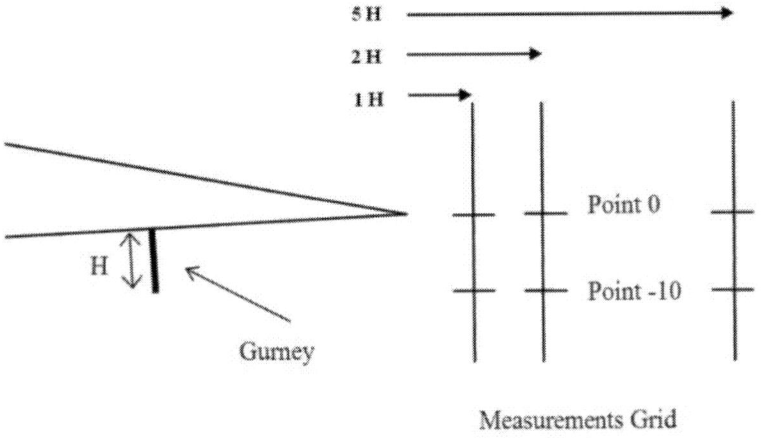

Figure 1. Wake measurement schema.

3. Experimental results

Below we show up some of the experimental results with the aim to explore qualitatively and quantitatively the particular fluid dynamic pattern, promoted by the mini-flap oscillations, in the near wake region, as a consequence of the active flow control of such device upon the airfoil´s aerodynamic characteristics. We observed the almost perfect matching between the mini-flap oscillating frequencies and the special wake structure, with a peak at the same frequency than the oscillating one and other peaks which are other structures, not harmonics because they aren´t multiples of the first (fundamental). In order to support such assumptions, we also showed the corresponding velocities spectra at some selected vertical points (Figure 3) and the horizontal (U) and vertical velocities (V) components on each "x" position (1h, 2h and 5h) for all the vertical points (mean free stream of 10m/s and 0^0 angle of attack):

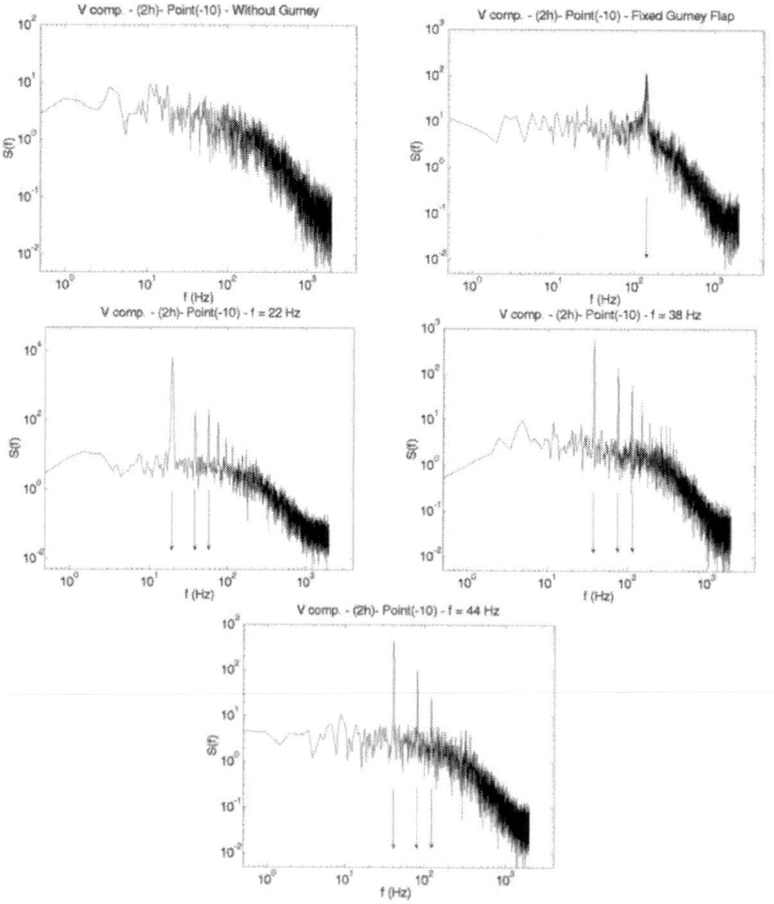

Figure 2. V component velocities spectra for the clean airfoil, fixed Gurney and moving one (three frequencies), 2h "x" position and vertical point -10.

We observe, above, the power density spectra evolution of the downstream vertical velocities, at the 2h "x" position in a vertical point located at the same horizontal level of the mini-flap´s trailing edge (point -10), for the clean airfoil (PS), the mini-flap deployed fixed (GF) and for the mini-flap oscillating at 22; 38 and 44 Hz. In all cases, the free stream upstream velocity was 10 m/s.

The spectra peaks were as follows. For 22 Hz (oscillating frequency), first peak at 22 Hz and successively (approximate) 39 Hz, 58 Hz, 74 Hz, 92 Hz and 110 Hz; for 38 Hz (oscillating frequency), the peaks were at 30 Hz, 78 Hz, 110 z, 124 Hz, 190 Hz; for 44 Hz (oscillating frequency), the peaks were 44 Hz, 82 Hz, 108 Hz. For the fixed GF condition, the peak was at 142 Hz. One could see how the oscillation of the mini-flap made important changes on the wake characteristics. The periodic (coherent) vortex street, generated by the oscillating mini-flap, had enough strength to overlap and diminish the intensity of the turbulent structures typical of the airfoil with the fixed mini-flap. This behavior is more significant as the oscillating frequency grows. In that way, the important changes in the wake, promoted by the oscillating mini-flap, will affect directly the general circulation around the airfoil. Thus, our main concern is to measure the near wake with detail.

Figure 4 shows the instantaneous velocities at those -10 points. The curves exhibit peaks, of course, in accordance with those power density spectra (Figure 3). We could see how the turbulent intervals between peaks, are reduced as the oscillating frequency grows. This is due, probably, to the fact that once we overcome some frequency step, the characteristics of the periodic structures, shed by the oscillating mini-flap, become almost independent of the frequency and, so, the near wake structure will be similar for frequencies above such step.

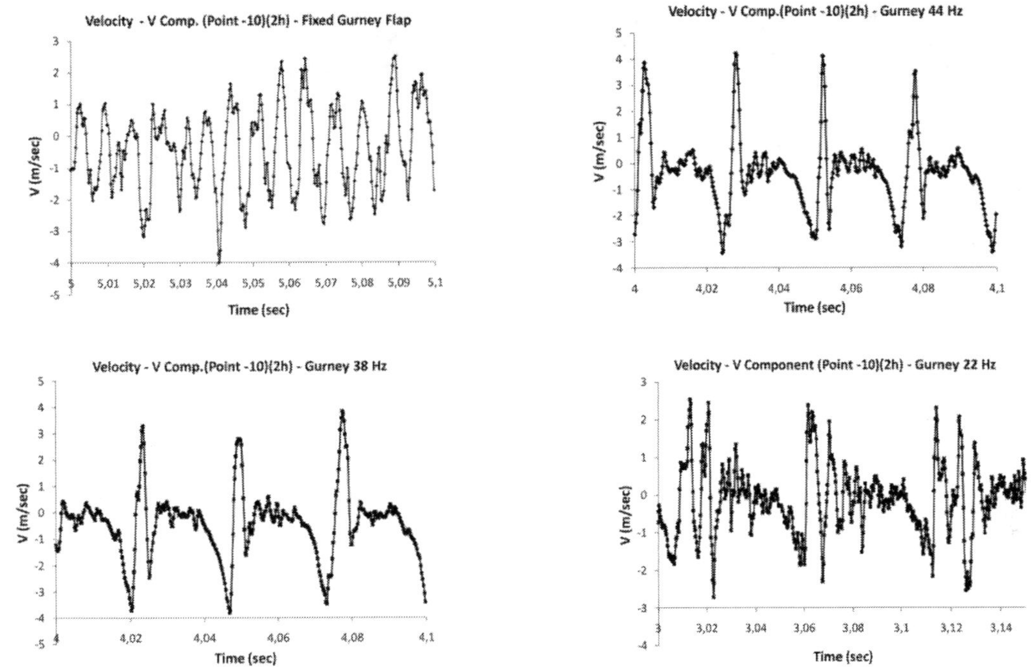

Figure 4. Instantaneous vertical velocities curves at -10 points for the wing with GF fixed and wing for the three mini-flap frequencies oscillation

Figure 5, 6 and 7 shows the horizontal and vertical velocities, for a free stream of 10m/s and 0^0 angle of attack, for all vertical points in each "x" position (1h, 2h and 5h), for the clean airfoil, the fixed deployed mini-flap and the oscillating condition for the three frequencies.

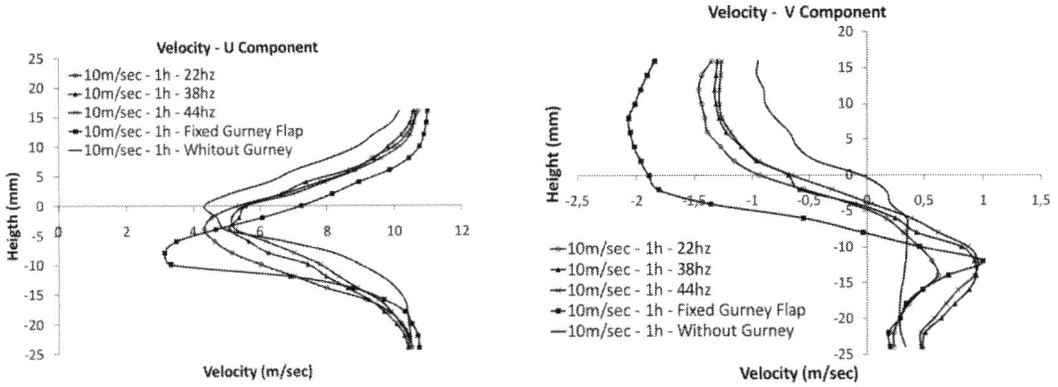

Figure 5. U and V component velocity distributions in the wake at 1h.

Regarding the Figure 5, we could conclude that the U-component has small variations between the different conditions (clean airfoil; fixed flap; etc), being always positive above and below the trailing edge, but with a reduction of its magnitude from the trailing edge level to the end of mini-flap level. The vertical velocities exhibit important differences, above the trailing edge, between the clean airfoil and the fixed mini-flap case. Respect the oscillating mini-flap, there are small differences between the vertical velocities for the three frequencies but, if we look close the vertical velocities at the mini-flap level and lower, their values are greater than the corresponding to clean airfoil or even the fixed mini-flap case. Qualitatively, the situation is similar for the 2h "x" position (Figure 6). Analyzing both Figures, it´s clear that we have an anticlockwise vortex behind the mini-flap. This is consistent with the results founded by other authors [1, 3, 4, 5, 8, 9, 10, 14, 15 and 17].

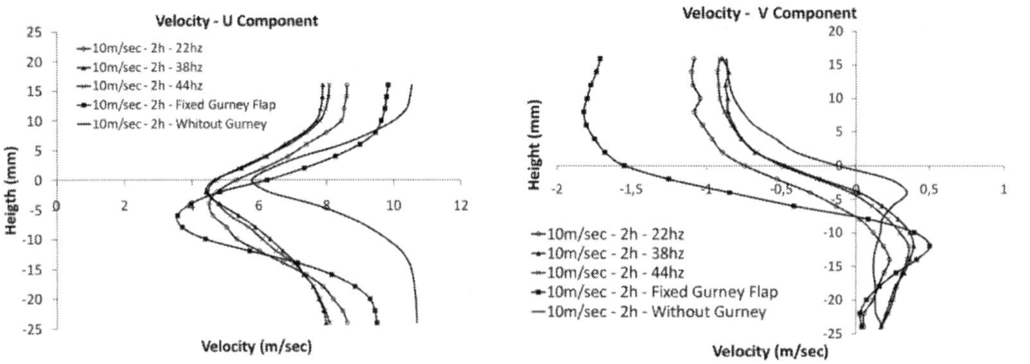

Figure 6. U and V component velocity distributions in the wake at 2h.

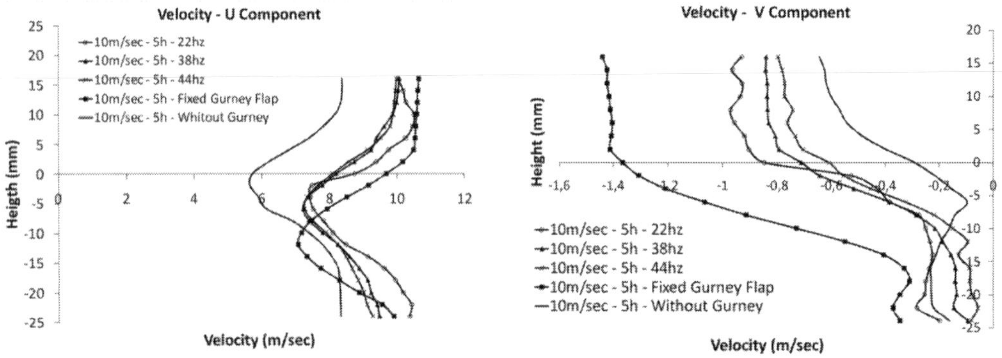

Figure 7. U and V component velocity distributions in the wake at 5h.

Following some of the ideas exposed by Tang et al [17], we selected two upstream points, one on the upper surface and the other on the lower surface (see Experimental procedure), to analyze the pressure time history. Our pressure taps were located at the "x" position 0.88c being the mini-flap location at 0.96c. Such pressure taps were designated Up (for upper surface) and Low (for lower surface). The main difference, regarding the procedure followed by Tang et al [17], was our election of the pressure taps location, upstream the perturbation device (mini-flap) location. Figures 8a and 8d show the C_p time history for 5^0 angle of attack (AOA) and 22Hz and 38Hz oscillating frequencies, respectively. Figures 8c and 8d are for those two frequencies but for 11^0 of angle of attack.

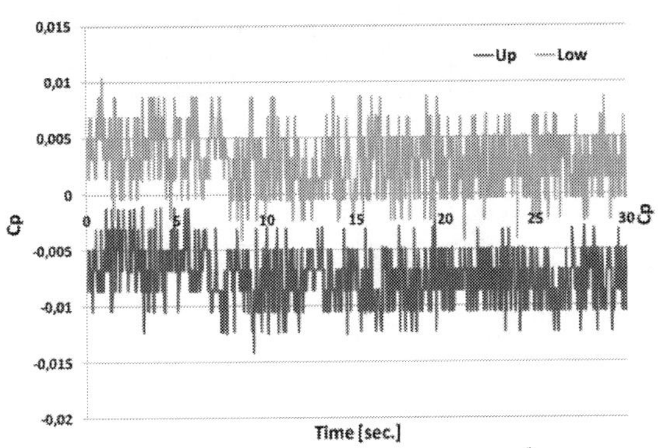

Figure 8a. Cp vs. time (22 Hz – 5^0 AOA)

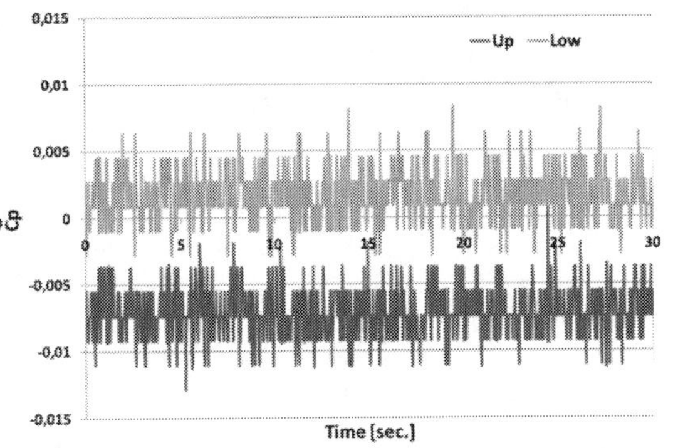

Figure 8b. Cp vs. time (38 Hz – 5^0 AOA)

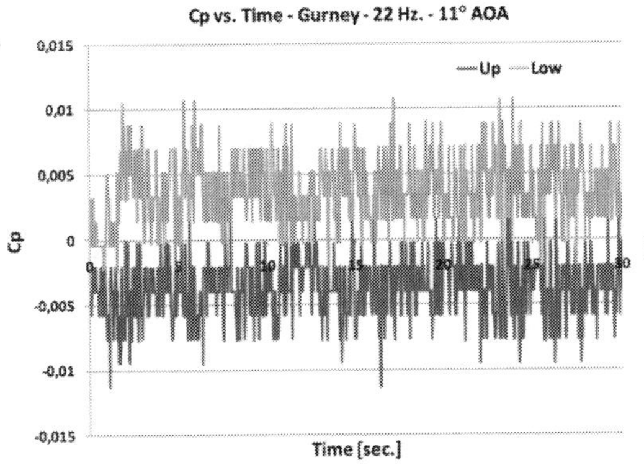

Figure 8c. Cp vs. time (22 Hz – 11^0 AOA)

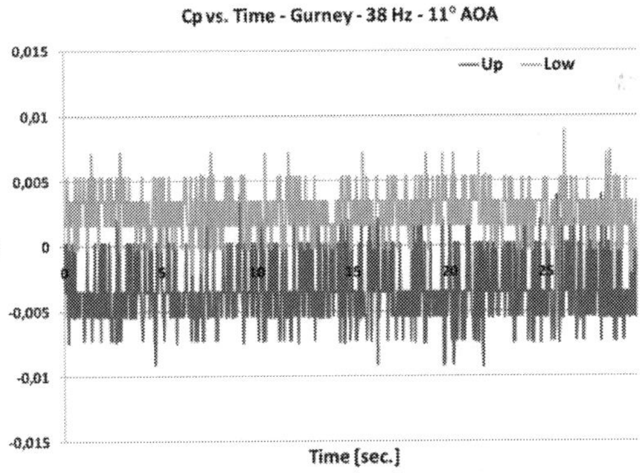

Figure 8d. Cp vs. time (38 Hz – 11^0 AOA)

Figure 8a show some irregularities in the pressure fluctuations, than Figure 8b. It seems that as frequency grows, the pressure fluctuations become similar in amplitude, both in the upper and lower surfaces. The difference between those times histories could be associated with the changes in the near wake as the frequency grows (see Figure 4). Although velocities spectra showed in Figure 4 corresponds to 0^0 of angle of attack, we could made a comparison between such results and the pressure time histories for 5^0 angle of attack, bearing in mind the similar qualitative behavior of the airfoil for 0^0 and 5^0 angles of attack. Moreover, such times history behavior is also associated with the pair of vortex structures in the near wake (described above, regarding Figures 5 and 6). If we observe carefully, for the same angle of attack and far away from the stall, as frequency grow, the upper C_p becomes more negative whereas the lower C_p becomes a bit less positive as the frequency grow. From an overall point of view we could conclude that as frequency grows the lift will enhance. Figures 8c and 8d show us the situation for 11^0 of angle of attack, exhibiting an overall increase of the pressure fluctuations, in comparison with the case for 5^0 of angle of attack, but with they seems to diminish the difference between the upper and lower C_p \`s. So, that could imply an small lift lowering, in comparison with the 5^0 angle of attack. Such behavior, on the upper surface, could be a result of the interaction of the external turbulent flow and the boundary layer near to stall and, in the lower surface, the interaction of the external flow and the fluctuations induced by the oscillating mini-flap.

Finally, looking to achieve an overall understanding of the whole phenomena, we prepared the Table 1 in order to compare the upper and lower C_p\`s, for the airfoil with the fixed mini-flap and the airfoil with the oscillating mini-flap, for the three frequencies. The first conclusion is that the C_p differences between the lower and upper surfaces, for three reference angles of attack (0^0, 5^0 and 11^0), are greater for the fixed mini-flap than the oscillating one. The second conclusion is that the corresponding C_p differences between the lower and upper surfaces diminish as the oscillating frequency grows, but in all cases (even for the minor frequency) the values are lesser than the fixed mini-flap case.

Table 1. Cp values at the trailing edge.

Device			Position	
			Cp Up	Cp Low
Deployed Gurney Flap		0°	-0.1673516	0.2247924
		5°	-0.413925	0.2588836
		11°	-0,2374264	0.2957263
Movable Gurney Flap	22 Hz.	0°	-0.1477534	0.1360552
		5°	-0.3680741	0.142911
		11°	-0.1969174	0.1875045
	38 Hz.	0°	-0.0647598	0.1507643
		5°	-0.3591227	0.0808359
		11°	-0.1732171	0.1346238
	44 Hz.	0°	-0.0308086	0.1724985
		5°	-0.3501714	0.0187608
		11°	-0.2016496	0.1066335

4. Conclusions

From the analysis of the power density spectra, vertical velocities spectra, horizontal and vertical velocities in the near wake and pressures time history, we could say that the oscillating mini-flap change the wake flow pattern, alleviating the near wake turbulence and enhancing the vortex pair near the trailing edge at the mini-flap level and below that level, magnifying the effect described first by Liebeck [1]. That effect is more evident as the oscillating frequency grows. Additionally, the wake alleviation probably affects also the far wake. All of these facts suggest us to continue with the experiments, trying to measure the pressure distribution around the airfoil, with an enough number of taps around it to find relations between the special near wake characteristics and the overall pressure behaviour around the airfoil.

References

[1] Liebeck, R. H., "Design of Subsonic Airfoils for High Lift," *Journal of Aircraft*, Vol. 15, No. 9, 1978, pp. 547–561.

[2] Roesch, P., and Vuillet, A., "New Designs for Improved Aerodynamic Stability on Recent Aerospatiale Helicopters," *Vertica*, Vol. 6, Jan. 1982, pp. 145–164.

[3] Neuhart, D. H., and Pendergraft, O. C., "A Water Tunnel Study of Gurney Flaps," NASA TM 4071, Nov. 1988.

[4] Katz, J., and Largman, R., "Effect of 90 Degree Flap on the Aerodynamics of a Two-Element Airfoil," *Journal of Fluids Engineering*, Vol. 111, March 1989, pp. 93, 94.

[5] Storms, B. L., and Jang, C. S., "Lift Enhancement of an Airfoil Using a Gurney Flap and Vortex Generators," *Journal of Aircraft*, Vol. 31, No. 3, 1994, pp. 542–547.

[6] Kentfield, J. A. C., and Clavelle, E. J., "The Flow Physics of Gurney Flaps, Devices for Improving Turbine Blade Performance," *Wind Engineering*, Vol. 17, No. 1, 1993, pp. 24–34.

[7] Ross, J. C., Storms, B. L., and Carrannanto, P. G., "Lift-Enhancing Tabs on Multielement Airfoil," *Journal of Aircraft*, Vol. 32, No. 5, 1995, pp. 1072–1078.

[8] Bloy, A. W., and Durrant, M. T., "Aerodynamic Characteristics of an Aerofoil with Small Trailing Edge Flaps,"*Wind Engineering*, Vol. 19, No. 3, 1995, pp. 167–172.

[9] Myose, R., Papadakis, M., and Heron, I., "Gurney Flap Experiments on Airfoils, Wings, and Reflection Plane Model," *Journal of Aircraft*, Vol. 35, No. 2, 1998, pp. 206–211.

[10] an Dam, C. P., Yen, D. T., and Vijgen, P. M. H. W., "Gurney Flap Experiments on Airfoil and Wings," *Journal of Aircraft*, Vol. 36, No. 2, 1999, pp. 484–486.

[11] Jeffrey, D., Zhang, X., and Hurst, D. W., "Aerodynamics of Gurney Flaps on a Single-Element High-Lift Wing," *Journal of Aircraft*, Vol. 37, No. 2, 2000, pp. 295–301.

[12] Li, Y., Wang, J., and Zhang, P., "Effects of Gurney Flaps on a NACA 0012 Airfoil," *Flow, Turbulence and Combustion*, Vol. 68, No. 1, 2002, pp. 27–39.

[13] Zhan, J. X., and Wang, J. J., "Experimental Study of Gurney Flap and Apex Flap on a Delta Wing," *Journal of Aircraft*, Vol. 41, No. 6, 2004, pp. 1379–1383.

[14] Gai, S. L., and Palfrey, R., "Influence of Trailing-Edge Flow Control on Airfoil Performance," *Journal of Aircraft*, Vol. 40, No. 2, 2003, pp. 332–337.

[15] Boldes, U.; Delnero, J.; Marañón Di Leo, J.; Colman, J.; Camocardi, M. & François, D. "Influencia en la sustentación, de los vórtices de la estela de un perfil con miniflap tipo Gurney". Actas 1er Congreso Nacional de Ingeniería Aeronáutica. La Plata. Dic. 2008.

[16] Matalanis, C.G. & Eaton, J.K. (2007). Wake vortex alleviation using rapidly actuated Segmented Gurney flaps. Report 102. Flow Physics and Computation Division, Department of Mechanical Engineering, University of Stanford. (2007).

[17] Tang, D. & Dowel, E.H. (2007). Aerodynamic loading for an airfoil with an oscillating Gurney flap. Journal of Aircraft, Vol. 44, Nr. 4, pp. 1245-57.

Journal of Physics: Conference Series **296** (2011) 012008 doi:10.1088/1742-6596/296/1/012008

Dam-break release of a gravity current in a power-law channel section

B M Marino[1] and L P Thomas
Facultad de Ciencias Exactas, Universidad Nacional del Centro de la Provincia de Buenos Aires, Pinto 399, B 7000 GHG Tandil, Argentina

E-mail: bmarino@exa.unicen.edu.ar, lthomas@exa.unicen.edu.ar

Abstract. The evolution of gravity currents generated in open channels which represent those present in estuaries and many hydraulics structures are studied analytically and experimentally. We are particularly interested in the effects of the cross-section shape on the dam-break flows that satisfy the shallow-water equations by considering an equivalent channel cross-section represented by a power law. A stratified two-layer system is assumed and an analytical model based on the balances of mass and momentum in the frontal region, whose results are successfully compared with those obtained by means of experiments, is proposed. The solutions describe the dynamics of the gravity currents with sharp fronts and can be applied to a wide range of natural situations where the turbulence is not dominant.

1. Introduction

Gravity currents are flows driven by a difference in density between the current and the ambient fluid into which it penetrates. As they are present in numerous natural, man-made and industrial events, the comprehension of its features is important for a variety of practical situations [1]. Gravity currents are affected by the topography over which they flow, the entrainment of the ambient fluid and, in many cases, the particulate matter they carry. Most part of the effort has been directed towards investigating the release of such flows in flat-bottomed rectangular tanks, in open horizontal or inclined channels, and considering lock-exchange or dam-break configurations. Also significant, the studies of gravity currents propagating along channels of different cross-section shapes have been addressed in the geophysics context [2-6] although these flows are considered to be fed by a constant flux rather than by a constant volume.

The release of a volume of fluid from behind a lock into an ambient fluid of a different density is a well known situation in hydraulics. Interest in dam-break problems arises as a result of governmental regulations to prevent dramatic events which may become dangerous to human beings or nature, as occurs in Italy among other mountainous countries. The dam-break flow is that produced by the instantaneous release of a liquid by means of the sudden removal of a barrier. It may be thought as the idealized situation of that occurring when the wall of a reservoir open to the atmosphere collapses and the water initially contained within spreads, for example, along a valley or a gorge. The mass of water then flows horizontally under gravity, representing the non-Boussinesq limit in which the air is the fluid of smaller density and, as a consequence, its resistance to the water motion may be neglected. The one-dimensional problem (i.e. for wide cross-section channels) was solved by St. Venant (see for

[1] To whom any correspondence should be addressed.

example [7]) in the shallow-water equations approximation, assuming that the vertical accelerations are small enough with respect to gravity g. This hypothesis is valid except immediately after the dam collapse, so that the eventual errors derived from it are not significant for long periods of time [8].

A rather simple form of the dam failure problem in a dry channel was first solved by Ritter [9], who used the Saint-Venant equations in the characteristic form, under the hypothesis of instantaneous failure in a horizontal rectangular channel without bed resistance. A description of the subsequent studies about this problem was summarized by Zanuttigh & Lamberti [10]. Geometric effects on sudden releases were first investigated by Shih-Tun Su and Barnes [11], who obtained water surface profiles and mean velocity distributions for rectangular, triangular, and parabolic cross sections, represented by a power-law variation of the area with the depth in the case of a dam break in a dry right channel. Zanuttigh & Lamberti [10] analytically investigated the cross-sectional shape effects on the initial discharge of a dam-break flow assuming an equivalent channel section represented by a power-law variation of the section width. They considered three parameters characterizing the section: the channel reference width (not necessarily the bank-full width), the bed elevation, and the power-law index assumed to be constant along the channel. Velocity and bore speed for some typical sectional shapes (rectangular, triangular, concave, and convex banks) were determined as functions of the relative flow depth difference and of the power-law index. Later, Zanuttigh & Lamberti [12] extended the Weighted Averaged Flux (WAF) method by E. F. Toro [13] from rectangular to generic regular cross sections, in order to represent the geometric effects of the cross-section on wave propagation along channels. Assuming a power-law variation of the channel width, the 1D conservative shallow water equations, their characteristic form and the shock propagation equations were presented. The exact Riemann solver was derived and applied to the dam-break problem in valleys with different shape in order to test its efficiency and to check the accuracy order of solutions obtained by approximating the real cross-section with an equivalent rectangle. More recently Biscarini *et al.* [14] presented numerical simulations of free surface flows induced by a dam-break comparing the shallow water approach to full three-dimensional simulations based on the solution of the complete set of Reynolds-Averaged Navier-Stokes (RANS) equations coupled to the Volume of Fluid (VOF) method. The methods assessment and comparison were carried out on a dam-break over a flat bed without friction, a dam break over a triangular bottom sill and a dam break flow over a 90° bend. Experimental and numerical literature data were compared to their results.

On the other hand, Monaghan *et al.* [15] investigated high-Reynolds numbers gravity currents of constant volume and propagating horizontally along a rectangular tank with a V-shaped bottom filled with a lighter fluid. They showed that a numerical simulation based on a box model with the Froude number of the head defined using the distance from the top of the current to the bottom of the valley predicts the evolution of the position of the head in close agreement with the experiments. The major differences between these gravity currents and those flowing along a flat bottom were indicated. Also Marino and Thomas [16-17] introduced a theoretical treatment describing the dynamics of the counter-flows, which only includes the effect of the cross-section shape. As the front condition for steady (surface and bottom) flows generated in non-rectangular cross-section channels is analytically derived independently for each current, the results can be used to study situations where the attention is focused on a single current as in geophysical and engineering flows. The theoretical findings were compared with those provided by laboratory experiments where lock-exchange flows in open and closed channels of rectangular, triangular and parabolic cross-section were generated.

We experimentally and analytically investigate the gravity currents generated by the removal of a barrier that initially separates two fluids of slightly different densities, and provide an analytical model based on mass and momentum balances in the shallow-water approximation that describes the evolution of the front velocity and height profile of the dam-break flows generated in open channels with cross-sections defined by an index related to geometry. It is found that the current dynamics is affected in different ways. The coefficients of the power-law that describe the evolution of the front position with time are greater in the cases in which the cross-section shape is farther from being rectangular and the relative density difference is smaller, while the height profile tends to be uniform. These differences between the flows can affect the estimates of erosion and sedimentation effects on river and estuaries beds. The experiments indicate that the mixing does not significantly modify the frontal motions in stratified environments. Thus, the analytical solutions may be applied to a huge range of natural situations in which turbulence and mixing are not dominant.

2. Analytical solution for a dam-break gravity current in a power-law cross-section channel

We consider a horizontal channel with a cross-section shape that is uniform in the x-direction. As usual, a power-law variation of a channel width in terms of the flow depth is assumed in the form

$$y = \tfrac{1}{2} b z^{\alpha}, \quad (1)$$

where z and y are the vertical and transverse coordinates of the channel cross-section, while α and b are constants. In particular, $b = 2w/h_0^{\alpha}$ is determined by the width w and the depth h_0 of the channel occupied by the fluid. Figure 1 illustrates the cross-section shape associated with different values of α.

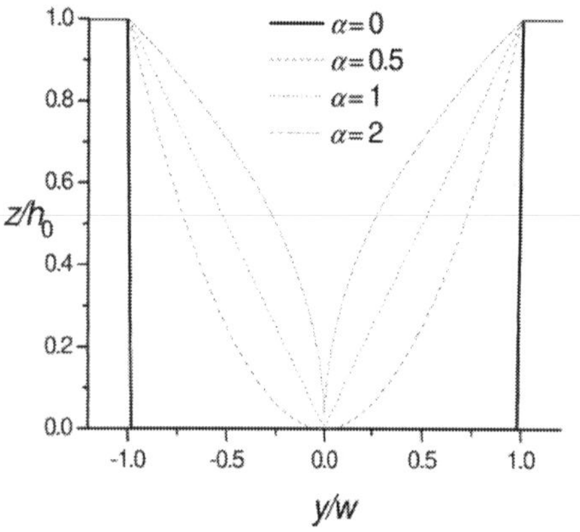

Figure 1. Cross-section shapes of a channel according to the values of α-parameter.

A fluid of density ρ_1 occupies the channel from the rear wall ($x = 0$) up to the position x_0 where a vertical barrier is inserted to separate it from another fluid of density $\rho_2 < \rho_1$. Both fluids fill their rooms up to a depth h_0 and are initially at rest. The volume V of the dense fluid is

$$V = \int_0^{x_0} \int_0^{h_0} 2y \cdot dz \cdot dx = \frac{b \cdot x_0 \cdot h_0^{\alpha+1}}{\alpha+1} \quad (2)$$

As soon as the gate is removed, the dense fluid collapses generating a layer that extends up to $x = x_f(t)$ where the front (or a height discontinuity) moves with velocity u_f. When inertial forces dominate, the equations of mass and momentum conservation in the shallow-water approximation for a channel of cross-section shape defined by (1) take the form [18]:

$$\frac{\partial h^{\alpha+1}}{\partial t} + \frac{\partial}{\partial x}\left(u h^{\alpha+1}\right) = 0 \quad (3)$$

$$\frac{\partial u}{\partial t} + \frac{\partial \left(g' h + u^2/2\right)}{\partial x} = 0 \quad (4)$$

where h is the current height, t is time, u is the component of fluid velocity in the x-direction and $g' = g(\rho_2-\rho_1)/\rho_1$ is the initial value of the reduced gravity (g: gravity acceleration).

We look for solutions of (3) and (4) in the form

$$h(x,t) = h_f(t) H(\eta) \quad (5)$$

$$u(x,t) = u_f(t)U(\eta) \tag{6}$$

where $\eta = x / x_f$ and the sub-index f denotes the values of the variables in the front. These are self-similar solutions since the dimensionless distributions of the current height and velocity have the same form at any time, that is $H(\eta)$ and $U(\eta)$ do not depend on t. Thus the time evolution is established only by the variation of the current length $x_f(t)$, that in turn determines $u_f(t)$ and $h_f(t)$.

The dense fluid volume is constant if mixing is negligible and, as the current profile is self-similar, it results that

$$V = \int_0^{x_f} \int_0^{h} 2y \cdot dz \cdot dx = \frac{b \cdot I \cdot x_f \cdot h_f^{\alpha+1}}{\alpha+1} \tag{7}$$

with the constant shape factor

$$I = \int_0^1 H^{\alpha+1} d\eta . \tag{8}$$

Differentiating (7) with respect to time, it is found that

$$\frac{dh_f}{dt} = -\frac{h_f u_f}{(\alpha+1) x_f} . \tag{9}$$

In addition, a local balance of mass and momentum in the front gives place to a constant densimetric Froude number

$$Fr = \frac{u_f}{\sqrt{g' h_f}} , \tag{10}$$

of the order of unity ($Fr \sim 1$). Integrating (10) in time with $u_f = dx_f/dt$ and using (7), the front position evolves according to

$$\frac{x_f}{x_0} = \xi \cdot \left(\frac{t}{t_c} \right)^{\beta} , \tag{11}$$

where

$$\beta = \frac{(2\alpha+2)}{(2\alpha+3)} , \tag{12}$$

$$\xi = \left(\frac{Fr}{\beta} \right)^{\beta} \left(\frac{1}{I} \right)^{1/(2\alpha+3)} \tag{13}$$

and

$$t_c = \frac{x_0}{\sqrt{g' h_0}} \tag{14}$$

is the characteristic time determined by the characteristic length x_0 and the velocity $\sqrt{g' h_0}$ of an internal wave propagating in shallow waters.

In this way the self-similar solutions are found. For $\alpha = 0$ the result agrees with that for gravity currents running in rectangular cross-section channels where $\beta = 2/3$. For $\alpha \neq 0$ the exponent β depends only on the channel cross-section shape, while the coefficient (13) depends also on the Froude number at the front. The released volume and the reduced gravity merely influence the characteristic distance x_0 and time t_c. In addition, the relationship between the width w and depth h_0 of the channel can be neglected under the assumed approximations since only the shape is of interest.

The current height profiles are determined by solving the equations (3) and (4). Differentiating (10), and employing (9), we obtain

$$\frac{du_f}{dt} = -\frac{1}{2(\alpha+1)}\frac{u_f^2}{x_f}. \tag{15}$$

On the other hand, by replacing (5) and (6) in (3) and (4), and using (9) and (15), it is found that

$$H^{\alpha+1}\left(\frac{dU}{d\eta}-1\right)+(U-\eta)\frac{dH^{\alpha+1}}{d\eta}=0 \tag{16}$$

$$-\frac{U}{2(\alpha+1)}+\frac{dU}{d\eta}(U-\eta)+\frac{1}{Fr^2}\frac{dH}{d\eta}=0. \tag{17}$$

The solution for any value of α and Fr results:

$$U=\eta \tag{18}$$

$$H=1-\frac{Fr^2}{4(\alpha+1)}\left(1-\eta^2\right) \tag{19}$$

Then, velocity increases linearly from $x = 0$ to $x = x_f$ independently of the cross-section shape. Figure 2 shows the distributions $H(\eta)$ given by (19). Again, for $\alpha = 0$ the result coincides with that of gravity currents in rectangular channels. For increasing α the distribution tends to be uniform but $H \approx 1$ only for large values of α.

The solutions found indicate that the height and velocity are analytical continuous functions in $0 \leq x < x_f$. At $x = x_f$, a height discontinuity exists as a step from h_f to $h = 0$ while the velocity changes from u_f up to $u = 0$.

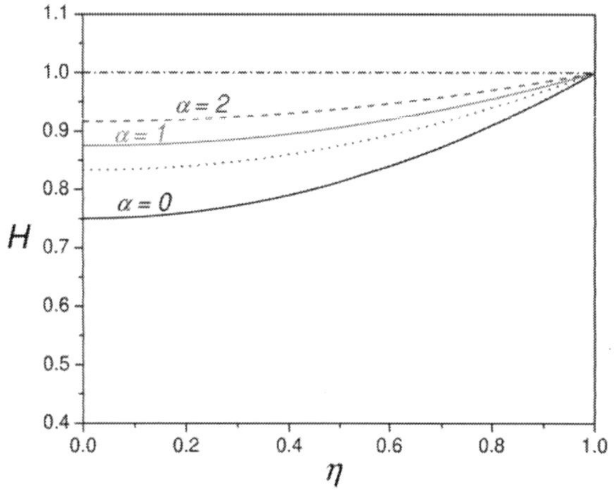

Figure 2. Height distributions between the origin $x = \eta = 0$ and the front $\eta = 1$ ($x = x_f$).

3. Experimental validation

3.1. Experimental description

Three sets of laboratory experiments were performed in Perspex channels with non-rectangular cross-sections which are considered as half the shapes defined by equation (2) with $\alpha = 0.44$ (concave), 1.00 (triangular) and 1.79 (convex). This experimental setup therefore includes a vertical wall at the centerline of the flow to reduce the optical distortion when it is observed laterally. Additional experiments confirm that the results can be extrapolated to the complete section by symmetry without losing validity. The channels are 2.00 m long, 0.20 m wide and 0.29 m high. In addition, one set of experiments was carried out in a rectangular cross-section channel where known results were reproduced. A gate located initially at 0.10 m from the rear of the channel separated two parts: the smaller one contained salt water while the bigger one was filled with fresh water, up to the same level. The experiment started when the barrier is suddenly removed giving place to a gravity current above the bottom. Each set includes four or five experiments with relative density difference $\Delta\rho/\rho_1 = 1-\gamma$ between 0.2% and 8%. Densities were measured using an Anton Paar 4500 densimeter with a precision of 5×10^{-5} g/cm^3. The non-rectangular channels were placed inside a greater rectangular cross-section tank filled with fresh water to reduce the optical distortion due to the non parallel or curved lateral walls.

The flows were observed laterally with a CCD video-camera located 6 m away from the tank. Experimental set-up and image processing are similar to those previously reported by Marino *et al.* [19]. An anamorphic lens was used in order to magnify the vertical scale and reduce the horizontal scale to improve the resolution of the images. Fluorescent strip lights and a light-diffusing screen located behind the tank provided nearly uniform back-lighting. A reference image of the initial intensity distribution of the tank containing only water was taken by averaging a sequence of images registered during 10 s. A calibrated quantity of dye was added to the salt water to provide visualization. After finishing the experiments, the images were processed considering the acquisition system response, converting the captured light intensity to absolute intensity. Then a correction compensating the small spatial variations of the back-lighting was performed and finally the decreasing of the light intensity with respect to the initial reference image was estimated for each pixel. The light attenuation due to the dyed water allows the cross-current averaged concentration to be measured from which the cross-current average density may be inferred. Thus, the intensity in the processed images is proportional to the cross-current average density.

For a rectangular channel the estimate of the density in excess with respect to the non-colored light fluid is direct [19]. For a non-rectangular cross-section, the intensity depends on two variables: mixing and variation of the channel width with depth. This complication is solved by applying an additional operation that uses another reference image of the channel filled with water and a known concentration of dye [16].

From the processed images, the front velocity and the height of the flow are estimated. With these values, Reynolds numbers ranging between 7000 and 35000 are found for the currents generated.

3.2 Results

The evolution of the front position with time for a rectangular cross-section channel is shown in figure 3a, while figures 3b, 3c and 3d show the cases corresponding to a concave, triangular and convex cross-section shape, respectively. In each case, the theoretical lines describing the *slumping* phase [19], characterized by a constant front velocity, and the self-similar solution given by equation (11) for long periods of time ($t/t_c > 20$) are included. It is important to note that the classical *slumping* phase of gravity currents evolving over flat bottoms is also present for $\alpha \neq 0$ cases. For a given value of α, the experimental points show the same trend, that is, for different initial relative density differences they collapse onto common curves with the slope depending only on the cross-section shape, thus indicating that the characteristic parameters x_0 and t_c are representative of the spatial and time scales, respectively, of the studied flows.

Figure 4a shows a typical lateral image of a gravity current evolving towards the right in a rectangular cross-section channel, processed as described in Section 2. The presence of instabilities generates mixing between the fluids and makes difficult to determine the height profile with precision. However, it is clear that the front region is higher than the following fluid and that the height decreases progressively up to the left vertical wall, consistently with the theoretical profile presented in figure 2.

Figures 4b, 4c and 4d show lateral views of the currents developed in parabolic, triangular and convex cross-section channels, respectively. Here, the intensity of each pixel is also proportional to the mean density, independently of the particular shape of the cross-section of the channel in which the flows evolve. Although it is observed that the general shape of the flows is the same, for increasing α ($\neq 0$) the frontal zone of breaking waves and intense mixing becomes larger. The current height decreases slightly behind the head in agreement with the theoretical model. The thickness of the gravity current increases with α since the fluid needs more space to flow in narrower channels.

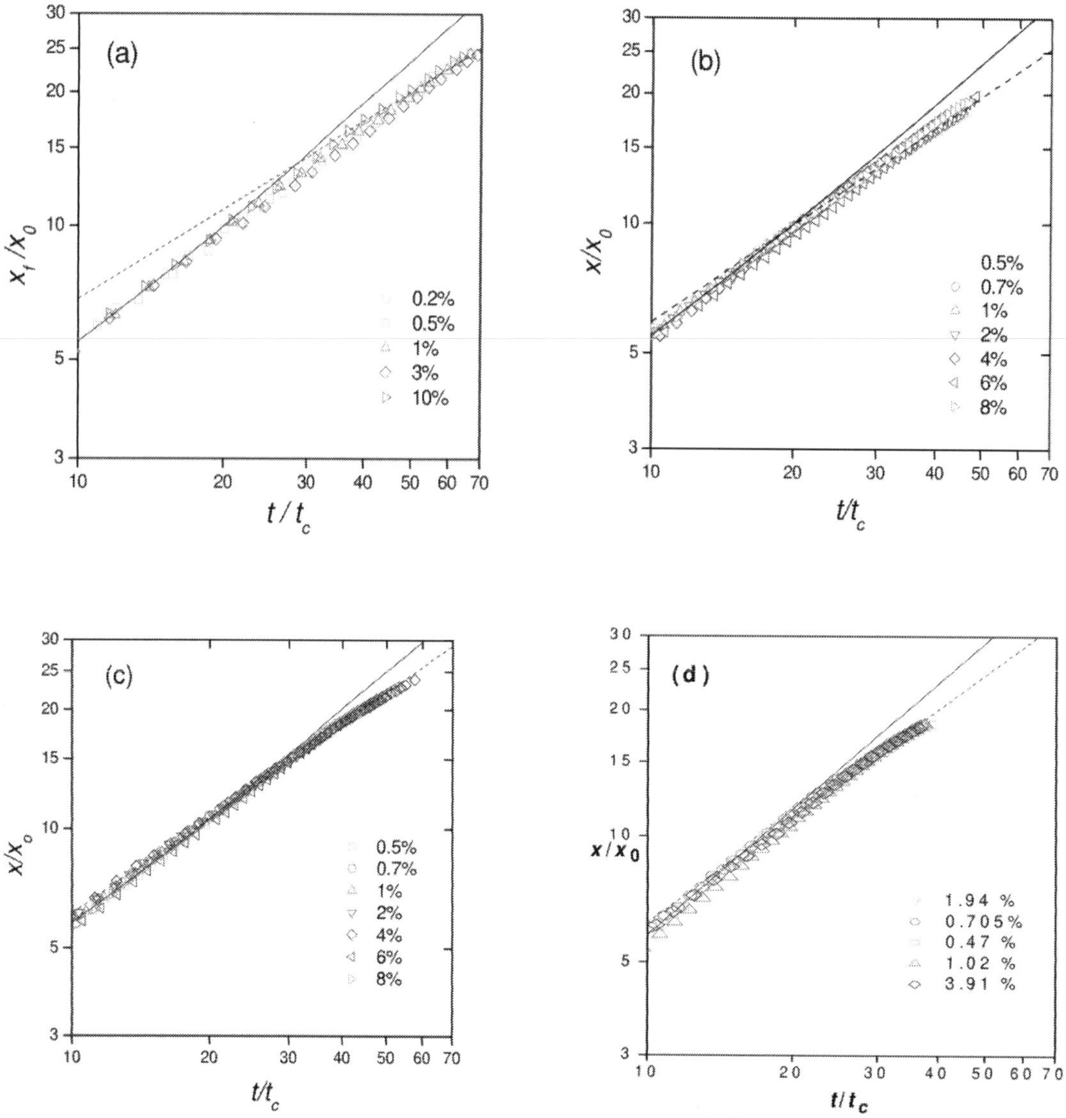

Figure 3. Time evolution of the front position of a gravity current evolving in (a) rectangular ($\alpha = 0$), (b) parabolic ($\alpha = 0.44$), (c) triangular ($\alpha = 1$) and (d) convex ($\alpha = 1.79$) channels. Solid lines represent the solution for the *slumping* stage and dashed-lines are the solution given by (11).

Figure 4: Density distributions of gravity currents in (a) rectangular ($\alpha = 0$), (b) parabolic ($\alpha = 0.44$), (c) triangular ($\alpha = 1$) and (d) convex ($\alpha = 1.79$) channels for $\Delta\rho/\rho \approx$ 4-6%. The lines are spaced 0.10 m.

4. Concluding remarks

We analytically and experimentally study the dam-break flows generated in open straight channels with cross-sections shape defined by a geometrical parameter that facilitates the analysis in the shallow water approximation. It is found a self-similar solution that describes the evolution of the front height and velocity for extended currents, that is, when the initial conditions are not important ($x_f >> x_0$). In this case the main characteristic magnitudes are the flow initial length x_0 and the velocity $\sqrt{g'h_0}$ of the internal waves propagating in shallow waters. Thus the space and time scales are determined by x_0 and t_c, independently of the cross-section.

After the classical *slumping* phase, the evolution of the front follows the power law given by equation (11) in which the similarity exponent only depends on the cross-section shape. The power law coefficient, however, also depends on the Froude number at the front. These facts are based on the total mass conservation, the self-similar height profiles shape [equations (5) and (6)], and the mass and momentum balance in the current front. In addition, the mass and momentum conservation equations in the shallow water approximation allow the determination of the height profiles of the flows and, with them, the calculation of the shape factor I and the coefficient ξ of the self-similar law. Thus, a relationship between the Froude number in the front and the respective height profile is obtained, which is different from that corresponding to a flow produced in a rectangular channel.

The experimental results obtained using channels with cross-section shape given by (1) agree with the theoretical treatment provided. The space and time scales found analytically are adequate to align the points on a single curve for different initial relative density difference between the fluids. The front evolution tends to be described by a self-similar solution for $x_f > 20x_0$ and the height profiles are consistent with the theoretical ones. In laboratory flows a non negligible mixing between the fluids exists, which is not considered by theory. Nevertheless, this does not notoriously modify the dynamics of the front since the proposed balances are equally verified. This is because mixing increases the volume and height of the current, but the dilution reduces the difference of density in an analogous proportion. As a result the front velocity, and consequently the Froude number, are not modified substantially (equation 10).

The self-similar solution represents an interesting achievement from theoretical-conceptual and practical point of views in the study of gravity currents evolving in channels with non-rectangular cross-section shape. Such solutions describe the dynamics of the flows with sharp fronts and are applied to a wide range of natural situations in which a well-defined stratification exists. It may also be applied to validate complex numerical simulations that consider more difficult analytical aspects.

Acknowledgments

This work was performed in the frame of PICT-ANPCyT 1185/06 and PIP-CONICET 54/2010.

References

[1] Simpson J E 1997 *Gravity currents: In the environment and the laboratory*, Cambridge University Press, Cambridge, United Kingdom

[2] Nunes R A and Simpson J H 1985 Axial convergence in a well mixed estuary *Estuar. Coast. Shelf Sci.* **20** 637-49

[3] Wong K C 1994 On the nature of transverse variability in a coastal plain estuary *J. Geophys. Res.* **99** 209-22

[4] Wong K C and Munchow A 1995 Buoyancy forced interaction between estuary and inner shelf-observation *Contin. Shelf Res.* **15** 59-88.

[5] Valle-Levinson A and Lwiza K 1995 The effects of channels and shoals on exchange between the Chesapeake Bay and the adjacent ocean *J. Geophys. Res.* **100** 18551-63

[6] Engquist A and Hogg A 2004 Unidirectional stratified flow through a non-rectangular channel *J. Fluid Mech.* **509** 83-92

[7] Whitham G B 1974 *Linear and nonlinear waves* John Wiley & Sons, New York, USA.

[8] Landau L D and Lifshitz E M 1987 *Fluid Mechanics (Volume 6 of theoretical physics)* Pergamon Press, Oxford, England.

[9] Ritter A 1892 Die fortpflanzung der wasserwellen *Zeitschrift des Vereines Deutschei Ingenieure* **36** 947-54 (in German)

[10] Zanuttigh B and Lamberti A 2001 Dam-break waves in power-law channel section *J. Hydr. Engrg.* **127** 322–26

[11] Shih-Tun Su A M and Barnes A H 1970 Geometric and frictional effects on sudden releases *J. Hydr. Div., ASCE,* **96** 2185-200

[12] Zanuttigh B and Lamberti 2002 A Exact Riemann Solution and Weighted Average Flux Method for Power-law Channel Section *International Journal of Computational Fluid Dynamics* **16** 155–69

[13] Toro E F 1992 Riemann problems and the WAF method for solving the two-dimensional shallow water equations *Philosophical Transactions Royal Society of London Series A* **338** 43–68

Toro E F 1997 *Riemann solvers and numerical methods for fluid dynamics*, (Springer, Berlin).
Toro E F 2000 *Shock-capturing Methods for Free Surface Shallow Fluids* (Wiley, New York).

[14] Biscarini C, Di Francesco S and Manciola P 2010 CFD modelling approach for dam break flow stud ies *Hydrol. Earth Syst. Sci.* **14** 705–18

[15] Monaghan J J, Mériaux C A, Huppert H E and Monaghan J M 2009 High Reynolds number gravity currents along V-shaped valleys *Eur. J. Mech. B Fluids* **28** 651–9

[16] Marino B M and Thomas L P 2009 Front condition for gravity currents in channels of non-rectangular symmetric cross-section shapes *J. Fluids Eng.* **131** 1201-6

[17] Thomas L P and Marino B M 2004 Lock-exchange flows in non-rectangular cross-section channels *J. Fluids Eng.* **126** 290-292

[18] Thomas L P, Peña C C, Barrenechea A L and Marino B M 2007 Solutions of shallow water equations in non-rectangular cross-section channels *Applied Mathematics E-Notes* **7**, 131-8

[19] Marino B M, Thomas L P and Linden P F 2005 The front condition for gravity currents *J. Fluid Mech.* **536** 49-78

Gas flow in a channel semi-obstructed by an array of wires

D Dalponte[1], N Silin[2], and A Clausse[1]

[1]CNEA-CONICET and Universidad Nacional del Centro, Pinto 399, (7000) Tandil, Argentina.

[2] CONICET-CNEA and Instituto Balseiro, 8400 Bariloche, Argentina.

E-mail: clausse@exa.unicen.edu.ar

Abstract. An experiment of a gas flow in a narrow channel partially obstructed by a regular array of wires is presented. The purpose of the study is to recreate ideal conditions of flow through easily penetrable media. The velocity profile along the channel was measured by means of a hot-wire anemometer, inside and outside the obstructed region. Flow instabilities were found in the free zone adjacent to the porous region, which were characterized by the standard deviation and the power spectrum of the signal of the anemometer.

1. Introduction

The study of flows passing complex obstructions is relevant in several engineering branches. In general this phenomenon involves a wide range of scales practically impossible to calculate numerically over the whole spectrum. For this reason, usually the analyses are based either on experimental measurements or on effective models that average the fluid behavior in space and time. A special case of interest of complex obstructions is the flow parallel to porous pervasive media. Natural examples are atmospheric currents over vegetation [1], flow in channels with aquatic plants [2], and flooding over cultivated lands [3]. This type of flows also occurs in technological applications, such as refrigerant flow over electronic components, fins, pin arrays and metallic sponges [4].

A common feature encountered in the boundary between the porous obstruction and the free flow area is an inflexion point of the velocity profile, similar to what happens in a planar mixing interface [5], [6] and compound channels [7]. This in turn produces local coherent structures of large scale, which enhance the momentum and scalar transversal diffusion [8], [9]. These structures can modify the concentration of scalars and the velocity profile not only at the interface where they are generated but also inside and outside the obstruction [10], [11].

The present study reports a controlled experiment at low Reynolds number of air flow in a narrow channel partially obstructed by an array of cylindrical wires. The small diameter of the wires is chosen to ensure that the local Reynolds number is sufficiently low to avoid the typical instabilities around obstacles such as vortex shedding. This experiment presents a 2D geometry that involves a porous medium homogeneous and isotropic in these two dimensions. The chosen geometry can be easily represented by a rectangular mesh and simple boundary conditions. Hence, the experimental results can be used to evaluate and validate CFD codes. In particular we are interested in developing a CFD code capable of simulating flows involving porous media under conditions where large scale structures occur, *i.e.* fluid-dynamic structures with sizes larger than the elements of the porous medium.

2. Experimental setup and method

The design of the test section involves the compromise between attaining a low Reynolds and a good spatial resolution while keeping the flow velocity at measurable values. The test section consists of a rectangular channel as shown in figure 1, the total height being 56mm and the width 100mm. The lowest velocities are found in the porous region, which compromises the measurement accuracy in this region.

The channel walls are made of transparent acrylic 5 mm thick. The cross section is divided in two halves, one free and the other crossed by an array of cylindrical copper wires. The diameter of the wires is 0.25mm, and the gap between wires is Δx = 5mm in the horizontal direction and Δy = 2.5mm in the vertical direction (figure 1, “A”). The array is appropriate to simulate a pervasive porous media with low Reynolds number.

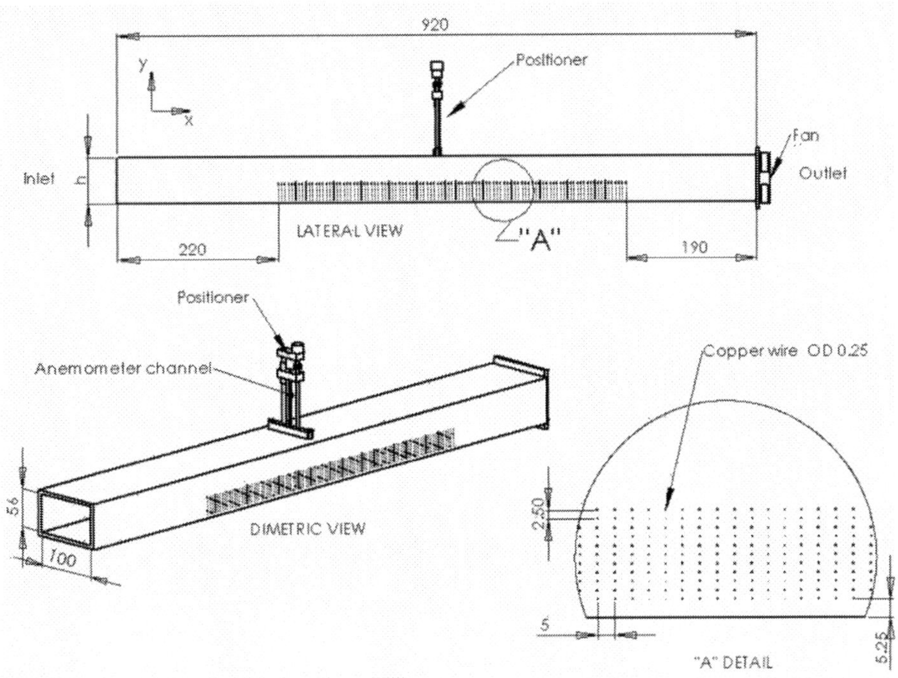

Figure 1. Test section (all dimensions in mm).

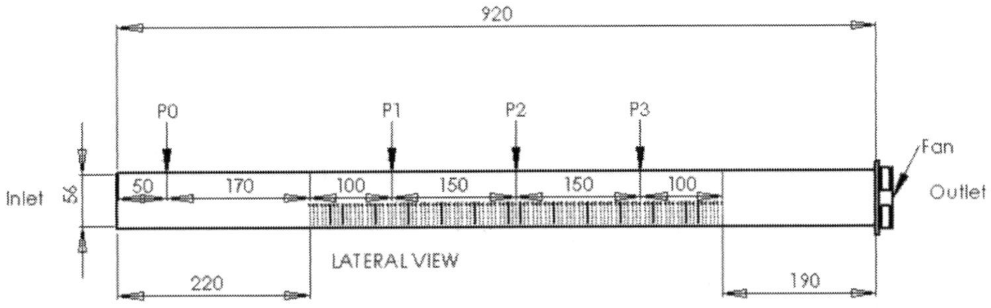

Figure 2. Measuring points P0, P1, P2 y P3 (all dimensions in mm).

The global Reynolds number is defined by:

$$\mathrm{Re} = \frac{\bar{U}_{in}\, h}{\nu} \tag{1}$$

where $\bar{U}_{in}$ is the mean inlet velocity, h is the channel width and ν is the fluid viscosity. The maximum Reynolds studied is approximately 1300.

The diameter and gap between wires was chosen taking into account the manufacture feasibility and that penetration length of the velocity profile in the porous region should be longer than the distance between wires. The volumetric resistance of the media can be calculated by averaging the drag force per unit length [3]. Taking into account that the separation between wires is 10 and 20 diameters in each direction, the drag of a wire can be calculated from correlations for low Reynolds numbers flows through a single cylinder, using the local Reynolds number:

$$\mathrm{Re}_d = \frac{U\, d}{\nu} \tag{2}$$

In our case, the maximum Re_d occurs at the front of the porous region, being smaller than 10, and it is substantially lower for most of the wires comprising the porous media. Under these conditions the drag force per unit length is given by $D' = 4\pi\,\mu\varepsilon U$ Hunner and Hussey [12], where μ is the dynamic viscosity and ε is a dimensionless coefficient. The volumetric resistance is then:

$$D''' = \frac{D'}{\Delta x\, \Delta y} = f\,\mu U$$

where f is a volumetric drag coefficient. The resulting coefficient for the experimental condition is approximately $f = 0.25\ \mathrm{mm}^{-2}$.

The flow is forced in the channel by means of an extractor fan located at the exit of the test section, which is controlled with a variable voltage power supply. Three velocities were studied, as shown in Table 1. The corresponding Reynolds numbers are lower than the critical value for rectangular channels [13], [14]. A honeycomb and a fine metallic mesh are located at the inlet in order to homogenize the entrance flow. Another honeycomb is placed 2 cm before the fan to avoid the introduction of vorticity upstream.

Table 1. Operating conditions.

Flow	$\bar{U}_{in}$	*Re*
W1	0.11 m/s	410
W2	0.24 m/s	900
W3	0.34 m/s	1270

The average velocity profiles were obtained by means of a constant-temperature hot-wire anemometer (CTA) guided by a mechanical device. The measurements were taken at positions P0, P1, P2 and P3 as shown in figure 2, located at 27.2, 42.2, 57.2 and 72.2 (±0.1) cm downstream from the inlet mesh. In each of these positions 28 measurements were made placing the probe along the vertical coordinate. Each measurement recorded 100,000 samples (20s, 5 kHz).

The CTA is basically a feedback resistive bridge that maintains the resistance (and hence the temperature) of the wire sensor constant. The probe consists of a Pt-Ir wire 15 μm thick 2 mm long, placed vertically, supported by two stainless steel prongs (figure 3). The vertical orientation of the sensible filament allows filtering the vertical component of the velocity. Since the horizontal velocity component normal the flow is negligible, as observed from smoke line visualizations, the signal carries information essentially of the axial velocity component U_x. The CTA was operated with 50% overheating. All measurements were performed with ambient temperatures between 19.9 and 20.8 °C.

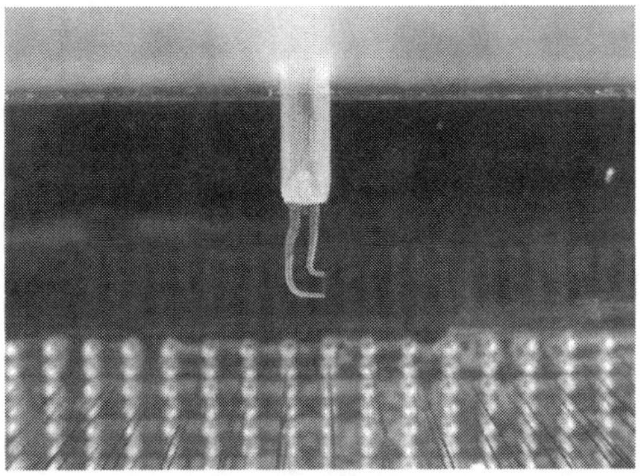

Figure 3. Hot wire sensor.

Figure 4. Photograph of a smoke line in the test section. The fluid flows from left to right.

The data was acquired by means of a PC with a National Instruments card PCI-6070E. Details of the electronics can be found in Osorio et al (2010). The CTA was calibrated, establishing the following correlation between the velocity and voltage:

$$U_x = 0.91(E - E_0)^{1.35} \quad (3)$$

where U_x is the axial velocity in m/s, E is the signal output from the CTA, and Eo is the stagnation voltage.

3. Results

The global operation of the test section was visualized injecting smoke at the inlet to track the stream lines. An image of the visualization is shown in figure 4. The curvature of the smoke lines at the inlet can be appreciated as well as the growth of undulation instabilities in the downstream region. No other type of instabilities was observed. It was also observed that, after the flow contraction brought about by the porous region, the stream lines rapidly become parallel to the channel walls. No significant deformation of the flow lines were observed in the lateral direction. The flow rate W2 corresponds to the visual observation of the onset of instabilities.

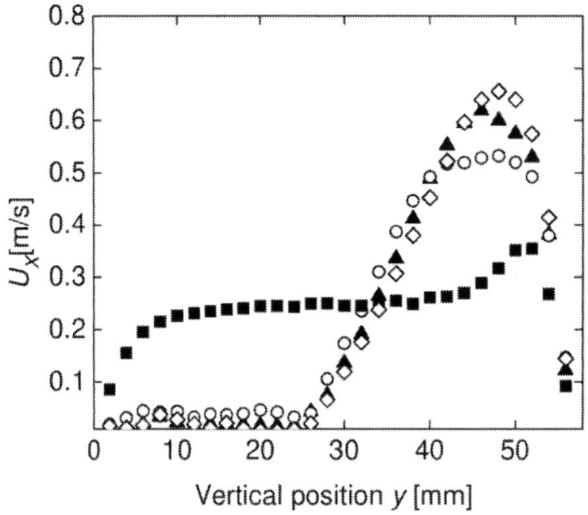

Figure 5. Velocity profiles at positions P0 (■), P1 (O), P2 (▲) and P3 (◊) for flow rate W2.

Figure 5 shows the mean velocity profile at positions P0, P1, P2 and P3 obtained when the flow rate W2 is imposed. The porous area spans from the channel wall at y= 0 to y = 28mm, while the free flow area spans from y = 28mm to y = 56 mm, which is the location of the opposite channel wall. It can be observed that the maximum development is reached at position P3, showing very low velocities in the porous region and a skewed parabolic profile in the free zone. The maximum velocity is found in the free area closer to the wall. An inflexion point can be distinguished in the transition between these two regions. The precise location of the inflection point is difficult to determine from the measurements due to the relatively large experimental error but lies between y = 30 mm and y = 40 mm for all cases.

Figure 6 shows the mean velocity and the standard deviation measured at P3 for each flow rate. As we have mentioned the lowest velocities exhibit relatively large errors. These errors are due to limitations of the thermal anemometry technique such as natural circulation, thermal perturbations from nearby objects (wires, walls, etc.), thermal drifts, etc., which are particularly important at low flow velocities. The maximum velocities are measured at the vertical positions y = 48 mm and 50 mm. The standard deviation shows for all the cases two clear peaks at y ~38 mm and y ~ 54mm. These positions are the locations of the highest derivative of the velocity profile. The latter is consistent with eddy diffusion brought about by instabilities. Furthermore, even though in the porous zone the errors are comparable with the measured values, for the highest flow rate W3 the fluctuations are relevant even in the porous zone, indicating that eddies penetrate into this area. It can be seen that even though the general aspect of σ_x profile is similar in all the flow rates, only for W2 and W3 there is a clear unstable behavior with strong velocity fluctuations. This is an indication that the critical flow rate is between W1 and W2.

In order to analyze the instabilities, the signals of the anemometer were processed to obtain the corresponding power spectrum densities. Figure 7 shows the power spectra at y = 38 mm for flow rate W2. A remarkable feature is the local minima appearing in the power spectra at P1, which coincide with local maxima at P2 and P3. The calculation was checked carefully to ensure that these features are not artifacts produced by the signal processing procedure. A possible cause of this singularity is turbulence reduction found in the contraction of stream lines, in this case at the beginning of the porous media [15].

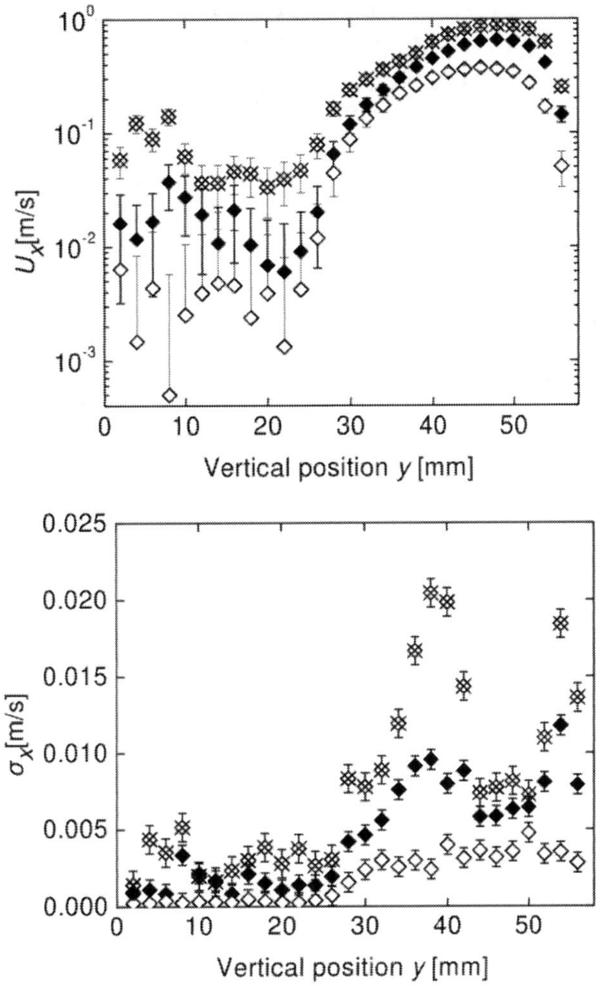

Figure 6. Mean velocity profile (up) and its standard deviation (down) at P3, for W1 (white symbols), W2 (black symbols) and W3 (crossed symbols).

4. Conclusions

An experimental study of the gas flow though a channel partially obstructed by a regular array of wires was presented. The velocity profiles, measured with a hot-wire anemometer at different positions along the channel, showed that the gas velocity decreases substantially in the porous region as a consequence of the resistance of the wires.

Flow instabilities were found in the free region adjacent to the wired region, which were quantified by means of the standard deviation and the power spectra of the signal provided by a hot-wire anemometer. It was found that the standard deviation presents two maxima in the free region, one adjacent to the porous media and the other to the wall. Both coincide with the maxima of $\partial \bar{U} / \partial y$.

For this geometry, the flow fluctuations quantified by the axial velocity standard deviation, shown in figure 6, are negligible for Reynolds number 400 and become significant for Reynolds numbers 900 and 1270. It was observed in figure 6 that the power spectrum of the measurements at y = 38 mm show a distinctive peak at a Strouhal value of approximately 0.6 for the positions P2 and P3, *i.e.* the fluctuations present in the flow have a characteristic frequency. Moreover, the low value of the

Strouhal number, smaller than 1, suggests that the fluctuations are originated by large scale structures. This is a relevant feature of the flow that might be used in the evaluation of numerical solutions to this kind of problems.

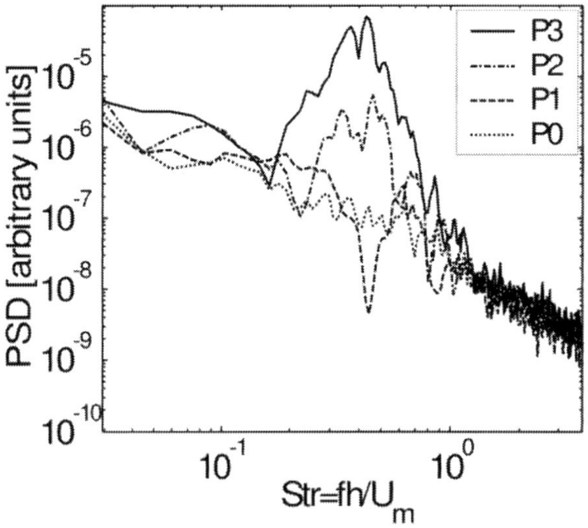

Figure 7. Power spectra at y = 38mm for the flow rate W2.

References

[1] Poggi V, Katul GG (2008) Micro and macro dispersive fluxes in canopy flows, Acta Geophysica 56, 778-800.

[2] López F., García MH (2001) Mean Flow and Turbulence Structure of Open-Channel Flow through Non-Emergent Vegetation, J. Hydraulic Eng. 127, 392-402.

[3] Gayev Y (2004) Variety of problems associated with Canopies. Flow and Transport Processes with Complex Obstructions I. Springer, The Netherlands.

[4] Beavers GS, Joseph DD (1967) Boundary conditions at a naturally permeable wall, J. Fluid Mechanics 30, pp 197-207.

[5] Raupach MR, Finnigan JJ, Brunet Y (1996) Coherent Eddies and Turbulence in Vegetation Canopies: The Mixing-Layer Analogy, Boundary-Layer Meteorology 78, 351-382.

[6] White BL, Nepf HM (2007) Shear instability and coherent structures in shallow flow adjacent to a porous layer, J. Fluid Mech 593, 1–32.

[7] Bousmar D, Riviere N, Proust S, Paquier A, Morel R, Zech Y (2005) Upstream discharge distribution in compound-channel flumes, J. Hydraul. Eng. 131, 408–412.

[8] Ozalp C, Pinarbasi A and Sahin B (2010) Experimental measurement of flow past cavities of different shapes, Experimental Thermal and Fluid Science 34, 505-515.

[9] Coppola G and Gomez A (2009) Experimental investigation on a turbulence generation system with high-blockage plates, Experimental Thermal and Fluid Science 33, 1037-1048.

[10] Ahmed A and Rifki R (2009) On the flow instabilities in the end-wall region of a surface mounted obstacle, Experimental Thermal and Fluid Science 33, 380-385.

[11] Finnigan JJ (2000) Turbulence in Plant Canopies, Ann. Review Fluid Mech. 32, 519-571.

[12] Huner B and Hussey RG, (1977) Cylinder drag at low Reynolds number, Phys. Fluids 20, 1211.

[13] Kao TW, Park C (1970) Experimental investigations of the stability of channel flows I: Flow of a single liquid in a rectangular channel, J. Fluid Mechanics 43, 145-164.

[14] Kundu PK, Cohen IM (2000) Fluid Mechanics, Academic Press, 339-340.

[15] Osorio O, Silin N, Converti J (2010) Fabrication of hot-wire probes and electronics for constant temperature anemometers, Latin American Applied Research 40, 233-239.

[16] Batchelor GK (1970) The theory of homogeneous turbulence. Cambridge Univ. Press, 55-75.

Flow rate estimation using acoustic field distortions caused by turbulent flows: time-reversal approach

A L Zimmermann, N Pérez and J C Adamowski

Department of Mechatronics and Mechanical Systems Engineering, University of São Paulo, São Paulo, SP, 05508-10, Brazil

E-mail: zimmer09@gmail.com

Abstract. A new acoustic technique for flow rate estimation is proposed here. This technique is based on the traditional ultrasonic cross-correlation flow meter, but instead of using a continuous wave or pulse trains in each transmitter-receiver pair, the acoustic time-reversal technique is applied. The system relies on the principle that a turbulent flow with multiple vortices will cause random distortions in a given acoustic field; hence, analyzing this noise caused in the ultrasound signal by the turbulence over time allows a "signature" or "tag" of the flow to be defined. In other words, the vortices modify the frequency response function of the flowing system uniquely, since the distortion is assumed to be random. The use of the time-reversal procedure in the cross-correlation flow meter provides improvements in several aspects: it simplifies the signal processing needed after the reception of the signals, avoiding the use of a demodulator to obtain the signature of the vortex; the signal is focused at the position of the reception transducer and; the sensitivity is also increased because the wave travels twice in the acoustic channel. The method is theoretically discussed showing its limitations and improvements. Experimental results in a laboratory water tank are also presented.

1. Introduction

Several methods of flow rate estimation using ultrasound have been proposed and tested over time. Since the early works, using transit time of an acoustic pulse [1], evolving to more complex models and techniques as Doppler shift detection [2], vortex shedding and the cross-correlation flow meters [3, 4], ultrasonic techniques are useful in flow measuring for different fluids and different regimens.

Ultrasonic techniques are no invasive, don't introduce pressure drops, and are suitable for flows with two phases or solids in suspension. The guidelines to select the most adequate type of flow meter depend on the application and the flow rate of the setup [5].

Ultrasonic flow meters can be classified in three different groups:

- Transit time flow meter; averages the difference in measured transit time between ultrasound pulses propagating together with and against the flow.
- Doppler effect flow meter; detects the frequency shift introduced in the original signal by reflectors travelling along the flow, such as air bubbles, to estimate the average velocity of the fluid.

- Cross correlation flow meter; the average velocity of the fluid is estimated from the cross correlation maximum of the signals measured in two different sections in the flow, upstream and downstream respectively.

Cross correlation flow meters are a robust solution to measure the average (bulk) velocity in the case of a non uniform flow. Compared to transit time flow meters and Doppler effect flow meters, the cross correlation flow meter presents the advantage of not depending upon the velocity of sound in the fluid, therefore, variations of the sound velocity caused by changes in the fluid temperature and other factors don't introduce measurement errors. Cross correlation flow meters can also acquire information from a larger portion of the velocity profile so that a more reliable estimate of the flow rate may be achieved [3].

Several devices are available in the market, including applications to measure ocean currents [6], blood flow [7] and mixture of gas-liquid flow [8]. Since this type of flow meter is dependent on the existence of perturbations within the flow, it fails in laminar flows in which there are no perturbations to correlate.

This work introduces an important modification in the cross correlation technique substituting the use of continuous wave and electronic demodulation of the signals for the time reversal of acoustic fields, thus using pulsed signals and avoiding problems associated with standing waves [9].

In section 2, the traditional cross correlation technique is reviewed. In section 3, the time reversal of the ultrasonic signals is introduced. Section 4 presents the experimental setup and the signal processing needed. Section 5 briefly explains the first experimental results obtained in laboratory and finally section 6 summarizes the conclusions.

2. Ultrasonic cross correlation flow meter

The basic arrangement of the cross correlation flow meter consist in two pairs of ultrasonic transducers axially distant, one upstream and the other downstream. Figure 1 shows a schematic diagram of the setup.

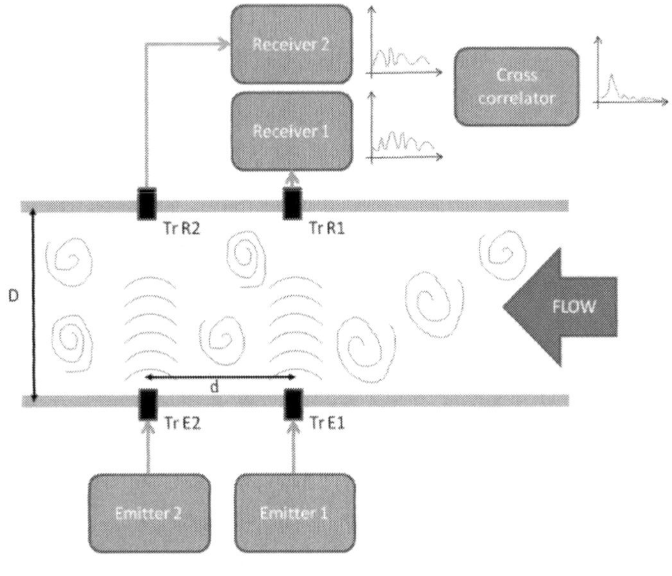

Figure 1. Ultrasonic cross correlation flow meter schematic diagram.

The proper operation of the flow meter depends on the presence of perturbations within the flow, such as, air bubbles, solids in suspension or differences in the density and local velocity produced by vortexes in the flow. These perturbations "modulate" the ultrasonic signal as it travels through the turbulent flow from the emitter transducers towards the receiver transducers. In the traditional version of the device, a continuous sinusoidal wave is used, and then the amplitude and phase of this wave are modulated by the perturbations in the flow [3]. At the first observation, the signals recorded at receptor 1, *Tr R1*, and at receptor 2, *Tr R2*, look like noisy signals with low similarity. But when the distance between the transducers pairs (marked as d in fig 1) is lower than a certain value, defined here as the "correlation length" of the perturbations, the cross correlation between these two signals displays a well defined maximum. The time position of this maximum, τ, is displaced from zero and represents the average time taken by a perturbation to travel axially from the position of transducer pair 1 (*Tr E1* – *Tr R1*) to the position of transducer pair 2 (*Tr E2* – *Tr R2*). The analytical formulation of this correlation R_{y1y2} can be expressed as

$$R_{y_1y_2}(\tau) = \frac{1}{T}\int_0^T y_1(t)y_2(t+\tau)dt \tag{1}$$

Here, $y_1(t)$ is the signal recorded at the pair 1 and $y_2(t)$ is recorded at pair 2. In practice the cross correlation is computed in the frequency domain as

$$R_{y_1y_2}(\tau) = \mathcal{F}^{-1}(\mathcal{F}(y_1).\mathcal{F}^*(y_2)) \tag{2}$$

Were the operator $\mathcal{F}$ is the Discrete Fourier transform and $\mathcal{F}^*$ is its complex conjugate. The average velocity $\langle v \rangle$ of the flow can be obtained from the correlation time τ as

$$\langle v \rangle = k\frac{d}{\tau} \tag{3}$$

where k is the calibration factor, usually adjusted empirically to take into account the non uniformity in the velocity profile. There are several research works focused on finding the relation between the correlation time τ and the average velocity $\langle v \rangle$ [10, 11, 12].

The use of continuous sinusoidal waves introduces stationary waves in the pressure field generated by the transducers. To avoid this problem a pulsed version of the cross correlation flow meter was proposed [8], where the estimation of the average flow velocity is made taking into account only the amplitude of the emitted pulse.

The selection of the transducers depends on the distance between the emitter and the receiver (marked as D in fig 1), where distance of several meters are required, like rivers and intakes of hydroelectric turbines, low frequency transducers form 20 KHz to 100 KHz are suitable, for small distances such as pipe diameters, high frequency transducers are more sensitive to small perturbations.

3. Time reversal of ultrasonic signals

3.1. Time reversal basis

The focusing of acoustic waves using time reversal in signal processing was developed by Mathias Fink and his collaborators during the late eighties [13]. In those first works, the main application scope was the improvement of acoustic images through inhomogeneous media. However, within the last years, several other practical applications using time reversal focusing of acoustic waves have been developed [14].

Time reversal of acoustic waves starts with the measure of the impulse response, $h(r,t)$, of a given system. In practice a sinc pulse (flat frequency spectrum up to the cut-off frequency) is inputted in the emitter transducers (*Tr E1* and *Tr E2*). The received signal at the receivers (*Tr R1* and *Tr R2*) is recorded and time reversed, generating $h(r,T-t)$ which is then retransmitted through the media, again from *Tr E1* and *Tr E2*, to *Tr R1* and *Tr R2*, respectively, where T is the temporal length of the signal. After this second emission, the signal is again received and recorded. This resultant signal can be obtained theoretically by the convolution between the time reversed signal and the impulse response of the system $h(r,T-t) \otimes h(r,t)$, assuming that the impulsive response h does not change from the first to the second shot, this operation is given by

$$\int_{-\infty}^{+\infty} h(r,\tau).h(r,t-T+\tau)d\tau \tag{4}$$

Changing the variable $t - T = t'$ the product (4) is equivalent to the autocorrelation function of the impulse response h, given by

$$\int_{-\infty}^{+\infty} h(r,\tau).h(r,t'+\tau)d\tau \tag{5}$$

Expression (5) presents a maximum at $t' = 0$, which corresponds to the signal's temporal length T. As these signals travel from *Tr E1* and *Tr E2* to *Tr R1* and *Tr R2* they are modulated by the perturbations and recorded in *Tr R1* and *Tr R2*. It is important to notice that in the frequency domain, the time reversal process is equivalent to the product of the transfer function $H(r,\omega)$ by its complex conjugate $H^*(r,\omega)$. This product cancels the phase of each component in the frequency response.

$$H(r,\omega).H^*(r,\omega) = |H|^2 \tag{6}$$

3.2. Application to detect changes in the media.
The time reversal maximum depends on the amplitude of each frequency component and also on the phase of each component. Every change in the frequency response affect this maximum, thus the time reversal maximum is sensitive to any change in H, both in amplitude and phase. As observed, the time reversal process is the autocorrelation function of the impulsive response; this fact was used to detect variations in the system's response for example introduced by temperature [15].

Arrays of ultrasonic transducers (Time reversal mirrors) have been used to characterize vortex and to estimate velocity profiles of various flows, showing the sensitivity of the time reversal technique to vortex patterns [16, 17, 18].

The application of the technique starts with the characterization of the impulsive response; this is achieved emitting a sinc pulse both in *Tr E1* and *Tr E2*. In this work the spectrum of the sinc is flat up to 5 MHz. In this application both emitters and receivers have central frequency 5 MHz.

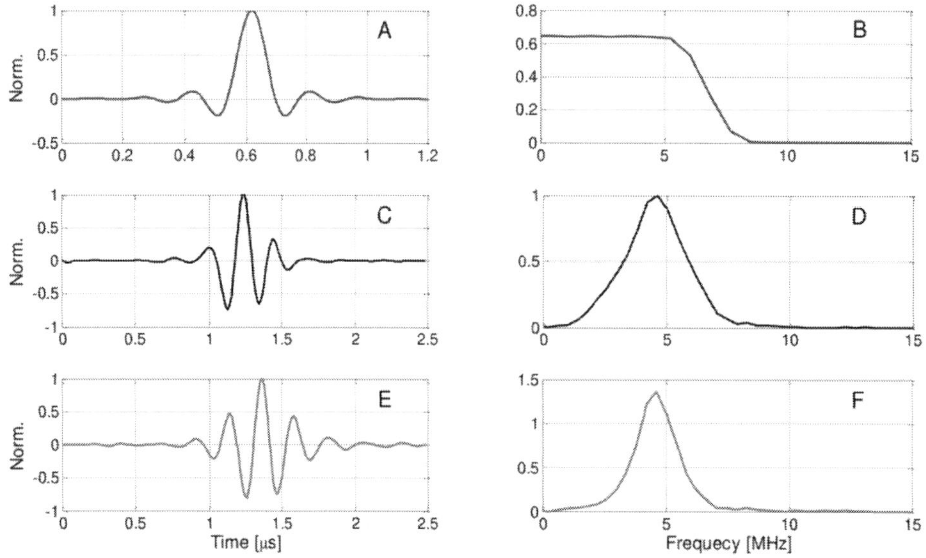

Figure 2. A) Sinc signal used to obtain the impulsive response, B) Spectrum of the sinc, C) Impulsive response, D) Spectrum of the impulsive response. E) Response to the time reversed impulsive response or ASIR (Autocorrelation of the System's Impulse Response). F) Spectrum of the signal E.

After recording the impulse response, figure 2C, the system is programmed to reverse it in the time domain, send it trough the emitter transducer, acquire the new response and record the maximum of the received signal. In this example, the acquisition window is 2.5 μs wide; this limits the maximum sampling frequency to 400 KHz, one point (the maximum of the acquisition window) for each window. However, the main limitation is the transmission time between the electronic device used to acquire the signals and the computer which lowers, in this case, the maximum sampling frequency to 50 Hz. Observe that the signal recorded after time reversal process, figure 2E, has a symmetrical shape. This is typical in an autocorrelation function, the presence of asymmetries shows that the impulsive response of the system has some differences in each shot. These differences are also reflected in the value of the correlation maximum.

Two noisy signals containing the information of the flow perturbations are then recorded; these signals are built as arrays whose elements are the maximums of each acquisition window at *Tr R1* (downstream) and at *Tr R2* (upstream). However the cross correlation of these signals shows a defined maximum, see figure 3C.

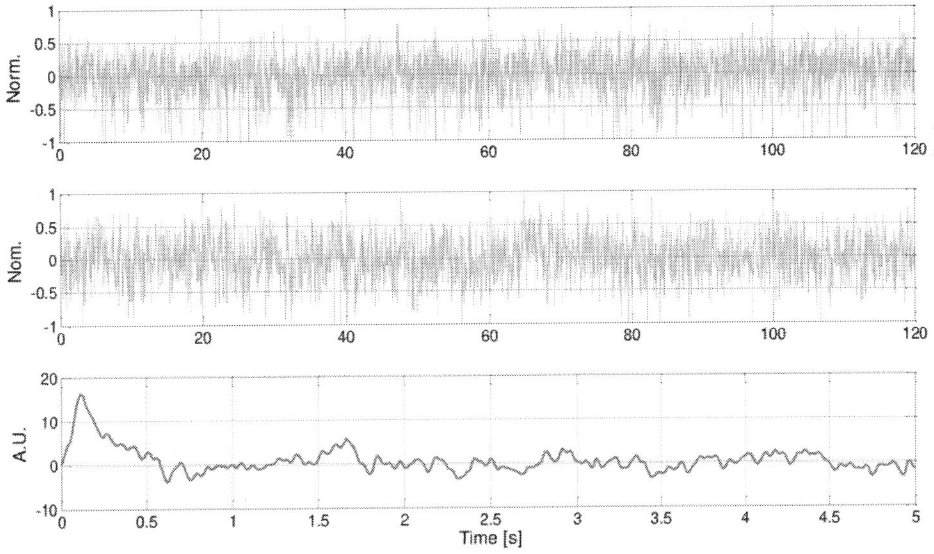

Figure 3. A) Maximum sequence at *Tr R1*, B) maximum sequence at *Tr R2*, C) Cross-correlation between the maximum sequences from *Tr R1* and *Tr R2*.

Results obtained for different flow rates and acquisition windows are presented in section 5.

4. Experimental setup

4.1. Hydrodynamic setup

The experiments were realized in a rectangular cross-section water channel, placed inside a water tank where the flow is generated by a submersible pump. The water channel is made of Plexiglas and its dimensions are: 1,000 mm length, 100 mm high and 100 mm width, being the axial distance between the two transmitter-receiver pair of 30 mm. The transducers stand 800 mm from the pump and the water depth is of 60 mm, measured from the base. The transducers are placed 30 mm high, measured from the base. Figure 4 shows the setup's schematic diagram.

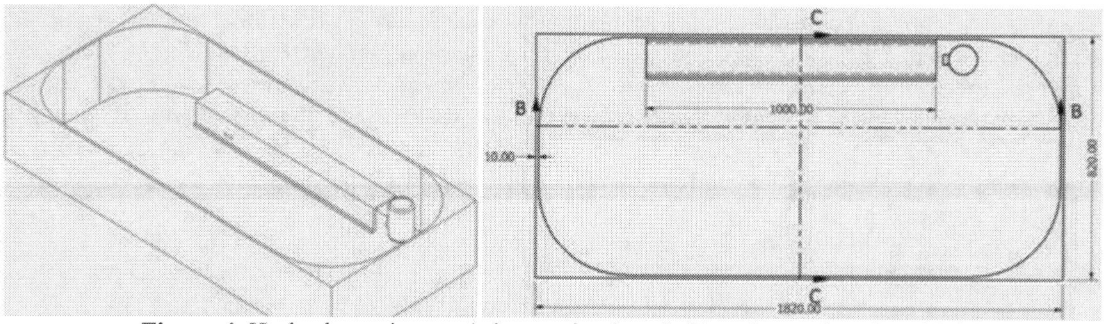

Figure 4. Hydrodynamic setup's isometric view (left) and superior view (right).

The tank was filled with water up to 60 mm from the base, leaving the water channel's cross-section not completely filled, since its high is of 100 mm. This partial filling combined with the pump placed close to the channel's entrance create an adequate level of turbulence for the tests, generating non-stationary vortexes.

The flow is generated by a calibrated DC pump and its flow rate can be well controlled. The transducers are placed inside holes in the water channel walls, as shown in figure 5, in order to reduce signal decay and multiple reflections in the Plexiglas. All the measurements and acquisitions were made during system's steady state operation.

Figure 5. The positioning of transducers into the walls of the water channel.

4.2. Electronic setup and signal processing

The experimental prototype of the cross-correlation flow meter consists of 2 transmitter channels and 2 receiver channels, all connected to piezoelectric ultrasonic transducers with central frequency of 5.0 MHz.

The electronic apparatus is implemented with *OPEN System*, a Multi channel ultrasonic device made by *Lecoeur Electronique*. *OPEN System* provides accessible parameters and its open architecture makes it suitable for research, along with its analog transmitters that allow the generation of arbitrary waveforms used in this work for the emission of sinc and time-reversed signals. It has also transmitting and receiving functions, which concentrate all the signal emission and reception channels and functions in a single configurable electronic device. For this experiment, the system was used with 4 active channels and sampling rate of 80 MHz, 12 bits A/D and D/A converters, input impedance of

50 Ohms in the 2 transmitter channels and gain of 10dB at the receiver channels. The *OPEN System*'s interface is implemented through a PC linked to the device using USB 2.0.

For the first step of sending a spike excitation through the two emitter channels, a sinc function of 1.25 μs was used, as shown in the figure 2A, being its formula given by

$$\mathrm{sinc}(2.\pi.f_0.t) \tag{7}$$

Where $f_0 = 2.0$ MHz. The signal has a cut-off frequency of 5.0 MHz in order to match the 5.0 MHz transducers used, as shown in the figure 2B. The sinc signal was built digitally with 100 samples for all tests, based in the sampling rate of 80 MHz. The signal is then treated with a Blackman window for smoothing its edges and is then normalized before being sent to the *OPEN System*.

The reception of the impulse response is implemented through A-scans with transducers *Tr R1* and *Tr R2*, using acquisition windows of 2.5 μs as shown in the figure 2C.

The cycle of emission of the sinc signal and reception of the impulse response is repeated 128 times for each transducer pair in order to estimate a statistical mean of the impulse response for both pairs. The calculated mean signals are then time reversed and loaded into the *OPEN System* through USB 2.0 to be emitted through transducers *Tr E1* and *Tr E2*.

The emission of these signals generates responses as the one shown in the figure 2E, which will be called ASIR (Autocorrelation of the System's Impulse Response) from now on. The ASIR signals are obtained at both receiver transducers, *Tr R1* and *Tr R2*, also with sampling frequency of 80 MHz and acquisition window of 2.5 μs. After the acquisition, the ASIR signals are then sent to the PC where a peak detection algorithm stores the value of the signal's maximum. This cycle is repeated continuously in order to build arrays with 90,000 samples of ASIR's maximums for pairs 1 and 2, forming the digital signals A1(t) and A2(t) respectively.

Although the ASIR signals are sampled with 80 MHz, the signals A1 (t) and A2 (t) are built with a sampling frequency of 50Hz, therefore, with a Nyquist frequency of 25Hz [19], due to the time spent in the USB 2.0 communications between PC and the *OPEN System*.

Offsets of A1(t) and A2(t) are removed and the signals are normalized. A1 (t) and A2 (t) are then interpolated using the algorithm 8.1 [20] to emulate a sampling rate 20 times higher, forming the signals AI1 (t) and AI2 (t), as the ones shown in the figures 3A and 3B. These interpolated signals are then cross-correlated using a PC-implemented digital algorithm based on the equation (2), then generating a signal as the one shown in the figure 3C.

As explained in section 3.2, the position of the maximum of the cross-correlation signal provides an estimation of the transit time (τ) of perturbations travelling from pair 1 to pair 2. As the distance between the two pairs of transducers is d = 30 mm, the bulk velocity of the flow can be calculated by equation (3).

5. Experimental results

The system was tested for flow rates ranging from 202ml/s to 391 ml/s, with Reynolds numbers from 4023 to 7793 respectively, being the Reynolds number given by [21]

$$Re = \frac{\rho.V.D_H}{\mu} \tag{8}$$

Where ρ is the water's density; V is the flow's average velocity; μ is the dynamic viscosity and D_H is the hydraulic diameter defined as

$$D_H = 4.A/P \tag{9}$$

Where A is the cross sectional area and P is the wetted perimeter [21]. Reynolds numbers were calculated assuming the following properties for the water: μ=1.002 x 10^{-3} N s/m^2, ρ=1,000 kg/m3.

The experiment was realized acquiring signals during steady state operation of five flow rates: 202 m/l, 254 ml/s, 307 ml/s, 351 ml/s and 391 ml/s. For each of the four flow rates, the samples of τ were acquired and saved continuously for 30 minutes, forming arrays with 90,000 samples.

The tests showed a high dependency of the cross-correlation results on the flow rate applied to the water channel, where the peak of the cross-correlation becomes narrower and closer from t=0s as the flow rate is increased. This is primarily expected since the reduction of flow rate brings the Reynolds number close to the boundary between turbulent and transition flows. Thus as the Reynolds number becomes smaller, the level of turbulence decreases and so the distortions in the acoustic field. Once the level of distortions decreases, the modulation on the signal's amplitude becomes smoother, making the cross-correlation peak broader and smaller until the point where it vanishes at the beginning of the laminar regimen, increasing the errors associated with the estimation.

Figure 6 shows the effect of the flow rate on the cross-correlation peak, with its maximums normalized.

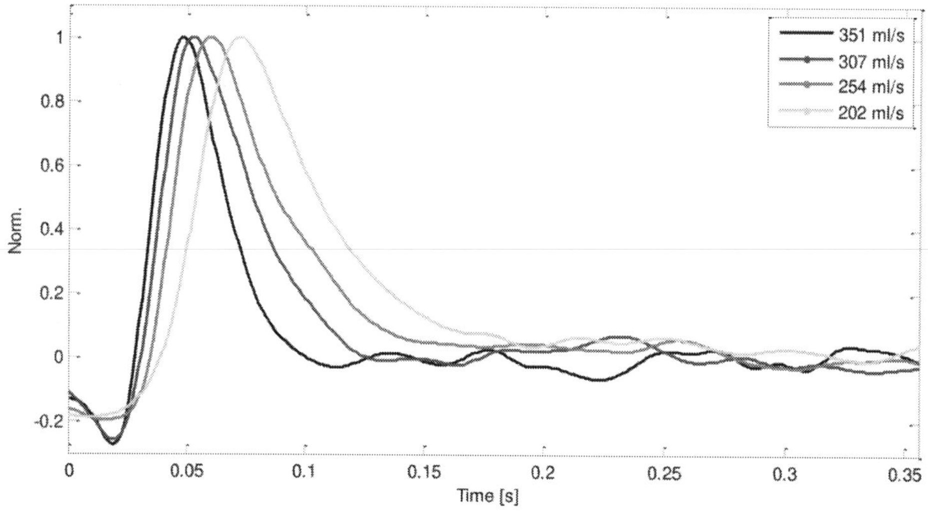

Figure 6. Normalized cross-correlation for the first four flow rates tested.

Another important factor concerning the cross-correlation peaks is the length of the signals AI1(t) and AI2(t) used in the calculations. Estimations of τ were made using signals with 90,000, 67,500, 45,000 and 22,500 samples. The tests showed that the length of signals (the length of the acquisition window) used for the calculation of the cross-correlation has a significant influence over results, where the longer is the acquisition window; the bigger is the peak in the cross-correlation which provides more reliable estimations. Figure 7 shows the influence of the acquisition window's length on the peak of the cross-correlation.

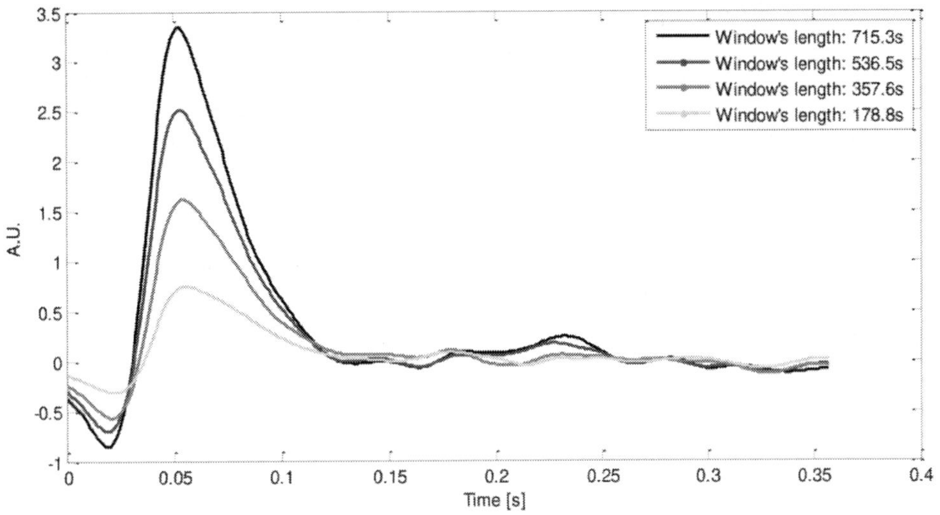

Figure 7. Dependency of the cross-correlation peak on the length of the acquisition window.

The dependency of the cross-correlation peak on the length of the acquisition window shows strong linearity, as expected by the theoretical model, for the reason that with longer signals, the integral represented by equation (5) will in fact be a discrete sum with more non-zero elements. Figure 8 shows the linear relation between the cross-correlation peak and the time length of the acquisition window.

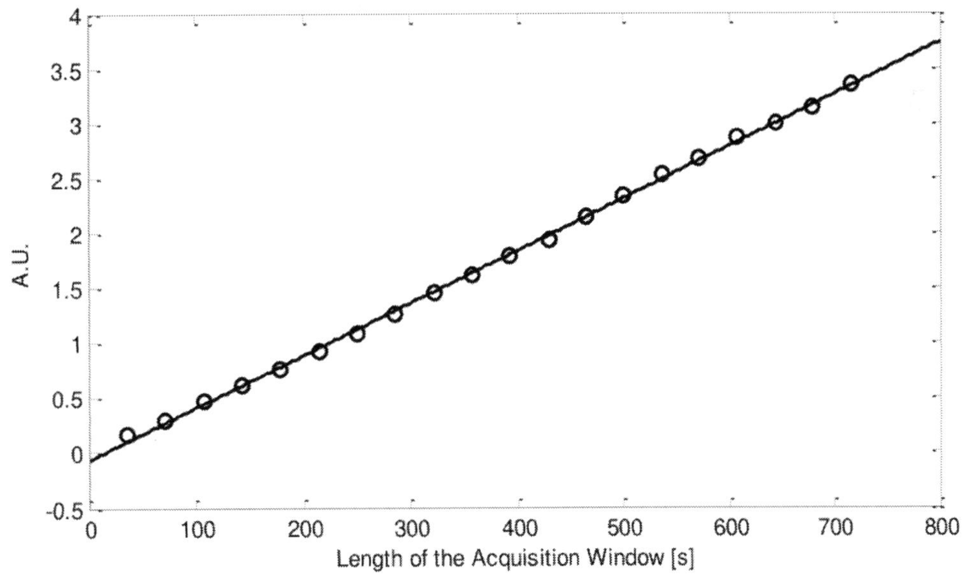

Figure 8. Value of the cross-correlation peak (Y axis) plotted against the time length of the acquisition window (X axis) and a linear regression curve, showing strong linearity.

The mean values of τ estimated from the acquisitions allow the estimation of the calibration factor for this experiment. The real mean velocity can be calculated as follows

$$\langle Vr \rangle = Q/A_{CS} \tag{10}$$

Where Q is the flow rate given by the pump and A_{CS} is the cross-sectional area of the flow. Substituting (11) in the equation (3) the calibration factor is given by

$$k = \frac{Q.\tau}{A_{CS}.d} \tag{11}$$

Table 1 shows the values of k estimated from equation (11) using the five mean values of τ estimated with the acquisitions.

Table 1. Estimation of the calibration factor.

Flow Rate [l/s]	τ	k
0.202	0.0972	81.6
0.254	0.0735	77.7
0.307	0.0654	83.6
0.351	0.0598	87.4
0.391	0.0505	82.2
	Average k:	**82.5**

Figure 9 shows the average estimated transit time of the perturbations (τ) between ultrasonic beams 1 and 2 as a function of the actual flow rate.

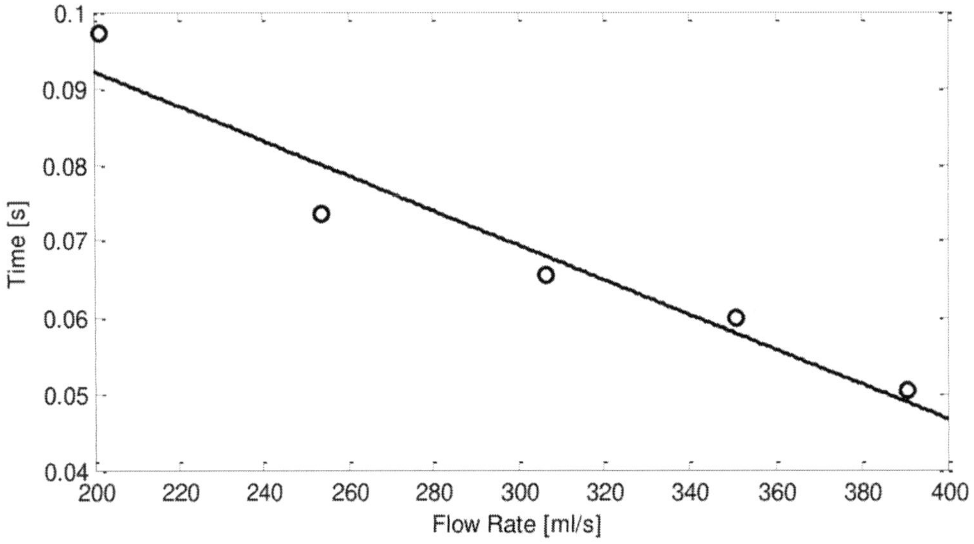

Figure 9. Average estimated transit time of perturbations between ultrasonic beams 1 and 2 (Y axis) plotted against the actual flow rate (X axis) and a linear regression.

After the calculus of the calibration factor, the flow rate estimation was then executed using the ultrasonic apparatus configured with the calculated k. Table 2 shows a comparison between the actual flow rate, estimated by measuring the time taken by the flow to fill a known volume, and the flow rate estimated by the ultrasonic system developed here.

Table 2. Comparison between the actual flow rate and the flow rate estimated with the ultrasonic system after calibration.

Actual Flow Rate [l/s]	Measured Flow Rate [l/s]	Error (%)
0,202	0,204	1,1
0,254	0,269	6,2
0,307	0,303	1,3
0,351	0,331	5,6
0,391	0,392	0,4

6. Conclusions

As table 2 shows, the initial results for the flow rate estimation in a pipe with rectangular cross section were considerably accurate, presenting errors smaller than 7%. An important point here is the characterization of the vortex, bubbles and other kinds of disturb generators, since it is important to know the frequency band width of these perturbations in order to develop suitable hardware for the detection of all the information carried across the flow.The implementation of hardware which does not require a high level of communication between the PC and de *OPEN System* is the next step in this research, intending to improve the system's sampling rate through the use of DSP directly in the *OPEN System*.

References

[1] Moore P, Brown G and Stimpson B 2000 *Meas. Sci. Technol.* **11** 1802
[2] Brody W and Meindl J 1974 *IEEE T. Bio-Med Eng.* **21** 183
[3] Coulthard J 1973 *Ultrasonics* **11** 83
[4] Beck M S 1981 *J. Phys. E: Sci. Instrum.* **14** 7
[5] Sanderson M L and Yeung 2002 *J. Flow Meas. Instrum.* **13** 1245
[6] Iorio D D, Lemon D and Chave R 2005 *J. Atmos. Ocean Tech.* **22** 1602
[7] Sisbot S 2005 *J. Flow Meas. Instrum.* **16** 27
[8] Xu L A, Green R G, Plasokowski A and Beck M S 1988 *J. Phys. E: Sci. Instrum.* **21** 406
[9] Jacobson S A, Denbigh P B and Naude D E 1985 *Ultrasonics* 128
[10] Lysak P D, Jenkins D M, Capone D E and Brown W L 2008 *J. Flow Meas. Instrum.* **19** 1
[11] Lysak P D, Jenkins D M, Capone D E and Brown W L 2008 *J. Flow Meas. Instrum.* **19** 41
[12] Schneider F, Peters F and Merzkirch W 2003 *Meas. Sci. Technol.* **14** 573
[13] Fink M 1992 *IEEE Trans. UFFC.* **39 (5)** 555
[14] Fink M 1997 *Physics Today*, **20** 34
[15] Tourin A, Derode A and M. Fink 2003 *World Congr. Ultrasonics* **1** 851
[16] Roux P and Fink M 1995 *Eurosphysics Lett.* **32** 25
[17] Manneville S, Maurel A, Roux P and Fink M 1999 *Europ. Phys. Journ. B* **9 (3)** 545
[18] Manneville S, Robres J H, Maurel A, Petitjeans P and Fink M 1999 *Phys. Fluids* **11 (11)** 3380
[19] Oppenheim A V, Schafer R W 1975 *Digital Signal Processing* (Englewood Cliffs: Prentice Hall)
[20] DSP Committee 1979 *Programs for Digital Signal Processing* (New York: IEEE Press)
[21] Munson B R, Young D F, Okiishi T H 2005 *Fundamentals of Fluid Dynamics* (New York: Wiley)

Numerical analysis of the immersed boundary method applied to the flow around a forced oscillating cylinder

L. C. Pinto[1], E. B. C. Schettini[2] and J. H. Silvestrini[3]
[1,2]Instituto de Pesquisas Hidráulicas, Universidade Federal do Rio Grande do Sul, Av. Bento Gonalves 9500, 91501-970, Porto Alegre - RS - Brazil
[3]Faculdade de Engenharia, Pontifícia Universidade Católica do Rio Grande do Sul, Av. Ipiranga 6681, 90619-900, Porto Alegre - RS - Brazil
E-mail: [1]leandroconp@gmail.com, [2]bcamano@iph.ufrgs.br, [3]jorgehs@pucrs.br

Abstract. In present paper, Navier-Stokes and Continuity equations for incompressible flow around an oscillating cylinder were numerically solved. Sixth order compact difference schemes were used to solve the spatial derivatives, while the time advance was carried out through second order Adams Bashforth accurate scheme. In order to represent the obstacle in the flow, the Immersed Boundary Method was adopted. In this method a force term is added to the Navier-Stokes equations representing the body. The simulations present results regarding the hydrodynamic coefficients and vortex wakes in agreement to experimental and numerical previous works and the physical lock-in phenomenon was identified. Comparing different methods to impose the IBM, it can be concluded that no alterations regarding the vortex shedding mode were observed. The Immersed Boundary Method techniques used here can represent the surface of an oscillating cylinder in the flow.

1. Introduction

The applications of the study of flow around oscillating bodies are diverse. The process of oil extraction from deep waters, for example, can be done through floating platforms, which are equipped with slender, cylindrical structures, called risers. Risers are responsible to conduct the oil from the bottom of the sea to the platform. Another important application is concerned to the study of the oscillating macrophyte vegetation in lakes and watercourses. Emerging macrophytes prominently in lakes can start a movement caused by external agents, as wind or waves, changing its morphology.

When a cylinder is oscillating, significant changes can be observed in the flow. Studies by [1] showed that, if the oscillation has well-defined frequency and amplitude displacement, the vortex street wakes presents characteristics shedding modes. Those authors mapped such modes, and introduced a symbolic code classification of letters and numbers, describing the combination of paired or simple vortex shedding during each forced oscillation cycle of the cylinder.

According the experiments by [2], when the oscillation frequency (f_o) is close to the frequency corresponding to the Strouhal number (f_s), a certain control in the vortex shedding is caused by the cylinder oscillation. It could be observed only for a range of f_o and up to a certain oscillation amplitude. This phenomenon is usually known as *lock-in*.

In this paper, results of hydrodynamic coefficients and instantaneous vorticity fields were compared to identify which numerical technique best represents the solid boundaries of the oscillating cylinder in the flow.

2. Numerical Methodology

The incompressible Navier-Stokes and the continuity equations:

$$\frac{\partial \mathbf{u}}{\partial t} = -\nabla p - \frac{1}{2}[\nabla(\mathbf{u}\otimes\mathbf{u}) + (\mathbf{u}\cdot\nabla)\mathbf{u}] + \frac{1}{Re}\nabla^2\mathbf{u} + \mathbf{f}, \tag{1}$$

$$\nabla\cdot\mathbf{u} = 0, \tag{2}$$

are directly solved on a computational grid in non-staggered configuration. In this equations, $\mathbf{u}$ and p are the velocity and pressure fields, respectively, and Re is the non-dimensional Reynolds Number. The present computational code, called `Incompact3d`, uses sixth order compact difference schemes to solve the spatial derivatives, while the time advance was carried out through second order Adams-Bashforth accurate scheme. In order to represent the obstacle in the flow we used here an Immersed Boundary Method (IBM), which is based on the addition of a forcing term in the flow governing equations to model the presence of the body. The direct forcing term $\mathbf{f}$ can be expressed as:

$$\mathbf{f} = \varepsilon\{\frac{1}{2}[\nabla(\mathbf{u}\otimes\mathbf{u}) + (\mathbf{u}\cdot\nabla)\mathbf{u}] - \frac{1}{Re}\nabla^2\mathbf{u} + \nabla p + \frac{\mathbf{u}_o - \mathbf{u}}{\Delta t})\}, \tag{3}$$

with $\varepsilon = 1$ in the body, and 0 everywhere else. The no-slip condition in the moving boundaries is obtained when:

$$\mathbf{u}(D/2, \theta, z, t) = \mathbf{u}_d, \tag{4}$$

where $\mathbf{u}_d$ is the forced oscillating cylinder velocity. The target velocity $\mathbf{u}_o$ is calibrated by the external velocity field and an artificial flow is created inside the cylinder. According to [3], a formal expression to $\mathbf{u}_o$ can be given as:

$$\mathbf{u}_o(r, \theta, z, t) = \mathbf{u}_d + [\mathbf{u}_d - \mathbf{u}(D - r, \theta, z, t)]g(r), \tag{5}$$

with

$$g(r) = sin(\frac{2\pi r^2}{D^2}). \tag{6}$$

The principle of direct forcing is to preserve the no-slip boundary condition avoiding discontinuities in the first derivative at the cylinder surface [4].

The oscillating cylinder velocity $\mathbf{u_d}$ is defined from the cylinder displacement, which can be expressed by $y_c = y_0 + A/Dcos(2\pi f_o t)$, where y_0 is the initial coordinate cylinder and A/D the non-dimensional displacement amplitude based on the diameter of the cylinder (D).

The equations (1) and (2) are solved in a computational domain $L_x \times L_y$ discretized on a Cartesian mesh of $n_x \times n_y$ nodes. Free-slip boundary conditions are assumed at the lateral boundaries ($y = 0$, $y = L_y$) while in the streamwise direction (inflow/outflow boundary) are imposed as velocity Dirichlet conditions. At the inlet ($x = 0$), an streamwise uniform profile (U) is prescribed while at the outlet ($x = L_x$), an one-dimensional convection equation is solved.

Regarding the validation, studies as [4, 5, 6, 7] using `Incompact3d`, presented both qualitatively and quantitatively results for conventional and complex flow geometries around fixed obstacles. According such studies, the code is a very attractive tool for solving flows in the context of Direct Numerical Simulations (DNS). Here, it was used in two-dimensional configuration.

3. Results

Simulations for oscillating cylinder with the frequency ratio $F = f_o/f_s$ varied from 0.8 to 1.2 were performed using the computational grid n_x, n_y = 361, 217 (and 325) and the computational domain L_x, L_y = $20D$, $12D$ (and $18D$). In all simulations, the parameters $Re = 185$ and $A/D = 0.2$ were constants. A fixed time step $\Delta t = 0.005D/U$ is used in order to ensure both convective (Courant-Friedrich-Levy) and diffusive criteria.

3.1. Hydrodynamic coefficients

The results of the mean drag $<C_D>$ and root mean square lift C_{Lrms} coefficients varying the frequency ratio $F = f_o/f_s$ are compared in figure 1 with other authors [8, 9].

The numerical simulations by [8] were done with the computational grid n_x,n_y = 1201, 1201 and the computational domain L_x, L_y = $24D$, $24D$. The IBM was implemented using the feedback forcing technique based on an artificial term that can "freezes" the fluid in the body region through a damping oscillation process.

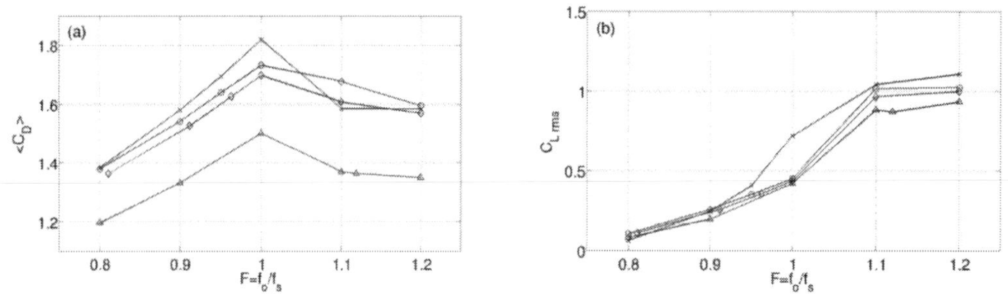

Figure 1. Force coefficients versus frequency ratio F. (a) $<C_D>$, (b) C_{Lrms} . × - Results by [8] using feedback forcing; O - Present work (n_x, n_y = 361, 217; L_x, L_y = $20D$, $12D$); ◊ - Present work (n_x, n_y = 361, 325; L_x, L_y = $20D$, $18D$); △ - Numerical simulation by [9].

The results in the figure 1 shown that the shape curves of the coefficients is quite similar. All simulations presented a pick for $<C_D>$ when F is close to one and an increasing C_{Lrms} characterizing the lock-in phenomenon. Despite the use of low grid resolution, this study presented results for C_{Lrms} very close to the reference [9]. It could be observed a difference in the $<C_D>$ of approximately 0.2 between our results and those obtained in [9]. Numerical simulations by [10], found similar discrepancies for this hydrodynamic coefficient. They attribute these discrepancies to the $<C_D>$ obtained for simulations with the fixed cylinder. Simulation for the fixed cylinder were accomplished and its results for $<C_D>$, C_{Lrms} and f_s are compared in table 1. The results for $<C_D>$ shown that the discrepancy between the present work and [9] for fixed cylinder is approximately 0.2. This value is near to the differences observed in the drag curves for the oscillating cylinder (Figure 1). Numerical and experimental simulations for fixed cylinder at low Reynolds number are compared in [11]. Such authors presented a $<C_D> \approx 1.4$, at $Re \approx 190$, which is close to our results.

Comparing simulations with different computational domain, it was observed that as the vertical domain length L_y increases, the drag and lift coefficients decreases. These results indicate the existence of the blockage effect as expected.

Table 1. Comparative of the hydrodynamic coefficients $<C_D>$, C_{Lrms} and the Strouhal Number f_s for fixed cylinder.

Simulations	$<C_D>$	C_{Lrms}	f_s
IBM using feedback forcing by [8]	1.420	0.492	0.192
Present work (n_x, n_y: 361, 217; L_x, L_y: 20D, 12D)	1.488	0.516	0.199
Present work (n_x, n_y: 361, 325; L_x, L_y: 20D, 18D)	1.469	0.507	0.196
Results by [9]	1.287	0.443	0.195
Results by [11] at $Re \approx 190$	1.4	-	-

3.2. Vortex wakes

Instantaneous vorticity fields of the oscillating cylinder for $F = 0.8$ are shown in figure 2. Regarding the street wakes, significant alterations were not observed. The vortex shedding mode 2S was identified here for all simulations, corresponding to the shedding of two counter-rotating simple vortex at each oscillation cycle.

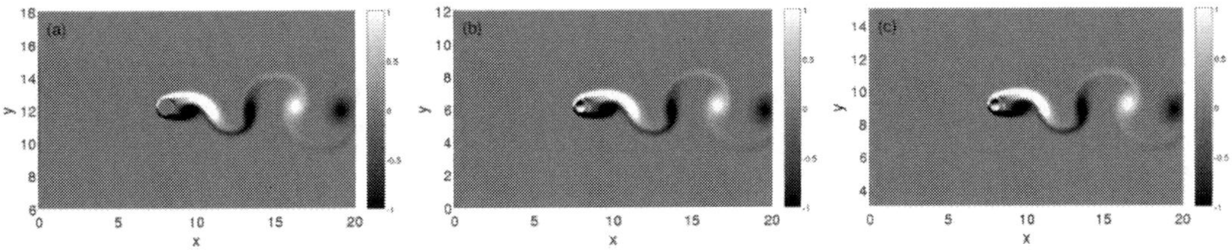

Figure 2. Instantaneous vorticity fields for simulations with $F = 0.8$: (a) IBM using feedback forcing by [8], (b) Present work (n_x, n_y: 361, 217; L_x, L_y = 20D, 12D), (c) Present work (n_x, n_y = 361, 325; L_x, L_y = 20D, 18D).

The figure 3 presents a comparative of the region next to the cylinder. In present simulations degraded results were expected due the lower grid resolution. However, the vicinity region of the cylinder presents quite similar results as shown by the isocontours of vorticity.

4. Conclusions

The results obtained shown that IBM technique used in the present work can represent the surface of a forced oscillating cylinder. The lock-in phenomenon was represented and the hydrodynamic coefficients are in agreement to other works in the current literature. Despite the use of lower grid resolution, this study presents results for the drag and lift forces very close to the others references (see, for instance [9] and [11]), and the vorticity fields are in agreement to others simulation of higher grid resolution.

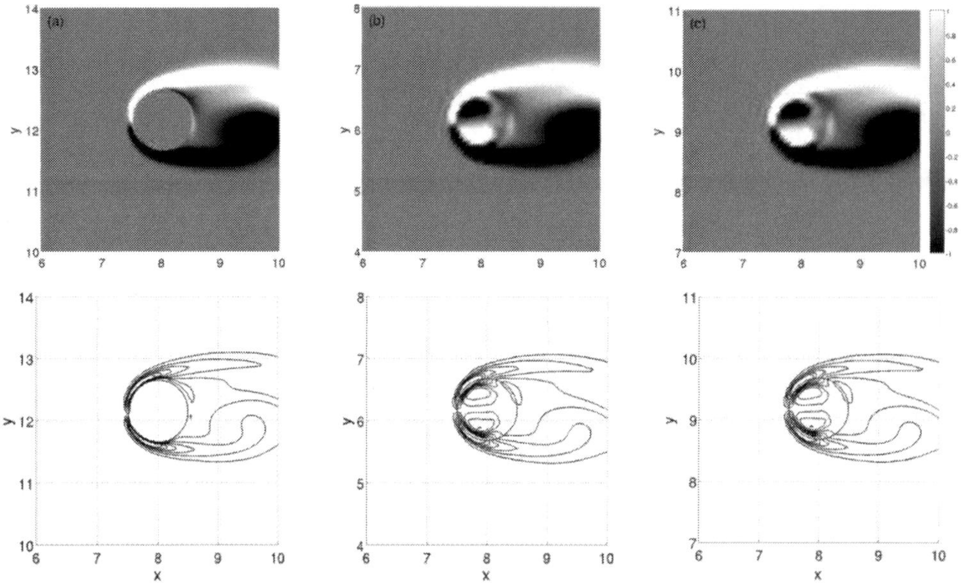

Figure 3. Instantaneous fields and isocontours (isolevels from -6 to 6 at each 1) of vorticity for simulations with $F = 0.8$: (a) IBM using feedback forcing by [8]; (b) Present work (n_x, n_y: 361, 217, L_x, $L_y = 20D$, $12D$); (c) Present work (n_x, n_y = 361, 325, L_x, $L_y = 20D$, $18D$).

References

[1] Williamson C H K and Rohsko A 1988 *J. Fluid. Struct.* **2** 355
[2] Bishop R E D and Hassan A Y 1964 *Proc. R. Society* **277** 51
[3] Hoby S 2005 Master's thesis Université de Poitiers
[4] Parnaudeau P, Carlier J, Heitz D and Lamballais E 2008 *Phys. Fluids* **20** 085101
[5] Parnaudeau P, Lamballais E, Heitz D and Silvesrini J H 2003 *Proc. DLES-5* (Munich)
[6] Laizet S and Lamballais E 2009 *J. Comput. Phys.* **228** 5989
[7] Laizet S, Lamballais E and Vassilicos J C 2010 *Comput. Fluids* **39** 471
[8] Pinto L C, Schettini E B C and Silvestrini J H 2010 *Proc. BBVIV-6* (Italy) p 369
[9] Guilmineau E and Queutey P 2002 *J. Fluid. Struct.* **16** 773
[10] Placzek A, Sigrist J F and Hamdouni A 2009 *Comput. Fluids* **38** 80
[11] Sheard G J, Hourigan K and Thompson M C 2005 *J. Fluid Mech.* **526** 257

Experimental and numerical studies on the mixing at the intersection of millimetric channels

F. Etcheverry[1], **J. M. Gomba**[2], **C. Wolluschek**[3], **M. Cachile**[1]

[1] LIA-Grupo de Medios Porosos, Facultad de Ingeniería, UBA, Buenos Aires, Argentina.
[2] Instituto de Física Arroyo Seco, UNCPBA, Tandil, Argentina.
[3] Mecánica de Fluidos e Ingeniería Térmica, Centro tecnológico Cemitec, Noáin, Navarra, Spain.

E-mail: mcachil@fi.uba.ar

Abstract.
In this work, experimental and numerical results on the effect of diffusion and geometrical dispersion on the mixing of confluent flows are presented. Two channels with an internal diameter $D_h = 4mm$ intersect with an angle $\alpha = 30, 60, 90, 120, 150, 180°$. The experimental setup allows to accurately control the flow rate and assures a constant flow at both inlets. The mixing properties are studied by injecting pure water in one inlet and colored water at the other. The effects of the inlet flow and the intersection angle on the diffusion of ink is analyzed. We observed that the mixing by convection is only important for $\alpha = 180°$. For other angles, diffusion is the main mechanism for mixing.

1. Introduction

In both theoretical and experimental research, porous media are modeled in different ways such as obstacles distribution, network of channels of different sizes, granular media and artificial fractures just to mention a few [1–8]. An additional approximation made is to study the flow using Hele-Shaw cells, that is, to employ two-dimensional configuration in order to simplify the theoretical analysis or, in experiments, to optimize the flow visualization [9, 10].

A subject, closely related to the presented here, which has recently received a great interest deals with the mixing at the micrometer scale. Its study is essential to improve our understanding of the transport phenomena in channels as well as in network of channels. When the Reynolds numbers are small, as usual in the flows in porous media, the main mechanism of interaction is through molecular diffusion [1, 15]. Due to this fact, the diffusion time is a key parameter to be taken into account when designing micromixers. Nevertheless, the importance of the geometry has been recognized and, recently, many geometrical configurations have been explored [11–14], all of them trying to reduce the mixing length and mixing time.

Despite the large number of works on this field, a systematic study on the impact of the geometry parameters on the mixing process has been seldom studied. Here, we show results in the investigation of the effect of the crossing angle α on the mixing of the species. This has been achieved by analyzing the behavior of the flow and its dispersion in the intersection of two channels in the stationary regime.

2. Experimental setup

A schematic view of the experimental setup is shown in figure 1. The liquids are injected into two crossing channels at the same flow rate by using a double-syringe pump (Harvard Apparatus [16]). The millimetric pipes cross with angles, defined as the angle between the injection channels, of $\alpha = 30°$, $60°$, $90°$, $120°$, $150°$ and $\alpha = 180°$. The inner diameter of the channels is $D_h = 4mm$, constructed by drilling a solid block of transparent plexiglass, as shown in figure 2. The external and internal surfaces of the channels were carefully polished in order to obtain an optimal visualization of the flow. The drainage pipes are identical in length and diameter and to ensure the same impedance at both outlets, they are submerged in a vessel at the same depth.

The injected liquids are (a) pure distilled water and (b) a mixture of distilled water with blue ink (water blue). The density and viscosity of both liquids are the same, given the low concentration of ink needed to obtain a reasonable intensity of color. The diffusion constant for the ink in water is $D = 10^{-11} m^2/s$. The blue color of the ink allows to analyze only the red channel of the RGB images, where a darker level indicates a higher concentration of blue tonality. The channels were illuminated by transmission using a uniform light source, the illumination being controlled and kept constant during all the experiences. The analysis of the images is carried out after subtracting each image with ink from the image without ink.

The study presented here corresponds with stationary state regime, so the images are taken after an injection of about 10 times the volume of the mixing zone. We do not quantify the amount of ink in the experiments but only qualitatively compare the images with those obtained from the simulations.

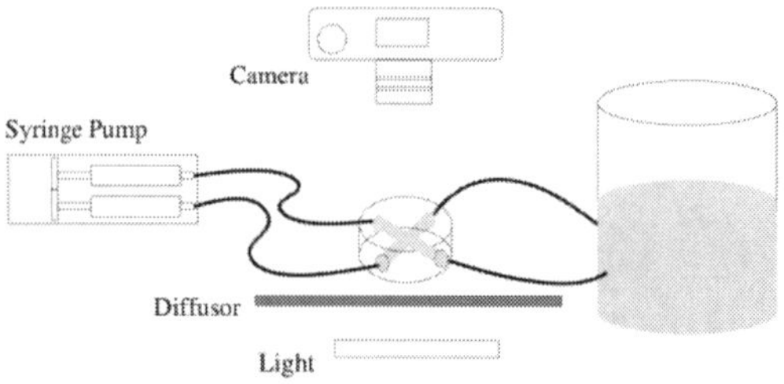

Figure 1. Experimental setup for the study of the flow in the X mixer.

Figure 2. Top view of the X millichannel.

3. Formulation and numerical issues

The flow is described by the Navier-Stokes and continuity equations:

$$\rho\frac{\partial u}{\partial t} - \eta\nabla.(\nabla u + (\nabla u)^T) + \rho(u.\nabla)u + \nabla p = 0 \quad (1)$$

$$\nabla.u = 0 \quad (2)$$

where ρ is the density, u is the velocity, η is the viscosity and p the pressure. The considered fluid is water with $\eta = 1\times10^{-3}N.s/m^2$ and $\rho = 1\times10^3kg/m^3$. The convection-diffusion process is governed by the following equation:

$$\frac{\partial c}{\partial t} = -\nabla.(-D\nabla c) - u.(\nabla c) = 0, \quad (3)$$

where $D = 10^{-11}m^2/s$ represents the diffusion coefficient and c is the ink concentration. The boundary conditions at the inlets are $c = 0$ and $c = 1mol/m^3$, similar to the experiments. Considering this low concentration value, it is licit to assume that the solute molecules only interact with water molecules (without interacting with themselves) and thus, it is possible to use Fick's law to describe the diffusion process. We also assume that density and viscosity of the water is not modified by the presence of the ink. Under these assumptions, it is legitimate to first solve the Navier-Stokes equations at steady state and then use its solution as a base flow to solve the steady convection-diffusion equation. Nevertheless, in order to check the validity of the numerical solutions and explore some experimental issues not shown in this work, we solved the complete partial differential equation given in (3) until the steady state is reached. The convergence criteria employed to assume a stationary solution are that the concentration at the outlets and the velocity at several control points vary less than 0.5% with time.

In order to simplify the simulation and to exploit the symmetry of the problem, we have studied only one half of the volume, as shown in figure 3(a). The geometry is meshed using tetrahedral elements, as schematized in Figure 3(b). The maximum elements size are $5\times10^{-4}m$ at the intersection volume and $8\times10^{-4}m$ at the outlets channels. In total, we have a mesh of about 10600 finite elements, which corresponds to approximately 68100 degrees of freedom. As solver we have chosen the direct PARDISO. For the diffusion equations, the model includes an isotropic artificial diffusion as stabilization procedure. Briefly, the method increases the effective value of D. This increment, proportional to the cell size and local velocity, is kept as small as possible to obtain meaningful and realistic solutions. The artificial diffusivity is iteratively reduced down to a value that assures a stationary solution without spurious oscillations. The artificial isotropic diffusion is necessary in order to keep the local Peclet number near 1 (Pe = vh/2D, where v is the convective velocity, h is the mesh element size and D is the diffusion coefficient). In our case, due to the computer memory limitations it is not possible to reduce h. Instead, we need to adjust the coefficient diffusion as explained above.

4. Results

The working regime of mixers as well as micromixers is usually characterized in terms of the Reynolds number, $Re = UD_h/\nu$, and the Peclet number, $Pe = D_hV/D$, where, V is a typical velocity, D_h is the hydraulic diameter and ν the kinematic viscosity. Nevertheless, both dimensionless numbers are related to each other through the Schmidt number, $Sc = Pe/Re = \nu/D$. For a given fluid , Sc is a constant, in our problem $Sc = 10^5$. Thus, in what follows, both numbers are indistinctly employed to describe the flow.

Figure 4(a) shows the streamlines for a case with $Re = 40$. The flow is symmetric with respect to a bisecting plane that is not crossed by the streamlines. The velocity at the central

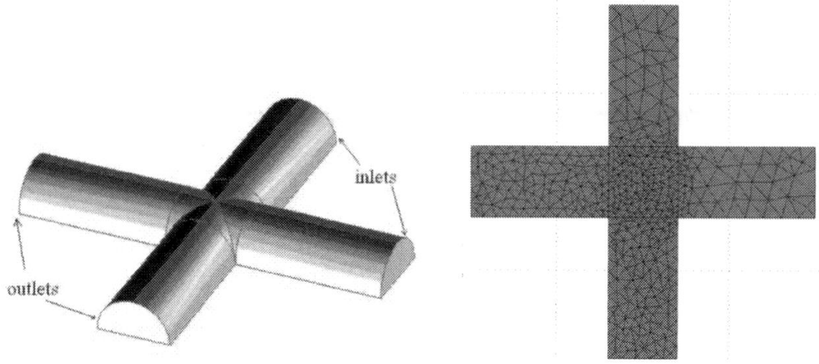

(a) (b)

Figure 3. (a) Sketch of the numerical domain and position of the inlets and outlets. (b) Size distribution of the tetrahedral elements.

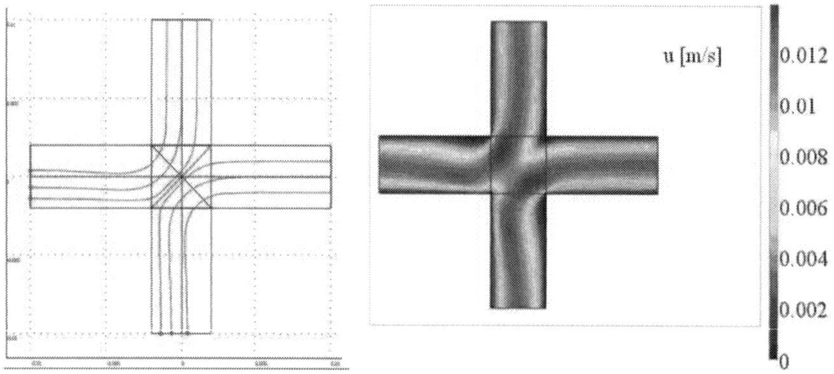

(a) (b)

Figure 4. Case with $Q = 6ml/min$ (Re=40), $\alpha = 90°$. (a) Streamlines are symmetric with respect to a bisecting plane. (b) Velocity: the highest value occurs at the mixing zone. In both figures, inlets are at right and top channels.

zone, Figure 4(b), is increased from $0.01m/s$ (velocity at the inlets) to $0.0139m/s$, an increment that exactly corresponds to a section narrowing that occurs in the central zone.

Figure 5(a) shows the ink distribution for an experiment with $\alpha = 90°$ and an inflow $Q = 6ml/min$. The numerical results are in good agreement with the experiments, as shown in Figure 5(b). In this experiment, ink is not detected at the outlets while in the simulation the ink concentrations at the outlets are $0.93mol/m^3$ and $0.06mol/m^3$, respectively. This slight difference between the experiment and the simulation can be explained in terms of the isotropic artificial diffusion employed to solve the convection-diffusion equation, which induces diffusion of a small amount of ink towards the opposite outlet.

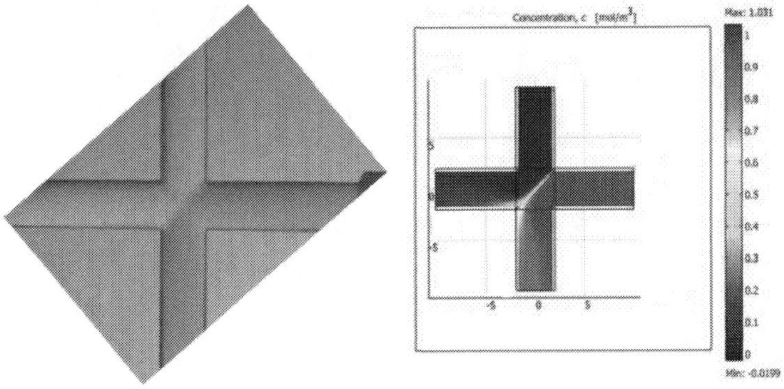

(a) (b)

Figure 5. Concentration of blue ink for case with $Q = 6ml/min$ and $\alpha = 90°$. (a) Experiment. The image was rotated for an easier comparison with numerical result. (b) Simulation. Here, the red color represents the maximum of ink concentration, the blue color is pure water. In both figures, inlets are at right and top channels.

In general, experiments show that diffusion is almost negligible. This observation is in agreement with numerical results and can be explained in terms of the large values for the Peclet numbers ($Pe > 10^3$). Effectively, Pe is the ratio between the diffusion time, $\sim D_h^2/D$ to the characteristic transit time int the mixing zone, $\sim D_h/V$, V being a characteristic velocity. Figure 6(a) shows that for an inlet flow of $Q = 1ml/min$, there is a presence of ink at the water outlet (upper right corner of the figure). When Q is increased, diffusion is less important (see figure 6(b)) and, for example for the large inflow $Q = 20ml/min$, ink can not be detected at the water outlet.

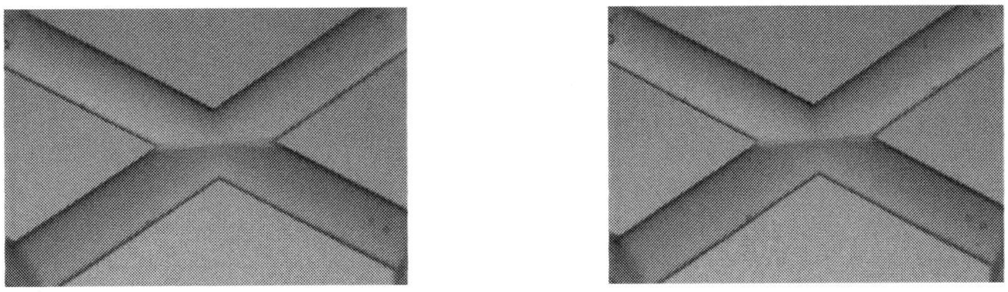

(a) (b)

Figure 6. Flow for $\alpha = 60°$. (a) $Q = 1ml/min$: Notice that a small amount of ink is diffused to the water outlet. (b) $Q = 10ml/min$: The amount of ink diffused to the water outlet is reduced with respect to the case with $Q = 1ml/min$.

We analyzed the effect of the crossing angle α on the diffusion of the ink. Although several angles were studied, only two typical values (60° and 90°), for the sake of simplicity are presented. For the range of $5 < Re < 150$, given Q, we did not appreciate significant differences in the amount of ink in the water outlet when α is changed. The exception is the case with $\alpha = 180°$, for which the ink equally convects to both outlet pipes, as shown in figure 7. Nevertheless, it is worth to point out that the geometry for $\alpha = 180°$ differs from other cases in that the outlets channels are not aligned to the inlets and, thus, the ink is equally splitted to the outlets.

Figure 7. Flow for the case with $Q = 10ml/min$ and $\alpha = 180°$. On the contrary of what happens for other values of α, here the ink convects toward both outlet pipes. Here, the inlets are on the left, outlets on the right.

5. Conclusions

The convection-diffusion behavior of a colored liquid injected in a X millichannel is characterized. Numerical results show that the flow is symmetric with respect to a bisection plane that is not crossed by the streamlines. The velocity is increased in the confluence zone due to a reduction of the channel volume. The effect of the crossing angle on the diffusion of the ink towards the pure water outlet is negligible except for the case $\alpha = 180°$, for which the ink equally convects to both outlet channels.

Acknowledgments

M.C. and J.M.G. acknowledge CONICET and ANPCyT for funding support.

References

[1] J. Bear, Dynamics of fluids in porous media, Elsevier, New York, 1972.

[2] M. Blunt, Curr. Op. in Coll. and Int. Sci.,6, 197-207, 2001.

[3] V. Kumar, M. Aggarwal, K. Nigam, Mixing in curved tubes, Chem. Eng. Sci., 61, 5742-5753, 2006.

[4] F. Teruel, Rizwn-uddin, Characterization of a porous medium employing numerical tools: permeability and pressure-drop from Darcy to turbulence, Int. J. Heat and Mass Transf. 52, 5878-5888, 2009.

[5] E.Villermaux, J. M. Chomez, Kluwer Eds, en *Mixing: Chaos and Turbulence*, pp. 307-342, 1999.

[6] J. Yang, W. Fang, K. Tung, Fluids mixing in devices with connected-groove channels, Chem. Eng. Sci., 63, 1871-1881, 2008.

[7] H. Wang, P. Iovinetti, E. Harvey, S. Masood, Optimizing layout of obstacles for enhanced mixing in microchannels, Smart Mater. Struct., 11, 662-667, 2002.

[8] H. Wang, W. Li, A novel 3D porous micromixer fabricated using selective ultrasonic foaming, J. Micromech and MicroEng. 17, 1835-1842, 2007.

[9] P. Gaganis, E. Skouras, M. Theodoropoulou, D. Tsakiroglou, V. Burganos, On the evaluation of dispersion coefficients from visualization experiments in artificial porous media, J. Hydrol., 307, 79-91, 2005.

[10] D. Córdoba, F. Gancedo, R. Orive, Analytical behavior of 2-D incompressible flow in porous media, J. Math. Phys. 48, 065296, 2007.

[11] N. Nguyen, Z. Wu, Micromixers- a review, J. Micromech. Eng., 15 R1-R16, 2005.

[12] C. Chang, Y. Yang, T. Yen, C. Chen, Numerical investigation into thermal mixing efficiency in Y-shaped channel using Lattice Boltzmann method and field synergy principle, Int. J. Ther. Sci, 48, 2092-2099, 2009.

[13] R. Liu et al., Passive mixing in three-dimensional serpentine microchannel, J. MicroElect. Sys., 9, 190- 197, 2000.

[14] P. Luo, Y. Cheng, Y. Jin, W. Yang, J. Ding, Fast liquid mixing by cross-flow impingement in millimiter channels, Chem. Eng. Sci., 62, 6178-6190. 2007.

[15] W. Jost; Diffusion in Solids, Liquids, Gases, Academic Press, 3rd Ed., NY, 1960.

[16] Double syringe pump, Model 55-2226. Harvard Apparatus, http://www.harvardapparatus.com.

Journal of Physics: Conference Series **296** (2011) 012013 doi:10.1088/1742-6596/296/1/012013

Lubrication theory applied to the convergent flows of two stacked liquid layers

Julio Gratton[1] and Carlos Alberto Perazzo[2]

[1] INFIP–CONICET, Dpto. de Física, FCEyN, UBA, Ciudad Universitaria, Pab. I, 1428 Buenos Aires, Argentina.

[2] Universidad Favaloro and CONICET, Solís 453, 1078 Buenos Aires, Argentina.

E-mail: jgratton@tinfip.lfp.uba.ar, perazzo@favaloro.edu.ar

Abstract. With the aim of describing the mountain building process, we have previously applied the lubrication approximation to obtain the evolution equations of the problem of two stacked layers of viscous fluids with different densities and different viscosities. The lubrication approximation is a perturbation method where the small parameter is the aspect ratio (thickness/lenght) of the current. This approximation is widely used to study the slow flow of one layer of a viscous fluid, but it is not well known under which conditions it can be applied in more general settings. Here we analyze in detail the assumptions needed to apply the lubrication theory to study the flow of two stacked viscous fluid layers. We employ the same perturbation method and we found that, besides the usual conditions (low Reynolds number and gentle slope), we must require that the viscosity and density ratios are of the order of unity. These requirements determine the range of validity of the equations of our model of the mountain building.

1. Introduction

Mountain ranges are one of the most striking features of the Earth and their origin and evolution have been investigated for a long time. It is known that the lithosphere (the outer solid layer of the Earth) is a two layer structure in which the crust rests on the denser lithospheric mantle (see Fig. 1), being separated by the Mohorovičić discontinuity (called Moho). The lithosphere is divided into several approximately rigid plates that rest on the hotter and more fluid asthenosphere. The relative motion of these plates is the cause of mountain building, because of the shortening and consequent thickening of the crust that occurs when two continental plates collide or when an oceanic plate is subducted beneath a continent. On the timescale of the orogenic processes the lithosphere is in local hydrostatic equilibrium (a condition called isostasy) that implies that the visible regional topography is accompanied by a corresponding anti-topography (called root) of the Moho (see Fig. 1).

Clearly mountain builiding is an important problem that involves many disciplines and interests a broad range of scientists. To attempt a realistic and detailed theoretical description of mountain building is an exceedingly complex task because across the lithosphere there are large variations of the temperature, density and rheological parameters as well as other properties (many of which, to compound the issue, are poorly known). To this should be added the complications due to the geometry and the time dependence of the motion of the plates.

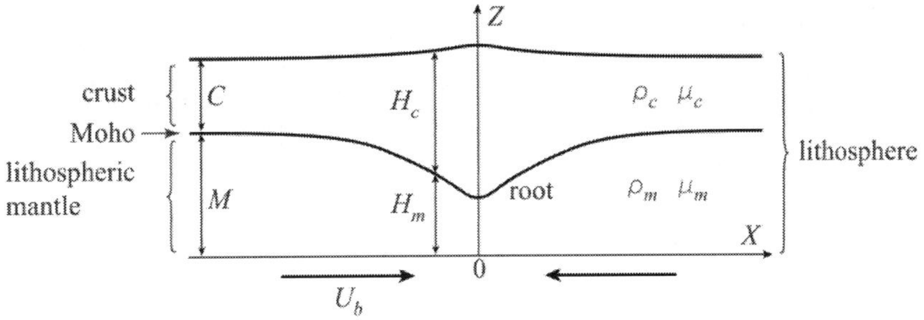

Figure 1. Formation of a ridge due to the convergent motion of a two–layer system.

The basic phenomena that govern the large scale evolution of mountain belts are the spreading flow at the depth of the roots together with isostasy and crustal shortening. The profile of the ridge is determined by the dynamic balance between buoyancy and viscous forces. Based on these ideas we have recently developed a model to describe the process of mountain building which showed a very good agreement when it was compared with real topographical data from approximately straight segments of several mountain belts [1]. In this model we describe the lithospheric plates as two stacked layers of Newtonian liquids. The upper layer, which is the lightest, represents the crust, and the lower one represents the lithospheric mantle. The motion of the plates is driven by the basal traction that is assumed it is exerted by the astenosphere at the bottom of the lithospheric mantle. We introduced the basal traction in our model as a boundary condition which states that below the plates there is a prescribed horizontal velocity. When the motion due to the basal traction is convergent, the crust is shortened and thus its thickness increases in the region of convergence producing a ridge. We also assume in our model the isostasy condition, which is introduced assuming that at the bottom of the lithospheric mantle the pressure is uniform. From these hypothesis and using the lubrication approximation we derive a partial differential equation for the thickness of the crust.

The lubrication approximation is a well known and established treatment to study the viscous slow flow of a thin current whose typical length is much larger than its thickness. Depending on the problem under study, the driven forces can be the gravity, centrifugation, surface tension, Marangoni effect, Van der Waals forces, thermocapillarity, condensation and evaporation, etc. It was used to study various problems with very different spatial and temporal scales, from lava flows to tear films in the eye (see for example the reviews by Oron *et al.* [2] and Myers [3]). However, in most of the works where the lubrication approximation was applied, the problem involves only one layer of liquid over a rigid substrate. Much less is known about the lubrication approximation when it is applied to study the flow of more than one layer of fluid, as it happens in our model of mountain building. Here we discuss the conditions and hypotheses needed to apply the lubrication approximation to study the flow of two stacked viscous layers under the action of gravity.

This paper is organized as follow. In the next Section we state the problem and the full system of equations that describes it. In Section 3 we reduce the mathematical problem by means of a set of simplifications which are the bases of the lubrication approximation. In Section 4 we particularize the problem to derive the equations of our two–layer model of mountain building [1]. Finally in Section 5 we present the conclusions.

2. The complete system of equations

We consider a system which consists of two layers of Newtonian liquids with different viscosities and densities. The upper layer has viscosity μ_c and density ρ_c, and the lower one has viscosity

μ_m and density ρ_m. For simplicity we assume planar symmetry, and X and Z are the horizontal and vertical coordinates respectively (see Fig. 1). The domain of the two–layer system is $Z \geq 0$. The thickness of the upper and lower layers are H_c and H_m respectively, so that the system has two interfaces, one at $Z = H_c + H_m$ and the other at $Z = H_m$. For $H_m \leq Z \leq H_m + H_c$ the Navier–Stokes equation takes the form

$$\rho_c \left(\frac{\partial U_c}{\partial T} + U_c \frac{\partial U_c}{\partial X} + W_c \frac{\partial U_c}{\partial Z} \right) = -\frac{\partial P_c}{\partial X} + \mu_c \left(\frac{\partial^2 U_c}{\partial X^2} + \frac{\partial^2 U_c}{\partial Z^2} \right), \quad (1)$$

$$\rho_c \left(\frac{\partial W_c}{\partial T} + U_c \frac{\partial W_c}{\partial X} + W_c \frac{\partial W_c}{\partial Z} \right) = -\frac{\partial P_c}{\partial Z} + \mu_c \left(\frac{\partial^2 W_c}{\partial X^2} + \frac{\partial^2 W_c}{\partial Z^2} \right) - \frac{\partial \Phi}{\partial Z}, \quad (2)$$

and for $0 \leq Z \leq H_m$

$$\rho_m \left(\frac{\partial U_m}{\partial T} + U_m \frac{\partial U_m}{\partial X} + W_m \frac{\partial U_m}{\partial Z} \right) = -\frac{\partial P_m}{\partial X} + \mu_m \left(\frac{\partial^2 U_m}{\partial X^2} + \frac{\partial^2 U_m}{\partial Z^2} \right), \quad (3)$$

$$\rho_m \left(\frac{\partial W_m}{\partial T} + U_m \frac{\partial W_m}{\partial X} + W_m \frac{\partial W_m}{\partial Z} \right) = -\frac{\partial P_m}{\partial Z} + \mu_m \left(\frac{\partial^2 W_m}{\partial X^2} + \frac{\partial^2 W_m}{\partial Z^2} \right) - \frac{\partial \Phi}{\partial Z}, \quad (4)$$

where T is the time, Φ is the potential of the gravitational force, P is the pressure, U and W are horizontal and vertical components of the velocity respectively, and the subscripts c and m indicate to which layer each magnitude belongs.

At the interface at $Z = H_m + H_c$ the usual kinematic condition holds

$$\frac{\partial}{\partial T}(H_M + H_c) + U_c \frac{\partial}{\partial X}(H_M + H_c) = W_c, \quad \text{for } Z = H_m + H_c, \quad (5)$$

and we assume that the normal and tangential components of stress vanish (we neglect surface tension effects)

$$\hat{n}_c \cdot \Sigma_c \cdot \hat{n}_c = 0, \quad \hat{t}_c \cdot \Sigma_c \cdot \hat{n}_c = 0, \quad (6)$$

where Σ_c is the stress tensor of the upper layer and $\hat{n}_c$ and $\hat{t}_c$ are the unitary vector normal and tangential to the upper surface.

At the interface between the two layers at $Z = H_m$ we have the kinematic condition

$$\frac{\partial H_m}{\partial T} + U_c \frac{\partial H_m}{\partial X} = W_c, \quad (7)$$

and we assume continuity of the components of the velocity and the stress

$$U_m = U_c, \quad (8)$$

$$W_m = W_c, \quad (9)$$

$$\hat{n}_m \cdot \Sigma_c \cdot \hat{n}_m = \hat{n}_m \cdot \Sigma_m \cdot \hat{n}_m, \quad (10)$$

$$\hat{t}_m \cdot \Sigma_c \cdot \hat{n}_m = \hat{t}_m \cdot \Sigma_m \cdot \hat{n}_m, \quad (11)$$

where Σ_m is the stress tensor of the lower layer and $\hat{n}_m$ and $\hat{t}_m$ are the unitary vector normal and tangential to the interface. We impose at $Z = 0$ a prescribed horizontal velocity U_b, so that

$$U_m = U_b, \quad \text{for } Z = 0. \quad (12)$$

Equations (1-12) together with the continuity equation and an additional condition at $Z = 0$ (usually this condition is $W_m = 0$, however we shall use a different one more appropriate for our problem) describe without any approximation the flow of the two layers of Newtonian liquids under the action of body forces with any initial conditions. However, this system of equations is difficult to solve, even numerically, so that in order to gain insight about a particular problem it is convenient to introduce some simplifying hypotheses and to reduce the system of equations.

3. The lubrication approximation

We are interested in flows which are slow and almost horizontal. Let Z_0 and X_0 be the characteristic vertical and horizontal scales, respectively. The main hypothesis of the lubrication approximation is that $X_0 \gg Z_0$, so that

$$\varepsilon \equiv \frac{Z_0}{X_0} \ll 1. \tag{13}$$

This means that both interfaces have gentle slopes. Calling W_0 and U_0 the characteristic vertical and horizontal velocities, form the continuity equation and (13) one has that

$$W_0 = \varepsilon U_0 \ll U_0, \tag{14}$$

so that the flow is predominately horizontal.

Now we introduce the following dimensionless quantities:

$$\begin{aligned} x &= \tfrac{X}{X_0} = \tfrac{\varepsilon X}{Z_0}, & z &= \tfrac{Z}{Z_0}, & t &= \tfrac{T}{T_0} = \tfrac{TU_0}{X_0} = \tfrac{\varepsilon T U_0}{Z_0}, & h_c &= \tfrac{H_c}{Z_0}, \\ h_m &= \tfrac{H_m}{Z_0}, & u_c &= \tfrac{U_c}{U_0}, & u_m &= \tfrac{U_m}{U_0}, & w_c &= \tfrac{W_c}{W_0} = \tfrac{W_c}{\varepsilon U_0}, \\ w_m &= \tfrac{W_m}{W_0} = \tfrac{W_m}{\varepsilon U_0}, & p_c &= \tfrac{P_c}{P_0}, & p_m &= \tfrac{P_m}{P_0}, & \phi &= \tfrac{\Phi}{P_0}. \end{aligned} \tag{15}$$

The scale of the pressure P_0 is determined from the understanding that the horizontal gradient of pressure is equilibrated by the vertical gradient of the viscous stress so that $\partial P/\partial X \sim \mu \partial^2 U/\partial Z^2$. Then we choose

$$P_0 = \frac{\mu_c U_0 X_0}{Z_0^2} = \frac{\mu_c U_0}{\varepsilon Z_0}. \tag{16}$$

Finally, we define the Reynolds number as

$$Re = \frac{\rho_c U_0 Z_0}{\mu_c}. \tag{17}$$

In terms of the dimensionless quantities the governing equations take the form

$$\varepsilon Re \left(\frac{\partial u_c}{\partial t} + u_c \frac{\partial u_c}{\partial x} + w_c \frac{\partial u_c}{\partial z} \right) = -\frac{\partial p_c}{\partial x} + \varepsilon^2 \frac{\partial^2 u_c}{\partial x^2} + \frac{\partial^2 u_c}{\partial z^2}, \tag{18}$$

$$\varepsilon^3 Re \left(\frac{\partial w_c}{\partial t} + u_c \frac{\partial w_c}{\partial x} + w_c \frac{\partial w_c}{\partial z} \right) = -\frac{\partial p_c}{\partial z} + \varepsilon^4 \frac{\partial^2 w_c}{\partial x^2} + \varepsilon^2 \frac{\partial^2 w_c}{\partial z^2} - \frac{\partial \phi}{\partial z}, \tag{19}$$

for $h_m \le z \le h_m + h_c$, and

$$\varepsilon Re \frac{\rho_m}{\rho_c} \left(\frac{\partial u_m}{\partial t} + u_m \frac{\partial u_m}{\partial x} + w_m \frac{\partial u_m}{\partial z} \right) = -\frac{\partial p_m}{\partial x} + \varepsilon^2 \frac{\mu_m}{\mu_c} \frac{\partial^2 u_m}{\partial x^2} + \frac{\mu_m}{\mu_c} \frac{\partial^2 u_m}{\partial z^2}, \tag{20}$$

$$\varepsilon^3 Re \frac{\rho_m}{\rho_c} \left(\frac{\partial w_m}{\partial t} + u_m \frac{\partial w_m}{\partial x} + w_m \frac{\partial w_m}{\partial z} \right) = -\frac{\partial p_m}{\partial z} + \varepsilon^4 \frac{\mu_m}{\mu_c} \frac{\partial^2 w_m}{\partial x^2} + \varepsilon^2 \frac{\mu_m}{\mu_c} \frac{\partial^2 w_m}{\partial z^2} - \frac{\partial \phi}{\partial z}, \tag{21}$$

for $0 \le z \le h_m$.

The boundary conditions at $z = h_m + h_c$ are given by

$$\frac{\partial}{\partial t}(h_m + h_c) + u_c \frac{\partial}{\partial x}(h_m + h_c) = w_c, \tag{22}$$

$$\begin{aligned} 0 = p_c \;&+\; \varepsilon^2 \left\{ \tfrac{\partial}{\partial x}(h_m + h_c)\left[p_c \tfrac{\partial}{\partial x}(h_m + h_c) + 2\tfrac{\partial u_c}{\partial z}\right] - 2\tfrac{\partial w_c}{\partial z} \right\} + \\ &+\; \varepsilon^4 \left\{ 2\tfrac{\partial}{\partial x}(h_m + h_c)\left[\tfrac{\partial w_c}{\partial x} - \tfrac{\partial}{\partial x}(h_m + h_c)\tfrac{\partial u_c}{\partial x}\right] \right\}, \end{aligned} \tag{23}$$

$$\begin{aligned} \tfrac{\partial u_c}{\partial z} &= \varepsilon^2 \left\{ \tfrac{\partial u_c}{\partial z}\left[\tfrac{\partial}{\partial t}(h_m + h_c)\right]^2 + 2\left(\tfrac{\partial u_c}{\partial x} - \tfrac{\partial w_c}{\partial z}\right)\tfrac{\partial}{\partial t}(h_m + h_c) - \tfrac{\partial w_c}{\partial x} \right\} + \\ &+ \varepsilon^4 \tfrac{\partial w_c}{\partial x}\left[\tfrac{\partial}{\partial t}(h_m + h_c)\right]^2 . \end{aligned} \tag{24}$$

The matching conditions at $z = h_m$ are given by

$$\frac{\partial h_m}{\partial t} + u_c \frac{\partial h_m}{\partial x} = w_c, \tag{25}$$

$$u_c = u_m, \tag{26}$$

$$w_c = w_m, \tag{27}$$

$$\begin{aligned} p_c &= p_m + \varepsilon^2 \left\{ (p_m - p_c)\left(\tfrac{\partial h_m}{\partial x}\right)^2 + 2\tfrac{\partial}{\partial z}\left[\tfrac{\partial h_m}{\partial x}\left(\tfrac{\mu_m}{\mu_c}u_m - u_c\right) + w_c - \tfrac{\mu_m}{\mu_c}w_m\right] \right\} + \\ &+ \varepsilon^4 2\tfrac{\partial h_m}{\partial x}\left[\tfrac{\partial h_m}{\partial x}\tfrac{\partial}{\partial x}\left(u_c - \tfrac{\mu_m}{\mu_c}u_m\right) - \tfrac{\partial}{\partial x}\left(w_c - \tfrac{\mu_m}{\mu_c}w_m\right)\right], \end{aligned} \tag{28}$$

$$\begin{aligned} \tfrac{\partial u_c}{\partial z} &= \tfrac{\mu_m}{\mu_c}\tfrac{\partial u_m}{\partial z} + \varepsilon^2 \left\{ \left(\tfrac{\partial h_m}{\partial x}\right)^2 \tfrac{\partial}{\partial z}\left(u_c - \tfrac{\mu_m}{\mu_c}u_m\right) - \tfrac{\partial}{\partial x}\left(w_c - \tfrac{\mu_m}{\mu_c}w_m\right) + \right. \\ &\left. + 2\tfrac{\partial h_m}{\partial x}\left[\tfrac{\partial}{\partial x}\left(u_c - \tfrac{\mu_m}{\mu_c}u_m\right) - \tfrac{\partial}{\partial z}\left(w_c - \tfrac{\mu_m}{\mu_c}w_m\right)\right] \right\} + \varepsilon^4 \left(\tfrac{\partial h_m}{\partial x}\right)^2 \tfrac{\partial}{\partial x}\left(w_c - \tfrac{\mu_m}{\mu_c}w_m\right). \end{aligned} \tag{29}$$

At $z = 0$ we have $u_m = u_b$.

Now we seek the solution as a perturbation series in powers of the small parameter ε:

$$u_c = u_{c0} + \varepsilon u_{c1} + \varepsilon^2 u_{c2} + \cdots, \tag{30}$$

$$w_c = w_{c0} + \varepsilon w_{c1} + \varepsilon^2 w_{c2} + \cdots, \tag{31}$$

$$p_c = p_{c0} + \varepsilon p_{c1} + \varepsilon^2 p_{c2} + \cdots, \tag{32}$$

$$u_m = u_{m0} + \varepsilon u_{m1} + \varepsilon^2 u_{m2} + \cdots, \tag{33}$$

$$u_m = u_{m0} + \varepsilon u_{m1} + \varepsilon^2 u_{m2} + \cdots, \tag{34}$$

$$u_m = u_{m0} + \varepsilon u_{m1} + \varepsilon^2 u_{m2} + \cdots. \tag{35}$$

To the leading order in ε the governing equations reduce to (after omitting the subscript 0)

$$\frac{\partial p_c}{\partial x} = \frac{\partial^2 u_c}{\partial z^2}, \tag{36}$$

$$\frac{\partial p_c}{\partial z} = -\frac{\partial \phi}{\partial z}, \tag{37}$$

$$\frac{\partial p_m}{\partial x} = \frac{\mu_m}{\mu_c}\frac{\partial^2 u_m}{\partial z^2}, \tag{38}$$

$$\frac{\partial p_m}{\partial z} = -\frac{\partial \phi}{\partial z}, \tag{39}$$

$$\frac{\partial}{\partial t}(h_m + h_c) + u_c \frac{\partial}{\partial x}(h_m + h_c) = w_c, \quad \text{for } z = h_m + h_c, \tag{40}$$

$$p_c = 0, \quad \text{for } z = h_m + h_c, \tag{41}$$

$$\frac{\partial u_c}{\partial z} = 0, \quad \text{for } z = h_m + h_c, \tag{42}$$

$$\frac{\partial h_m}{\partial t} + u_c \frac{\partial h_m}{\partial x} = w_c, \quad \text{for } z = h_m, \tag{43}$$

$$u_c = u_m, \quad \text{for } z = h_m, \tag{44}$$

$$w_c = w_m, \quad \text{for } z = h_m, \tag{45}$$

$$p_c = p_m, \quad \text{for } z = h_m, \tag{46}$$

$$\frac{\partial u_c}{\partial z} = \frac{\mu_m}{\mu_c}\frac{\partial u_m}{\partial z}, \quad \text{for } z = h_m, \tag{47}$$

$$u_m = u_b, \quad \text{for } z = 0. \tag{48}$$

It can be mentioned that to obtain this leading order approximation it is necessary to add the following condition

$$Re \lesssim \mathcal{O}(1), \text{ as } \varepsilon \to 0, \tag{49}$$

as usual in the lubrication approximation for a single fluid layer. In addition to this we must require that

$$\frac{\rho_m}{\rho_c} \sim \mathcal{O}(1) \text{ and } \frac{\mu_m}{\mu_c} \sim \mathcal{O}(1), \text{ for } \varepsilon \to 0. \tag{50}$$

These additional conditions are necessary due to the presence of the two layers.

Notice that the system (36–48) is much simpler than the original system of equations. In fact, some important results can be obtained immediately. First, from Eqs. (37) and (39) is clear that the pressure in both layers is hydrostatic, and with the conditions (41) and (46) the expressions for p_c and p_m can be completely determined. Second, once the pressure is found one can integrate (36) and (38) and use (42), (44), (47) and (48) to determine the horizontal velocities u_c and u_m. The velocity profiles thus obtained are parabolic in both layers. However, notice that the pressure and the velocity field in both layers are expressed in terms of h_m and h_c, which remain unknown.

Integrating in z the continuity equation from h_m to $h_m + h_c$, and using the kinematic conditions (40) and (43) it is possible to obtain

$$\frac{\partial h_c}{\partial t} = -\frac{\partial}{\partial x}\left(\int_{h_m}^{h_m+h_c} u_c \, dz\right), \tag{51}$$

which expresses the conservation of the mass of the upper layer. However, an analog expression for the lower layer can not be obtained until an additional condition at $z = 0$ be adopted. If the usual condition $w_m = 0$ is chosen, an expression for h_m equivalent to (51) will be obtained.

4. The two–layer model of mountain building

In the context of our model of mountain building the upper layer represents the Earth's crust and the lower one represents the upper mantle. We call C and M the initial thickness of the crust and lithospheric mantle respectively. We choose the characteristic scales as (see Ref. [1] for details)

$$Z_0 = C, \quad X_0 = (1 - \frac{\rho_c}{\rho_m})\frac{\rho_c g M C^2}{\mu_m U_0}, \quad U_0 = \max\left(|U_b|\right). \tag{52}$$

As it was said before, to close the system and derive evolution equations for h_c and h_m it is needed to add a boundary condition at $z = 0$. In our two–layer model of mountain building we introduce the isostasy by means of the condition

$$\frac{\partial p_c}{\partial x} = 0, \text{ at } z = 0. \tag{53}$$

From this is immediately obtained that

$$h_m = \frac{M}{C} + \frac{\rho_c}{\rho_m}(1 - h_c). \tag{54}$$

This simple linear relation between h_m and h_c allows us to reduce even more the problem because in the following the only unknown is h_c. On the hand, notice that the boundary condition (53) does not guarantee the mass conservation of the lower layer. In other words, flow of lithospheric mass trough the boundary $z = 0$ is allowed.

5. Conclusions

Here we applied the lubrication approximation to the flow of two stacked layers of Newtonian liquids. We conclude that the needed hypothesis are the same as in the usual case of only one layer: the ratio between the characteristic horizonal and vertical scales must be much lesser than one, the horizontal gradient of the pressure and the vertical gradient of the viscous stress are balanced, and the Reynolds number must be of the order of the unity or less. But, to apply the lubrication approximation to both layers we must require in addition that the viscosity and density ratios are of the order of unity.

Acknowledgments

We acknowledge support from Universidad Favaloro for participation in Fluidos 2010.

References

[1] Perazzo C A and Gratton J 2010 *Phys. Fluids* **22** 056603

[2] Oron A, Davis S H and Bankoff S G 1997 *Rev. Mod. Phys.* **69** 931

[3] Myers T G 1998 *SIAM Review* **40** 441

Influence of relative permeabilities on chemical enhanced oil recovery

M F Destefanis and G B Savioli

Laboratorio de Ingeniería de Reservorios, IGPUBA y Dpto. Ing. Química, Facultad de Ingeniería, UBA, Av. Las Heras 2214, 3° Piso, C1127AAR, Buenos Aires, Argentina

E-mail: fdestefanis@fi.uba.ar, gbsavioli@yahoo.com.ar

Abstract. The main objective of chemical flooding is to mobilize the trapped oil remaining after a secondary recovery by waterflooding. This purpose is achieved by lowering the oil-water interfacial tension and producing partial miscibility between both phases. The chemical partition among phases (phase behavior) influences all other physical properties. In particular, it affects residual saturations determining relative permeability curves. Relative permeabilities rule the flow of each phase through the porous medium, so they play an essential role in oil recovery. Therefore, in this work we study the influence of relative permeabilities on the behavior of a surfactant-polymer flooding for the three different types of phase behavior. This analysis is performed applying the 3D compositional numerical simulator UTCHEM developed at the University of Texas at Austin. From the examples studied, we conclude that the influence of relative permeabilities depends on the type of phase behavior, i.e., as microemulsion relative permeability decreases, oil recovery increases for Types II(+) and III while slightly decreases for Type II(-). Moreover, a better displacement efficiency is observed for Types II(+) and III, because they behave similarly to a miscible displacement.

1. Introduction

Statistically, after a waterflooding, more than half of the oil originally in place still remains trapped in the reservoir. The recovery of this trapped oil is the target of the Enhanced Oil Recovery processes (EOR), which implies the injection of materials not naturally present in the reservoir [1]. Among these materials, surfactants are injected to decrease the interfacial tension between oil and water in order to mobilize the trapped oil, and polymers are injected to reduce water mobility and improve the sweep efficiency. In a chemical flood a multicomponent, multiphase system is involved. The theory of multicomponent, multiphase flow has been presented by several authors [2]-[3]. Those studies led to the building of complex simulators, like the one developed at the University of Texas at Austin, UTCHEM [4]-[6], which accounts for most of the physical phenomena involved. For a Surfactant Pilot Simulation [7] those physical phenomena were described in terms of more than 70 parameters.

From a physical-chemical point of view, there are three pseudo-components: brine, oil and chemical. They are distributed between two or three phases: oleic, aqueous and microemulsion (phase behavior). The chemical has an amphiphilic character, making the oil partially miscible with water [8]. Phase behavior influences all other physical properties, and each of them, in turn, influences oil recovery. Relative permeability curves rule the flow of each phase through the porous medium, so they are essential in the oil recovery efficiency. These curves are represented by potential models depending on two parameters [9]. In this paper we apply UTCHEM to analyze the influence of relative permeability curves and phase behavior on cumulative oil production.

2. Mathematical model

UTCHEM is a 3D compositional simulator which considers up to 4 phases (a single component gaseous phase, "*g*", and three liquid phases: aqueous, "*w*", oleic, "*o*", and microemulsion, "*m*") and 19 components. It is based in the following partial differential equations:

- the mass conservation equation for component *k*,

$$\frac{\partial\left(\phi \tilde{C}_k \rho_k\right)_i}{\partial t} + \nabla.\left[\sum_{l=1}^{n_p} \rho_k \left(C_{kl}\, \vec{u}_l - \phi S_l \overline{\overline{D}}_{kl}.\nabla C_{kl}\right)\right] = \phi \sum_{l=1}^{n_p} S_l\, r_{kl} + (1-\phi)\, r_{ks} + Q_k \qquad (1)$$

where ϕ is porosity, ρ_k is the density of pure component *k*, n_p is the number of phases, C_{kl} is the mass concentration of component *k* in phase *l*, S_l is saturation of phase *l*, $\overline{\overline{D}}_{kl}$ is the dispersion tensor of component *k* in phase *l*, r_{kl} and r_{ks} are the reaction velocities of component *k* in phase *l* and in solid phase *s*, and Q_k is the injection o production rate per unit volume. The overall volume of component *k* per unit pore volume, $\tilde{C}_k$, is computed as,

$$\tilde{C}_k = \left(1 - \sum_{k=1}^{n_{cv}} \hat{C}_k\right) \sum_{l=1}^{n_p} S_l\, C_{kl} + \hat{C}_k\,, \qquad k = 1,\ldots, n_c \qquad (2)$$

where n_c is the total number of component, n_{cv} is the total number of volume-occupying components and $\hat{C}_k$ is the adsorbed concentration of species *k*.

- Darcy's equation for the flow of phase l, which defines Darcy velocity $\vec{u}_l$,

$$\vec{u}_l = -\overline{\overline{k}}\frac{k_{rl}}{\mu_l}\left(\nabla P_l - \gamma_l \nabla h\right); \quad l = 1,\ldots,n_p \qquad (3)$$

where $\overline{\overline{k}}$ is the absolute permeability tensor, *h* is depth, and k_{rl}, μ_l, P_l, are relative permeability, viscosity and pressure of phase *l*, respectively.

- Energy conservation equation. In this work we consider an isothermal reservoir, so this equation is not applied.

Summing up the mass balance equations over all volume-occupying components, substituting Darcy's law for the phase flux terms and using the definition of capillary pressure, the following equation, formulated in terms of aqueous pressure (*pressure equation*), is obtained:

$$\phi c_t \frac{\partial P_w}{\partial t} + \nabla.\overline{\overline{k}}.\lambda_{rTc}\, \nabla P_w = -\nabla.\sum_{l=1}^{n_p} \overline{\overline{k}}.\lambda_{rlc}\, \nabla h + \nabla.\sum_{l=1}^{n_p} \overline{\overline{k}}.\lambda_{rlc}\, \nabla P_{clw} + \sum_{k=1}^{n_{cv}} Q_k \qquad (4)$$

Finally, the system of differential equations consists on Eq. (4) (*pressure equation*) and Eqs. (2) (*concentration equations*). The basic boundary condition assumed in UTCHEM is that there is no flow through any boundary. The system is completed with equations representing physical properties of rock and fluids [5].

The system is solved using a block-centered finite-difference scheme. The solution method is implicit in pressure and explicit in concentrations (IMPES type). To increase stability and robustness, a flux limiter that is total-variation-diminishing (TVD) has been added [5].

3. Physical properties

The main physical properties involved in a chemical flooding are: phase behavior, interfacial tensions between fluid phases, residual phase saturations, relative permeabilities, phase viscosities, capillary pressures and adsorption. Phase behavior affects interfacial tensions, phase viscosities and adsorption. The main objective of surfactant injection is to lower oil water interfacial tension and thus mobilize the trapped oil. Therefore interfacial tension determines residual phase saturations. Besides, residual saturations affect capillary pressure and relative permeability curves. The relationship among physical properties is represented in Figure 1.

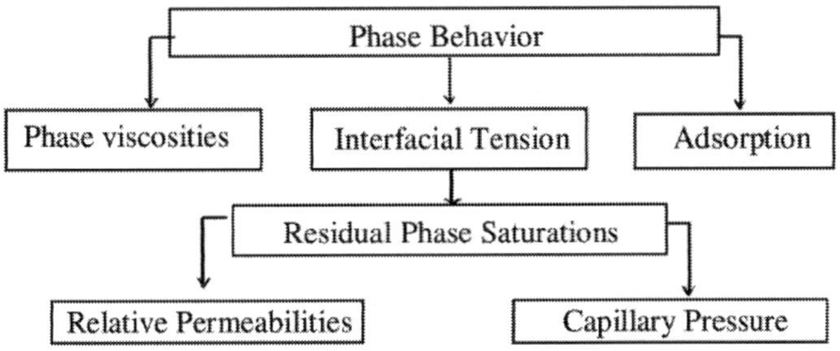

Figure 1. Relationship among physical properties.

In this work we analyze the influence of phase behavior and relative permeabilities, so only these properties are described.

3.1. Phase behavior
The brine-oil-chemical system is considered as pseudoternary, defined by 3 pseudo-components:

- *brine*: polar species (water and salts) and polymers.
- *oil*: non-polar species (hydrocarbons)
- *chemical*: amphiphilic species (surfactants and alcohols)

Three types of phase behavior, shown in Figure 2, may be considered:

- *Type II(-)*: surfactant is more soluble in aqueous-phase. There are two phases: *excess oil* (essentially pure oil) and *water-external microemulsion* (brine, surfactant and solubilized oil).
- *Type II(+)*: surfactant is more soluble in oleic-phase. There are two phases: *excess water* and *oil-external microemulsion* (most of the surfactant, oil and solubilized water).
- *Type III*: a third surfactant-rich phase is formed, the *microemulsion*. There are three phases: excess oil, excess water and microemulsion.

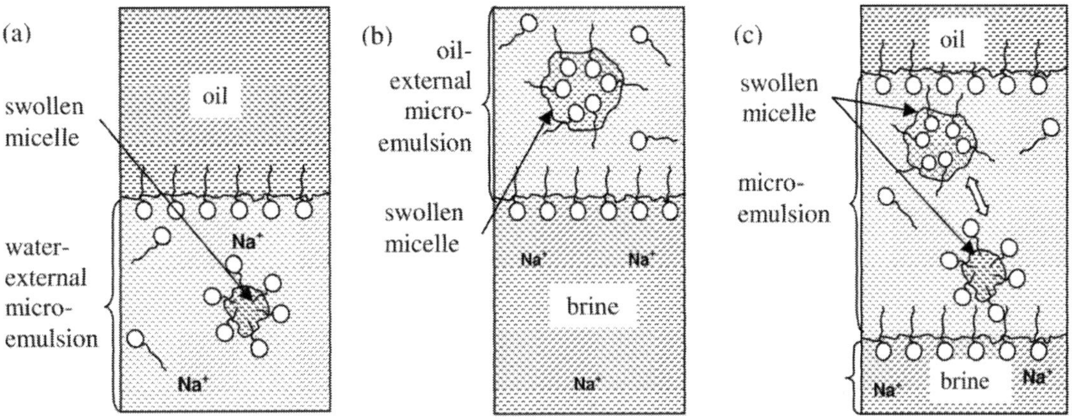

Figure 2. Phase behavior: (a) Type II(-) system, (b) Type II(+) system, (c) Type III system.

3.2. Relative permeabilities
Relative permeabilities are functions of saturation. In UTCHEM, they are represented by a potential model with 2 parameters: the endpoint (k_{rl}^{*}) and the exponent (n_l), which determines the curvature [9].

They also depend on residual saturations (S_r); therefore, the curves change as residual saturations decrease. This fact is illustrated in Figure 3: with superscript *high* and *low* we denote the curves for high and low S_r, respectively. It should be noted that the low microemulsion relative permeability has an atypical shape, it is concave ($n_m < 1$). This behavior was experimentally found by Delshad *et al.* [9].

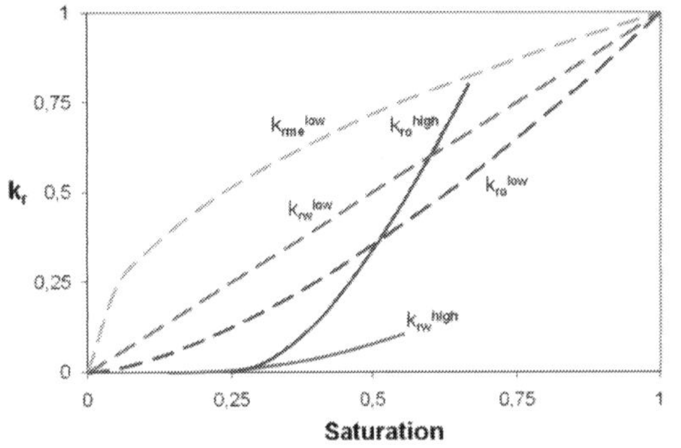

Figure 3. Water, oil and microemulsion relative permeability curves at high and low residual saturations.

4. Numerical Results

We consider a 1D linear portion of a 933 ft (284 m) length reservoir located between an injection and a production well. Porosity, permeability and cross-sectional area, are assumed constant (ϕ = *13.6 %*, *k* = *182 mD* = *1.82* 10^{-13} m^2, *A* = *65337* ft^2 = *6070* m^2). Two different cases are analyzed. The objective of **Case 1** is to describe the general behavior of a surfactant flooding, comparing the UTCHEM results with a simplified analytical solution developed by Larson [10]. The objective of **Case 2** is to perform a sensitivity analysis on the curvature of the microemulsion relative permeability curve on a polymer-surfactant flooding.

4.1. Case 1: Continuous surfactant injection

The injection fluid is a water solution with 1% surfactant and 0.54 mEq/mL salts. Initially the oil saturation is the residual saturation left behind after a waterflooding (S_{or}^{high}). The objective of surfactant injection is to mobilize the trapped oil, reaching a lower residual oil saturation (S_{or}^{low}). The injection flow rate is constant and equal to 1300 STB/d (STB=*stock tank barrel*=6.29 m^3).

Figure 4 shows the oil saturation as a function of dimensionless length corresponding to 300 days of injection. The pink curve represents the analytical solution and the grey one the results computed with UTCHEM. The analytical solution only considers two phases, so we choose a Type II(-) phase behavior. Besides, in UTCHEM, capillary pressure, dispersion, adsorption and chemical reactions have been neglected to satisfy other restrictions of the analytical solution. The remaining physical properties are taken from Han *et al.* [11]. Figure 4 illustrates the general behavior of a continuous surfactant injection. In the idealized analytical solution, an oil bank of approximate 55% saturation is formed. Behind it, there is a two-phase region that abruptly becomes a single phase with composition equal to that of the injected fluid. The UTCHEM solution approximates these results. The main difference is the time needed to reach $S_{or}^{low} = 0$. The analytical solution considers only two residual oil saturations (S_{or}^{high} and S_{or}^{low}) and only the microemulsion-oil relative permeability curves corresponding to the injected composition [10]. On the other hand UTCHEM is less restrictive, taking into account intermediate values of residual saturations and relative permeability curves. These intermediate values are computed using the reduced interfacial tensions [5]. As it can be seen in Figure 4, S_{or}^{low}=0 is obtained faster with the analytical solution than with UTCHEM. Besides, the use of intermediate relative permeabilities slightly delays the oil bank advance.

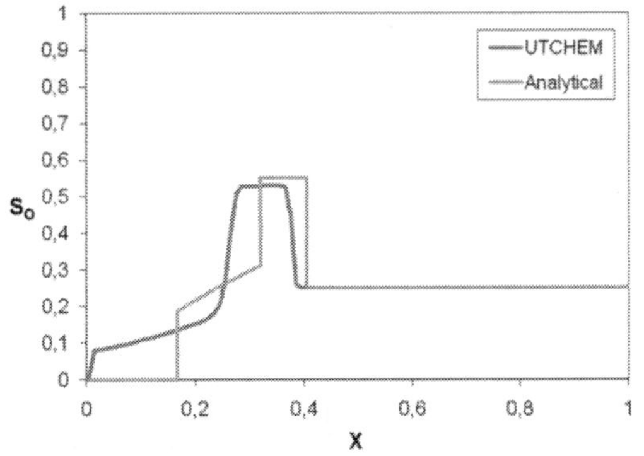

Figure 4. Oil saturation profile at t = 300 days of injection computed with UTCHEM and the analytical solution.

4.2. Case 2: Sensitivity analysis on the curvature of the microemulsion relative permeability curve

This example represents a pilot chemical flooding between an injector and a producer. From an economic point of view a continuous chemical injection is unreasonable. Therefore, chemical slugs are used, so a numerical solution is needed. Initially, there is 70% water and 30% oil. During the first 230 days, a *surfactant slug* (1% surfactant, 5% polymer and 0.54 mEq/mL salts) is injected, on the following 350 days a *polymer slug* (5% polymer and 0.54 mEq/mL salts) is injected, and on the following 2980 days water with the same salt concentration is injected. The injection flow rate is always 1300 STB/d (206.7 m^3/d). The three types of phase behavior are considered; they are determined controlling salinity parameters [11]. Capillary pressure and adsorption have been neglected. Other physical properties are taken from laboratory experiments (see references in Han *et al.* [11]).

The relative permeabilities are those shown in Figure 3. In this paper we only show the sensitivity analysis on the curvature of the microemulsion relative permeability curve (exponent n_m parameter). Sensitivity studies on other parameters can be found in Destefanis and Savioli [12]. We consider different curvatures: $n_m = 0.48$ (concave), $n_m = 1$ (linear) and $n_m = 1.5$ (convex).

The results obtained for the three types of phase behavior are shown in Figure 5. For Types II(+) and III (Figure 5 (b) and (c)), as n_m increases (or equivalently microemulsion relative permeability decreases) cumulative oil production (N_p) decreases. On the other hand, for Type II(-) (Figure 5 (a)), the behavior for different values of n_m is almost the same; in fact, as n_m increases, N_p slightly increases. Let us note that we have a water-external microemulsion in Type II(-) systems, therefore, the increasing of "water" mobility reduces oil mobility, and consequently, oil recovery.

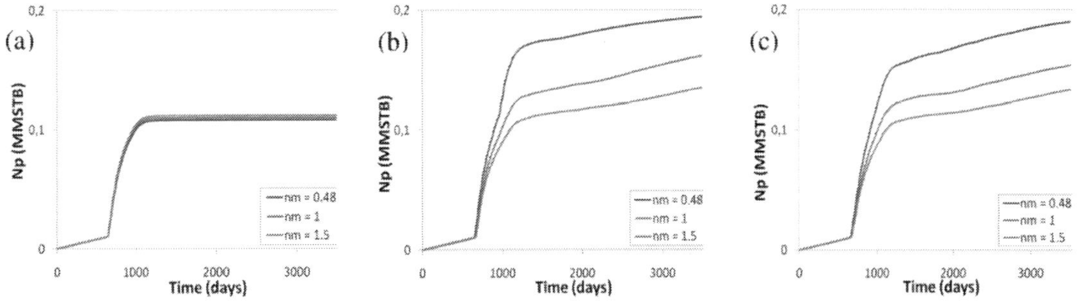

Figure 5. Phase behaviors: (a) Type II(-) system, (b) Type II(+) system, (c) Type III system.

In Figure 6, we compare the saturation profiles obtained with $n_m = 0.48$ at different times for the three types of phase behavior. Brine, oil and microemulsion saturation are represented in blue, brown and orange, respectively. Results for Type II(-) are shown in the second column of the plot, results for Type II(+) in the third column and results for Type III in the fourth one. Remember that Type II(-)

microemulsion is water external, Type II(+) microemulsion is oil external and Type III microemulsion constitutes a third phase. The typical characteristics of chemical displacement are seen in all figures, i.e., the formation of an oil bank (due to the reduction of interfacial tension and the partial oil-water miscibility) and the advance of the surfactant slug. In Type II(-) the displacement is practically immiscible, so no significant reduction of residual oil saturation is reached. On the other hand, Type II(+) and Type III displacements are miscible, oil is mostly removed and, consequently, recovery is higher.

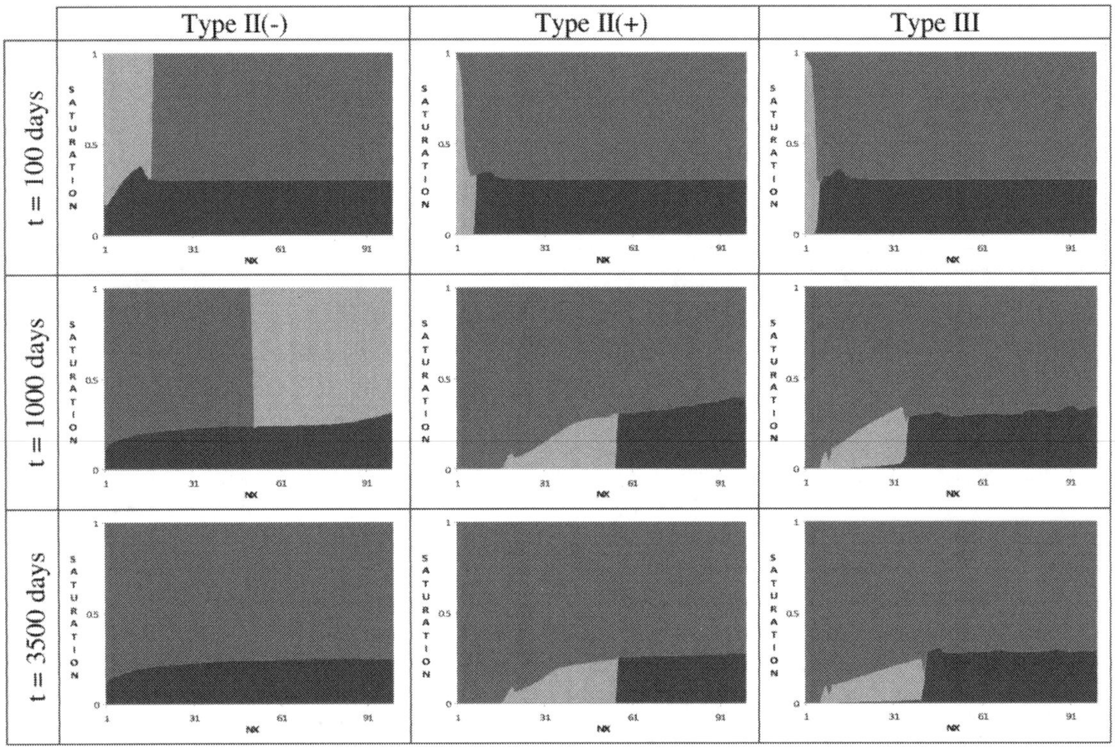

Figure 6. Saturation profiles at different times.

5. Conclusions

- The UTCHEM numerical results approach those of the idealized analytical solution, with a slower reduction of residual oil saturation.
- The influence of relative permeabilities depends on the type of phase behavior.
- For Types II(+) and III as microemulsion relative permeability exponent decreases, oil recovery increases.
- For Type II(-) the change in microemulsion relative permeability slightly affects oil recovery.
- Better displacements are obtained for Types II(+) and III because their characteristics are similar to that of a miscible displacement.

References

[1] Lake L W 1989 *Enhanced Oil Recovery* (Englewood Cliffs, New Jersey: Prentice Hall)
[2] Lake L W, Pope G A, Carey G F and Sepehrnoori K 1984 Isothermal, multiphase, multicomponent fluid flow in permeable media - part I: description and mathematical formulation *In Situ* **8** 1–40
[3] Fleming P D, Thomas C P and Winter W K 1981 Formulation of a general multiphase multicomponent chemical flood model *SPE Journal,* **21(1)** 63–76

[4] Camilleri D, Engelsen S, Lake L W, Lin E C, Ohno T, Pope G A and Sephernoori K 1987 Description of an improved compositional micellar/polymer simulator *SPE Reservoir Engineering* **2(4)** 427–32

[5] *UTCHEM 9-0, Technical Documentation*, 2000, Reservoir Engineering Research Program, Center for Petroleum and Geosystems Engineering, University of Texas at Austin

[6] Delshad M, Asakawa K, Pope G A and Sepehrnoori K 2002 Simulations of chemical and microbial enhanced oil recovery methods *SPE Paper #75237, Proc. SPE/DOE 13th Symp. on Improved Oil Recovery, (Tulsa, Oklahoma, 13-17 April 2002)* 1–13

[7] Saad N, Pope G A and Sepehrnoori K 1989 Simulation of big muddy surfactant pilot *SPE Reservoir Engineering* **4(1)** 24–34

[8] Bidner M S and Porcelli P C 1996 Influence of phase behavior on chemical flood transport phenomena *Transport in Porous Media* **24(3)** 247–73

[9] Delshad M, Delshad M, Pope G A and Lake L W 1987 Two- and three-phase relative permeabilities of micellar fluids *SPE Formation Evaluation* 327–37

[10] Larson R and Hirasaki G, 1978 Analysis of the Physical Mechanisms in Surfactant Flooding, Society of Petroleum Engineers Journal 18 (1) 42-58

[11] Han C, Delshad M, Sepehrnoori K and Pope G A 2007 A fully implicit, parallel, compositional chemical flooding simulator *SPE Journal* 322–38

[12] Destefanis M F and Savioli G B 2009 Modelización de la inundación química de reservorios *Mecánica Computacional* **XXVIII** 2445–65

Computation of unstable flows using system codes [1]

J C Ferreri [2]

Autoridad Regulatoria Nuclear, Av. Del Libertador 8250, (1429) Buenos Aires, Argentina

E-mail: jferreri@arn.gob.ar

Abstract. The prediction of unstable flows by means of so-called systems-codes is discussed in this paper. In the nuclear field, verification and validation (V&V) of codes applicable to predict the behavior of complex nuclear installations is a matter of huge collective effort presently. Multi-Physics–Multiple-Scales models, coupled with the needed wall laws is just one example if length scales go into wall layers or minute geometrical details. The validity of physical models as applicable to situations of interest in developing nuclear technology systems (e.g. natural circulation systems, super-critical conditions flows, very high temperature gas-cooled reactors) is now under consideration in the technical literature. This even applies to very simple problems, like cavity flows or one-dimensional (1D) systems. Some basic problems are presented for V&V that must correctly be solved by any systems code. Based on this, conclusions are drawn in relation to some present systems-codes, regarding the sensitivity to code options influencing its numerical approach and physical closure correlations.

1. Introduction

It must be recognized that the development of systems codes, like the ones usually utilized in the computation of the thermal-hydraulics of nuclear installation in relation to the nuclear safety evaluations impose huge efforts for a SYS-TH code developer group, in charge of its development, implementation and V&V. When adopted by a large technical community during decades, like in the case of RELAP5 [1], these codes are subject to the testing of practically all their options. Problems are continuously reported to the developers. These actions originate corrections and periodic updates. Needless to say, life is far easier for a SYS-TH code user (or one reporting on code capabilities and limitations) than for a code developer. Assessments constitute a combination of many aspects, involving standard problems of V&V ***and*** the code user, who gets results after setting up a discretization for the problem or installation under analysis. The present paradigm consists in the coupling of complex codes for analyzing complex systems. This increases the complexity of the problem, because of non-linearity of the models and the uncertainty related to code shortcomings and accuracy and also to possible implementation errors. Things become worse if different levels of approximations are used, like in the coupling of one-dimensional, time dependent systems codes with detailed nodalizations coming from computations fluid dynamics (CFD) codes. The present trend

[1] Revised, full text and updated version of a presentation to FONESYS: FORUM & NETWORK OF SYS-TH CODES IN NUCLEAR REACTOR THERMAL-HYDRAULICS, San Piero a Grado Nuclear Research Group headquarters, University of Pisa, Pisa, Italy, May 13-14, 2010

[2] Member of CIC, Conicet and Fellow of National Academy of Sciences of Buenos Aires, Argentina

consists in using CFD codes as a support to nuclear safety analysis, given the lack of enough detailed data to perform the V&V of CFD codes in situations of real practical interest in Nuclear Engineering.

CFD issues have been purposely left almost apart (only one example will be considered), because it would deserve much more than this presentation. In this case, discussion should be more focused into the physics and its associated equations because -from the numerical/computational point of view- the nuclear application of CFD codes benefits from well-established methods of many other branches of Engineering, such as Computational Mechanics

The validity of physical models as applicable to situations of interest in developing nuclear technology systems (e.g. natural circulation systems, super-critical conditions flows, very high temperature gas-cooled reactors) is now under consideration in the technical literature. This even applies to very simple problems, like cavity flows or 1D systems. In the nuclear field, V&V is a matter of huge collective effort presently. Multi-Physics/Multiple-Scales, coupled with the needed wall laws is just one example if length scales go into wall layers or minute geometrical details

The physical model of an installation implies the specification of a system of conservation partial differential equations (PDEs) and their initial and boundary conditions (I&BCs). The physical model may sometimes specify a mathematical problem with undesired properties (it is not a new here that an ill-posed system of PDEs may be accepted by a whole technical community). The mathematical analysis of the PDEs establish how the I&BCs must be formulated and specified. The numerical formulation relates this mathematical model to an algorithm allowing computing the solution. The numerical model may imply approximations that are not coherent with the mathematical properties of the system, even when said approximations are consistent and stable (implying solution convergence). I&BCs are usually coherent with the mathematical model despite the equations really solved (i.e. the algebraic system including the effects of truncation errors) by the computer implementation of the numerical model. As a consequence of the above considerations, a solution is "always" obtained (due to the so-called "code robustness"), which may show properties <u>different</u> from what must be obtained from the exact solution of the mathematical model, even more in the case of stabilization or posedness modification.

Differences may come from different causes, among them:

- Non detected errors in the code implementing a given numerical algorithm
- Artificial, numerical diffusion due to low order numerical approximation coupled with coarse nodalization
- Appropriate, stabilizing interphase correlations
- Use of steady-state correlations valid outside of the range of the code application (i.e. using friction laws, heat transfer correlations, etc., valid for steady-state, fully developed flow, in real life installation geometries)
- Inadequacy of component models, lack of appropriate closure laws
- User effects

Many other causes may not be easily discernible, like in the case of compensating errors originating in closure correlations uncertainties and problem representation specification.

The "User effects" is perhaps the most influencing source of differences between the solution of an exact code implementation of a numerical algorithmic representation of a physical model and an experiment. It has received great attention since many years ago. One example of this may be found in [2] and quite recently in [3]. In all these years, the formalization of the so-called "User effects" definition has consolidated and, again e.g. in [3], is stated as "the cumulative effect of (code) user community members producing a range of solutions for a well-defined problem with rigorously specified boundary and initial conditions". Contributions to these effects are: incorrect nodalization, wrong specification of the many input flags for a given system, use of different types of processors, etc. Ways to reduce these effects are the quality assurance of the input decks preparation and user training. User effects can not be eliminated completely.

In this paper, some results coming from a particular systems code (RELAP5/MOD3.3) and from in-house developed codes exercising are shown in the case of unstable single and two-phase flows. These few examples would hopefully illustrate some considerations to be taken into account for the simulation of this subclass of physical problems.

2. Single-phase, unstable, natural circulation flows in closed loops

This Author and his colleague at the University of Pisa, Prof. W Ambrosini, have studied the effects of nodalizations, closure correlations and their interaction in the prediction of flow behavior in single-phase, unstable, natural circulation flows since many years ago. One prototypical problem with known stability map boundary has been considered, namely the so-called Welander's loop [4], which may be considered a verification case. This loop consists of two parallel adiabatic tubes with a heat source at bottom and a heat sink at top. It has been represented with "real life" dimensions in our studies, namely two pipes 10 m height and 0.1 m in diameter and "pointwise" heat source and sink (0.1 m in length). In [5] the effects of different order nodalizations and number of nodes have been extensively discussed and illustrated, showing how the stability map of this system changes. In reference [6] three different geometries were considered, namely toroidal, square and again the Welander's loop, showing the effects of friction laws. In [7] all these results have been reviewed and a particular way of sensitivity analysis to parameters was introduced in this reference and a new one in [8]. Just to give an example of the coupled effects of nodes number and friction in this prototypical problem, the following may be mentioned: consideration of a numerical first order fully upwind approximation for the equation of energy and 90 nodes gives a correct unstable flow solution showing flow reversals. If 66 nodes are used, the flow becomes neutrally stable. Considering again 90 nodes and a friction factor magnified 1.43 times, the flow becomes again neutrally stable. This is just an example of the well known damping effects of numerical viscosity that, in this case, may be simply assimilated to friction. One interesting example of non-conservative results (i.e. results that show stability for a unstable flow situation) using a generalized 1D, single-phase code [9] for natural circulation in loops was documented in [10], showing how a physically fully unstable behavior map may have some stability island obtained by computational models due to the effects of adopting the standard law for friction, as implemented in systems codes. It must be remembered that a unstable flow with reversals traverses the zone of transition from laminar to turbulent flow. Fortunately in this case, the experimental map is known and adopting a smooth transition for friction (from Poiseuille to Blasisus friction laws) as suggested by the experimenters avoids the problem. It may be difficult to discern what law must be used in an industrial case, when dealing with unstable, cyclical flow oscillations and non fully-developed flow in relatively short pipes with bends and large flow area transitions. In a natural circulation flow, the driving forces originate in fluid buoyancy. Then, it is tempting to artificially diminishing by a known factor (simulating an error in a correlation) the heat input to the system. Again, using 90 nodes and multiplying the driving force by 0.6, the neutrally stable oscillations are obtained. Since flow rate variation comes from the balance between buoyancy and friction increasing the friction 1.43 times, the buoyancy by 1.66 and keeping 66 nodes nearly preserves this artificially neutrally stable condition. This simple, stereotyped example shows how errors may compensate.

In dealing with natural circulation in 1D loops, some problems may be found when the flow tends to stagnate. This case has been discussed in [11]. A square loop formed by tubes of 0.027 m of internal diameter and 2.2 m high and 1.4 wide, was heated with 100 W at its bottom horizontal pipe and this power was removed by a horizontal heat exchanger in its top horizontal pipe. The flow is unstable in almost all the range of power investigated (>105 - 600 W). Only for a heating power less than 105 W the flow was stable and one directional. This case was investigated using three codes, two 1D (TRANLOOP [9] andRELAP5[3] [1] and a CFD code (FLUENT 6.2 [12]). Both the 1D codes for 100 W arrived to a point of flow stagnation after a period of flow reversals that remained up to nearly 3600 s. The CFD code predicted the oscillation behavior in a correct way. The stopping of the flow with the 1D cods was due to the appearance of 1 hot fluid pocket at the top lasting for ever in the case

[3] The use of this particular systems code is due to its availability to the present Author and his colleagues. The same peculiarities of the solutions may also be found using other codes.

of TRANLOOP because of the assumption of the Boussinesq approximation for the fluid. In the case of RELAP5, the situation persisted up to the point in which the temperature of the fluid in the heater zone reached 273 K and, due to boiling, minute perturbations in the fluid allowed flow bursting, fluid cooling and a new period of stagnation. The CFD code succeeded to predict the instability because fluid stratification and sliding of hot fluid at the top of the horizontal bottom pipe precluded full stagnation. Then, the lessons learned were: a) 1D codes may fail to predict flow instability due to inherent limitations in the physical model or in the dimensionality; b) a detail code (like a CFD one) may be important to discern whether a model (the Boussinesq approximation) may still be useful due to dimensionality problem avoidance and c) these conclusions can not be generalized to any situation.

3. Two phase unstable flows in heated channels

Two-phase unstable flows may be studied to test the prediction capabilities of a code in order to perform V&V evaluations in the range that covers normal operating conditions in a nuclear installation or to reproduce the behavior of some experimental loop in controlled laboratory conditions, once again aiming to validate the computational predictions. The posedness of the governing equations for two-phase flows has been a subject of discussion. As mentioned before, most system codes solve an ill-posed system of PDEs. However it can not asserted that predictions are wrong, because there is always a method that allows circumventing the problem that consists in the introduction of numerical viscosity, like Von Neumann and Richtmyer did in 1950, or explicit physical viscosity in the equations. Up-winding is the usual technique. RELAP5 includes a term not affecting the steady-state solution. It is accepted that the least contribution of these terms regularize the system. Also, there are numerical ways to deal with ill-posedness, e.g. matrix pre-conditioning affecting only the time dependent terms. In this way, the steady-state solution is not influenced. This is important too when using pseudo-transients to get steady-state solutions.

3.1. A theoretical boiling channel simulation

In [13], the results of a computational study on boiling channel stability have been discussed. The study compared the information obtained by two programs, developed in previous work for the linear and the nonlinear stability analysis of boiling channels based on a simplified flow model, with the predictions of RELAP5. The phenomena, as described by the results of the latter, adopted with different flow models and numerical methods, have been discussed in a systematic way with the aid of the description obtained by the simplified code which, although limited in generality, permits a valid analysis in the case considered. The effects on the results obtained with RELAP5 were analyzed using different options like equilibrium and non-equilibrium fluid mixture, different numerical discretization schemes and variable number of nodes, evaluating their influence on the predicted stability boundaries.

In what follows, the results and conclusions that were arrived in [13]will be considered briefly. The adopted channel geometry and working conditions were as follows: channel diameter 0.0124 m; channel length 3.66 m (i.e., 12 ft.); inlet plenum pressure 7.0 MPa; pipe roughness 2.5×10-5 m; input and output concentrated loss coefficients Ki = 23 Ke =5; pressure drop between input and output was 0.11 MPa (imposed) and variable number of nodes (12, 24 and 48). Concerning the working fluid (water and its vapour), the homogeneous equilibrium model (HEM, in the following also named Equal Velocities Equal Temperatures, EVET) and the non-homogeneous non-equilibrium model (Unequal Velocities Unequal Temperatures, UVUT) were activated by choosing the related options in the "fvcahs" and the "tlpvbfe" codes. Regarding the numerical approaches, both the "semi-implicit" and "nearly-implicit" numerical methods were selected by choosing the related flags in the time step advancement cards.

The detailed discussion on the physics and computational results in [13], can not be considered here for the sake of brevity. The following conclusions may, anyway, be mentioned: a) the semi-implicit numerical method is considerably less prone to calculation problems than the nearly-implicit one in the application to boiling channel instability studies, showing a known code limitation; b) the latter method, as expected, introduces greater numerical diffusion, though this is evident only at low subcooling number, while at larger subcoolings the behaviour is less clear; c) the UVUT model is more "robust" than the EVET model and d) the ratio of the fluid transit time to the oscillation period at the stability boundary can be considerably different from 0.5, especially at high subcooling.

3.2. Simulation of a boiling channel in a closed loop

The computational simulation of the experimental results of [14] will be considered in the following. The code used was RELAP5 and a more detailed explanation will be included, since these results have not been published elsewhere. Figure 1 shows a schematic diagram of the two-phase forced circulation loop. It consists of a heated test section with a large bypass, a riser, an open tank and a downcomer. The fluid circulates by means of a centrifugal pump. Most of the flow is derived to the bypass, in order to maintain a constant pressure drop across the test section. In this aspect, the behavior of the heated channel would bi similar to the previous example. The downcomer, the upper and lower horizontal major sections, the bypass and the raiser are made of 50-mm inner diameter stainless steel pipes. The plena are cylinders 150 mm I.D. and 100 mm in height. Both plena and its corresponding exit tubes (1-inch i.d.) are made of brass. The valve at the inlet of the test section is 25 mm (1 inch nominal) I.D and the valve at the riser is 51 mm (2 inch nominal) I.D. The test section consists of an annular heated channel. It is connected to system through 25 mm I.D. tubing. The heated channel is a 29 mm I.D. transparent Pyrex glass tube with a coaxial electrical heater (9.5-mm O.D., 1000-mm length). A D.C. power supply unit provides the electrical power.

The experiment used water as working fluid, and the system was subject atmospheric pressure. The fluid temperature at the inlet and exit of the heated channel was measured with thermocouples. The temperature in the inlet plenum was also recorded. The flow rate in the channel was measured at the inlet of the heated section by means of an orifice plate and a fast response differential pressure transducer, variable reluctance type. The pressure drop and temperature signals were acquired by computer means. The sample interval was 0.25 seconds. The position of the valves was fixed in all experimental runs. The total loss coefficients upstream and downstream of the test section were 18 ± 5 and 8 ± 4, respectively. In this paper, the results corresponding to gravity driven oscillations will be considered, so the valves were completely open.

The experiments in [14] were carried out maintaining constant power (3400±100 W). Starting with high inlet subcooling (T_i = 68 °C) the inlet temperature was slowly increased (at approximately 0.1 °C/min.). At a certain point the stability margin is reached and the flow begins to oscillate. Further increasing the inlet temperature produced the transient excursions of the boiling boundary out of the channel, leaving the channel completely in liquid phase. Afterwards, the inlet temperature was forced to decrease until the system returned to stable conditions. Figure 2, taken from [14] shows the records of the flow measurements at different inlet temperatures. In this figure the oscillations start at T_i = 72.8 °C and it may be observed that the instabilities continue even if the temperature decreases below that value. The vertical bars are the standard deviations of the flow signal, and can be taken as indicators of the oscillation amplitude.

This experiment has been simulated using RELAP5 and some results will be shown in what follows. The heated channel was discretized using 52 nodes. Coherently with previous experience this would be an appropriate one. This was simply verified by increasing the nuber of nodes without success. Figure 3 shows the results obtained heating the system after getting steady state with the power turned off. The heating ramp, the plateau and the cooling down ramp allowed arriving to T_i = 80.2 °C. As may be observed, at this temperature the flow underwent constant amplitude oscillations. The code options were the ones corresponding to default conditions. The period of the oscillations was 18 s, the same that the period in the experiment and also coincident with the fluid transit time. These aspects are illustrated in figure 3 that shows these satisfactory results. Also, the expected hysteresis was reasonably represented. Most unfortunately, explicit activation of the UVUT or the EVET options led to non satisfactory results in the detection of the threshold temperature for the beginning of the flow oscillations and their amplitude. The causes for this behavior remain in the code and are uncertain.

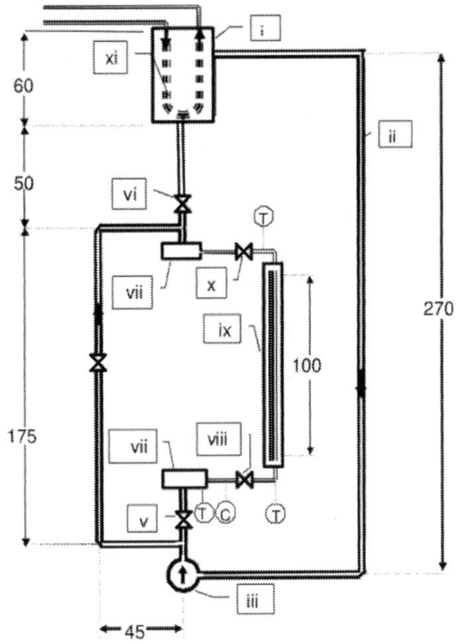

Loop components	
i	Upper tank open to atmosphere
ii	Downcomer
iii	Main pump
iv	Lateral bypass with a valve
v	Main input valve at pump exit
vi	Main output valve upstream of tank
vii	Lower and upper plena with connections to the heated channel
viii	Test section input valve
ix	Heated test chnnel
x	Test section output valve
xi	Secondary cooling system with a heat exchanger in the upper tank

Figure 1. Schematic loop representation and nomenclature

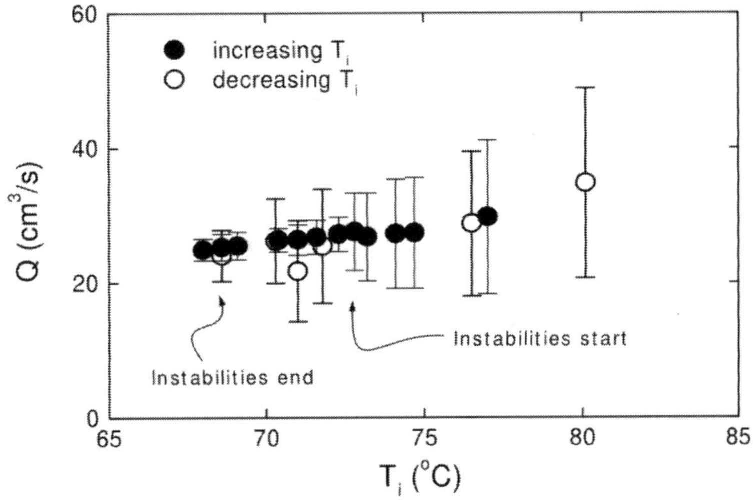

Figure 2. Volumetric flow rate in the heated channel vs input fluid temperature and oscillations ranges for gravity driven oscillations. Experimental data from [14]

3.3. Implications of the results shown on future research and development relevant to codes V&V

In the preceding subsections some limitations of present day's systems codes have been exemplified. Some of them are due to the so-called user effects. Many other are not so evident and are at the root of the need for uncertainty evaluation of computational results. The present trend, given the huge work done in the past on V&V of systems codes, seems turning to V&V of CFD codes. The international community is aware of the lack of detailed and reliable experimental information on two-phase flows in the vicinity of walls and is also aware of the need of advanced measurement techniques that permits obtaining said data. Complex flows (like two-phase flows, heterogeneous media flow like debris flows, localized heat transfer effects models, etc.) are progressively incorporated as CFD code capabilities. These type of codes (FLUENT, STAR-CD, ANSYS-CFX and many others) are typically considered as capable of giving support to safety evaluation of nuclear installations. As a consequence, V&V of CFD codes used in the nuclear industry constitutes nowadays a vast field for people dealing with the main topics treated here and need to be validated. A very valuable source of information on the needs and the current status of the activities may be found in the Nuclear Energy Agency site, namely at: http://www.nea.fr/html/nsd/csni/cfd/index.html

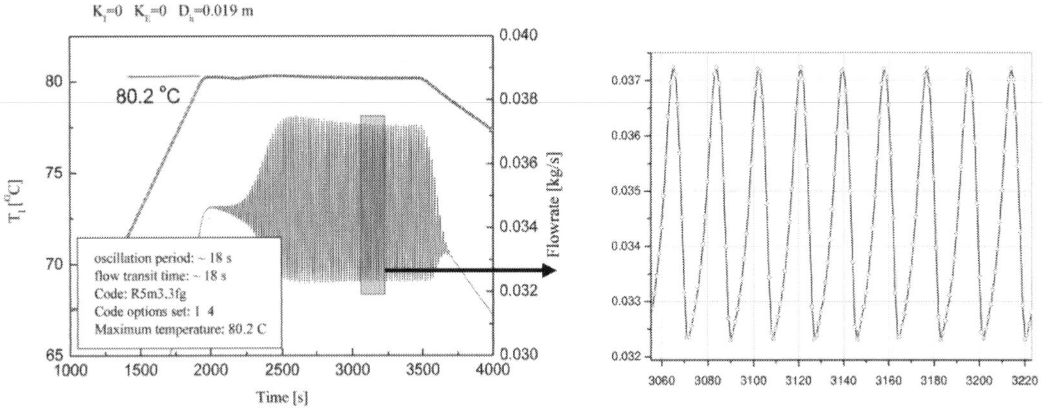

(a) (b)

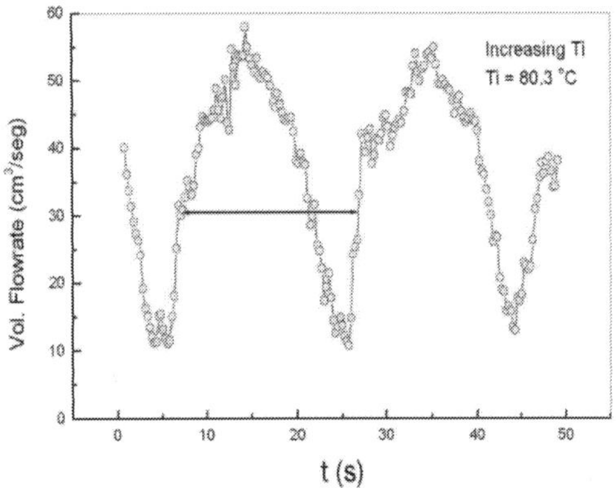

(c)

Figure 3. Comparison of predicted and experimental two-phase oscillating flow in the heated channel of [14] showing coincidence of experimental and predicted oscillation period.

a) Time variation of the computed flow rate and fluid inlet temperature

b) Zoom of the flow rate variation in figure 3a. Oscillation period is 18 s

c) Time variation of experimental [14] volumetric flow rate variation. Oscillation period is 18 s.

4. Conclusions and recommendations

In the previous sections, part of the experience gained along fifteen years of use of system codes and from the development of computational codes to deal with natural circulation flows in loops has been reviewed. Additionally, the simulation of two cases of forced two-phase flow has been presented. The one corresponding to an experiment showed good agreement, although the code used suffered from some limitations. In this case, user effects may also affect the accuracy of the predictions and can not be dismissed.

On the basis of Author's experience, there is a unique opportunity for researchers considering the physics and measurements of fluid flows in complex geometries and for the development of closure laws with a degree of detail compatible present technological demands. In the next decade, at least in the nuclear field, closure laws in two-phase bubbling flows considering multi-field approximations will be needed. This implies enhancing present measurement techniques near to walls to consider non-macroscopic space scales and the simultaneous consideration of non-isothermal flow regimes. Simultaneously, 1D system codes will be applied supported by coupled CFD detailed calculations adding complexity to nuclear safety evaluations. Then, the need for research in the physics of fluids in complex flows will be a challenge to face in the near future.

References

[1] Carlson K E et al., RELAP5/MOD3 Code Manual Volume I: Code Structure, System Models and Solution Methods, 1990 US NRC NUREG/CR-5535

[2] Ferreri J C 1998 IAEA *Specialists Meeting on User Qualification Effects for the Safety Analysis of Nuclear Power Plants, (Vienna, Austria, Aug 30th –Sept 3rd 1998)*

[3] Bucalossi A, Petruzzi A, Kristof M and D'Auria F 2010 *Nuclear Technology* **172** 29-47

[4] Welander P, On the oscillatory instability of a differentially heated fluid loop 1967 *J. Fluid Mech.* **29** 17-30,

[5] Ambrosini W and Ferreri J C 1998 *Nuclear Eng. and Design* **183** 53-76

[6] Ambrosini W and Ferreri J C 2000 *Nuclear Eng. and Design* **201** 11-23

[7] Ferreri J C and Ambrosini W 2002 *Nuclear Eng. and Design* **215** 153-170

[8] Ferreri J C and Ambrosini W 2001 *Int. J. Computational Fluid Dynamics* **16**, 277-281

[9] Ambrosini W and Ferreri J C 2003 *Annals of Nuclear Energy* **30** 1505-37

[10] Ambrosini W, Forgione N, Ferreri J C and Bucci M 2004 *Annals of Nuclear Energy* **31** 1833-65

[11] Pilkhwal D S, Ambrosini W, Forgione N, Vijayan P K, Saha D and Ferreri J C 2007 *Annals of Nuclear Energy* **34** 339-355

[12] FLUENT Inc., FLUENT 6.2 User's Guide 2003 Centerra Resource Park, Lebanon, NH USA

[13] Ambrosini W and Ferreri J C 2006 *Proceedings of ICONE 14, 14th International Conference on Nuclear Engineering (Miami, Florida USA July 17-20 2006)* ICONE PAPER 14-89863

[14] Juanico L, Identification of dynamic basins in boiling flows, 1997 *PhD Thesis* Inst. Balseiro UNCuyo Argentina

FLUIDOS 2010: XI Meeting on Recent Advances in the Physics of Fluids and their Applications IOP Publishing
Journal of Physics: Conference Series **296** (2011) 012016 doi:10.1088/1742-6596/296/1/012016

Harmonic vibration on a reactive fluid at boundary layer regime modifies the Turing instability

Jacobo Guiu-Souto[†], Jorge Carballido-Landeira, Alberto P. Muñuzuri

Group of Nonlinear Physics, Fac. de Fisicas, University of Santiago de Compostela. E-15782 Santiago de Compostela, Spain

[†] jacobo.guiu@rai.usc.es

Abstract. Studies of Turing instability in a reactive fluid, known as BZ-AOT microemulsion, under external harmonic vibration, show modifications on pattern formation. The characteristics of the reactor and forcing prevent flows to occur, and therefore the transport phenomena can only be due to microscopic diffusion. In order to understand this problem, we combine the reaction-diffusion model with a mesoscopic model. The performed linear stability analysis and numerical simulations show, on the one hand, that the forcing increases the wavelength of the patterns and, on the other hand, that the system is unable to exhibit any pattern for sufficiently large forcings thus yielding to a state of homogeneous concentration.

1. Introduction

Extensive studies have been performed in order to understand the formation mechanisms of the elegant patterns that are ubiquitous in Nature [1, 2], i.e. striped tiger skin or the sea-shells forms. Turing instability [3] represents a good mechanism to reproduce the pattern formation from a physicochemical perspective. Turing structures are stationary patterns that may appear when the propagating chemical waves are combined with the effects of diffusion in a medium. In this paper, we will consider the numerical and theoretical aspects of Turing instability obtained in a compartmented reaction-diffusion system (BZ-AOT system), by keeping in mind the location of the problem within the experimental context where these results can be applied.

The Belousov-Zhabotinsky reaction [4, 5] can be confined inside a water-in-oil microemulsion using AOT (sodium bis(2-ethylhexyl) sulfosuccinate) as the stabilizing amphiphile (BZ-AOT system). This closed system demonstrates a rich variety of dynamics, which are impossible to obtain by the classical BZ reaction [6, 7], such as Turing patterns. The BZ-AOT reaction involves a lot of elementary reactions [8]; however, the system dynamics is related to two main characteristics; on one hand, the antagonistic and competitive behaviour between some species of the reaction (known as activator and inhibitor), and on the other hand, their different diffusion rates. This difference in the diffusion coefficients lies in the different location of these antagonistic species. While the activator resides within the AOT microemulsion and diffuses as a micelle, the inhibitor is able to diffuse through the oil phase due to its nonpolar nature.

Nevertheless, in Nature, pattern formation usually occurs in presence of external perturbations [9-13]; thus, it is necessary to develop a more general description, in order to understand the transport phenomena and their impact in pattern formation mechanisms. In this work, we develop a numerical study in which the dynamics of pattern formation will be modified by introducing a harmonic temporal perturbation vertically to a two-dimensional BZ-AOT system [14, 15], so that, convection is not allowed due to physical constrains. This paper is organized as follows: in Sec. 2 we present the reaction-diffusion model, as well as a discussion establishing the kind of transport mechanism within our system. Sec. 3 deals with the mesoscopic treatment of the harmonic perturbation and how it affects the diffusion processes. The characteristics of Turing pattern formation are described in Sec. 4, by a linear stability analysis of the reaction-diffusion equations. In Sec. 5 we present numerical simulations of the pattern formation showing how the harmonic perturbation modifies the characteristics of the Turing structures. Finally, in Sec. 6 we present the conclusions and discussion.

2. Model equations

The problem we want to address is schematically depicted in Fig. 1. We consider a liquid pattern-forming system (BZ-AOT system) located within a closed reactor with a small vertical dimension (marked with L in figure). The whole system is bounded by glass boundaries that actually play a crucial role in determining the fluid dynamics as we will discuss below. The system described is periodically forced (up to 2.4 cm in amplitude and 100 Hz in frequency) in the vertical direction thus, the actual gravity affecting the reaction is periodically modulated with time.

The phenomenology of pattern formation including Turing instability can be studied through the Reynolds theorem. Turing structures are due to singular space-temporal evolution of the concentration of each species. By using this theorem we obtain the variation on the species concentration per unit volume in the presence of sources and flows:

$$\frac{\partial c_i(\overline{x},t)}{\partial t}+\nabla\cdot\overline{\Phi}_i - S_i(c_1,...,c_n)=0; \quad \overline{\Phi}_i=-D\nabla c_i + c_i\overline{v}_i \tag{1}$$

where $i=1,2,\ldots,n$ is the specie index and D is the diffusion tensor; c_i, S_i, $\boldsymbol{\Phi}_i$ and $\boldsymbol{v}_i$ are the terms associated with the concentration, the sources, the flows and the velocity field of the i specie, respectively.

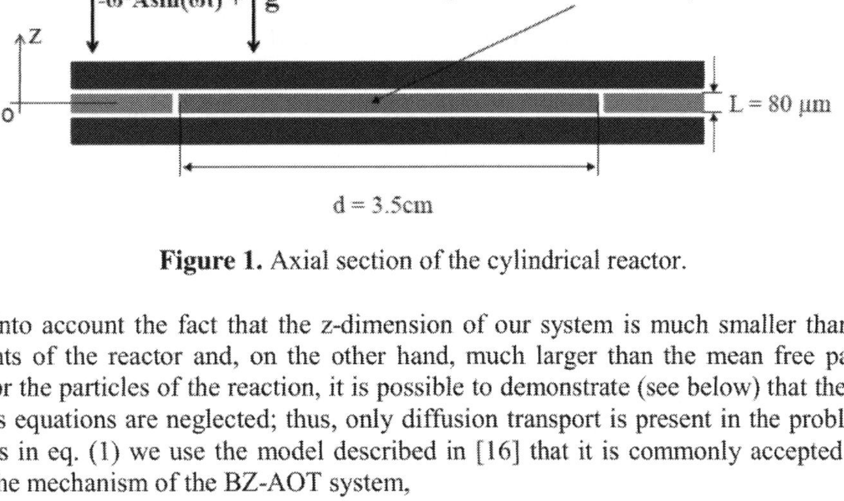

Figure 1. Axial section of the cylindrical reactor.

Taking into account the fact that the z-dimension of our system is much smaller than the global displacements of the reactor and, on the other hand, much larger than the mean free path between collisions for the particles of the reaction, it is possible to demonstrate (see below) that the convection terms in this equations are neglected; thus, only diffusion transport is present in the problem. For the source terms in eq. (1) we use the model described in [16] that it is commonly accepted to describe accurately the mechanism of the BZ-AOT system,

$$\frac{\partial c_1}{\partial t}=\frac{1}{\varepsilon}\left[fc_2\frac{q-c_1}{q+c_1}+c_1\frac{1-mc_2}{1-mc_2+\varepsilon_1}-c_1^2\right]+D_1\nabla^2 c_1 \tag{2}$$

$$\frac{\partial c_2}{\partial t}=c_1\frac{1-mc_2}{1-mc_2+\varepsilon_1}-c_2+D_2\nabla^2 c_2 \tag{3}$$

where c_1 and c_2 are the concentrations of the main species competing in the pattern formation mechanism known as activator and inhibitor, respectively, and D_1, D_2 are their corresponding diffusion coefficients. The parameters f, m, q, ε_1 and ε depend on the reaction rates and they appear as a result of applying the mass action law on the chemical equations and a posteriori dimensionless analysis [17]. Appropriated values to model the BZ-AOT systems are $f=1.85$, $m=10$, $q=0.004$, $\varepsilon_1=0.02$ and $\varepsilon=0.4$ [15-17]. In addition, it is necessary to consider, a difference of one order, at least, between the inhibitor and activator diffusion coefficients in order to produce Turing structures in this model, i.e. $D_1 = 1$ and $D_2 = 30$. This difference can be understood because of the different locations of both species; the activator remains inside of AOT nanodroplets and diffuses with the micelle. Meanwhile, the inhibitor (a nonpolar molecular compound) is able to diffuse through the oil phase and this process is much faster than that of the activator.

Furthermore, to complete the model we add the following Dirichlet and Newman conditions:

$$\begin{aligned} c_1(\bar{r},t)\big|_{t=t_0} &= c_{1,0};\ \hat{n}\cdot\nabla c_1\big|_{\partial\Omega}=0 \\ c_2(\bar{r},t)\big|_{t=t_0} &= c_{2,0};\ \hat{n}\cdot\nabla c_2\big|_{\partial\Omega}=0 \end{aligned} \tag{4}$$

Let us analyze now in detail the deep mechanism of the problem here presented that justify the absence of convection in our system. First, we proceed to estimate the boundary layer thickness, δ. Initially, we need to know the Reynolds number of the flow (in case it would exist). By supposing that a flow is present in the system, we may estimate the Reynolds number, Re:

$$Re = \frac{\rho\omega AL}{\mu} < 7 \tag{5}$$

here ρ ($854.3\ kg/m^3$) and μ ($0.002\ Pa{\cdot}s$) are the density and the dynamic viscosity of the BZ-AOT microemulsion, respectively; $v=\omega A$ ($0.2\ m/s$) is the typical velocity of oscillating flow at maximum forcing and L ($80\ \mu m$) is the thickness of the reactor [7, 15]. The low value of Re demonstrates that the flow is laminar. Thus, we may estimate the boundary layer thickness by using the Blasius boundary layer model [18]:

$$\delta = \frac{4.9d}{\sqrt{Re}} \approx 3.58\ cm \tag{6}$$

where d ($3.5\ cm$) is the diameter of the reactor (see figure 1). The boundary layer thickness in the worst case is several orders of magnitude larger than the thickness of the reactor ($80\ \mu m$). Consequently, it is appropriate to consider that the flows are heavily influenced by the boundaries, so convective phenomena are strongly inhibited. And, thus, our system is within the boundary layer regime and we can assume that the transport mechanisms are only due to diffusive processes.

3. Mesoscopic description

As no convective flows are allowed in the system due to the mechanical restrictions, as described in the previous section, the only transport mechanism for the molecules is molecular diffusion, thus, the continuum hypothesis can be applied [19]. We developed a microscopic description, based on the statistical mechanics, in order to analyze the influence of the vertical forcing on the molecular diffusion coefficients.

The particle position is composed by the sinusoidal motion of the forcing plus the motion of the particles inside of the reactor, where the BZ-reaction takes places. Therefore, the position vector within the laboratory reference frame is:

$$\bar{r} = x\hat{x} + y\hat{y} + z\hat{z} = x'\hat{x} + y'\hat{y} + \left(z' + A\sin(\omega t)\right)\hat{z} \tag{7}$$

where x', y', z' is the position of the particles inside of the reactor. In the xy-plane, there are only thermal fluctuations, meanwhile in the z-direction there is also a sinusoidal forcing. We want to note that as the forcing amplitude ($A \sim 10^{-2} m$) is much greater than thermal fluctuations, i.e. $z' << A\sin(\omega t)$), we assume that $z \sim A\sin(\omega t)$. Thus the resulting Hamiltonian, by neglecting variations in the gravitational contributions and introducing the vibration energy, is given by

$$H_{1,2} = \frac{\overline{p}_{1,2}^2}{2m_{1,2}} + p_{z_{1,2}}\omega\sqrt{A - z_{1,2}^2} + \frac{1}{2}m_{1,2}\omega^2 A^2 \tag{8}$$

where $\mathbf{p}_{1,2}$ and $m_{1,2}$ are the momentum and masses of the activator and inhibitor; ω and A are the frequency and amplitude of the applied forcing. In addition, we assumed that the collisions are only between identical particles, i.e. *1-1* or *2-2*. The molecules of different species, *1-2*, do not collide with each other due to their different sizes and charges. Therefore, the canonical partition function associated with our system is:

$$Z_{1,2} = \frac{V}{h^3}\left(\frac{2\pi m_{1,2}}{\beta}\right)^{3/2} \sum_{n=0}^{\infty} \frac{\left(-\beta m_{1,2}\omega^2 A^2\right)^n}{2^n(2n+1)n!} \tag{9}$$

here V is the volume of the reactor, h is the Planck constant and $\beta = k_B T$ with k_B as the Boltzmann constant and T the temperature of the reactor. Thus, the momentum of our particles can be estimated and it happens to increase with the forcing parameters (ω and A) and with the mass of the particles ($m_{1,2}$):

$$\left\langle p_{1,2}^2 \right\rangle = \frac{m_{1,2}}{\beta}\left\{3 + \beta m_{1,2}\omega^2 A^2 \left[1 - \frac{\displaystyle\sum_{n=0}^{\infty} \frac{\left(-\beta m_{1,2}\omega^2 A^2\right)^n}{2^n(2n+3)(n!)}}{\displaystyle\sum_{n=0}^{\infty} \frac{\left(-\beta m_{1,2}\omega^2 A^2\right)^n}{2^n(2n+1)(n!)}}\right]\right\} \tag{10}$$

It is important to note that the momentum domain is always positive for all ranges of forcings, and also at zero forcing we recover the classical expression for free particles.

By statistical procedures [20], we estimate the collision number and the mean free path of our particles. With these results we obtain the following expression for the diffusion coefficients:

$$D_{1,2} = \frac{D_{1,2}^0(\beta, m_{1,2})}{3\sqrt{3}}\left(3 + \beta m_{1,2}\omega^2 A^2\right)^{3/2} \tag{11}$$

where $D^0_{1,2}$ is the diffusion coefficient of the free particles given by:

$$D_{1,2}^0(\beta, m_{1,2}) = \frac{1}{8N_{1,2}\sigma_{1,2}}\sqrt{\frac{3}{\beta m_{1,2}}} \tag{12}$$

here $\sigma_{1,2}$ is the total cross section and $N_{1,2}$ is the number of particles per unit of volume for each species. It is important to note that to obtain simplified expressions in the calculations of the diffusion coefficients, we consider a first order approximation ($O(\omega^2 A^2)^1$) in the expansion of forcing term of the summations (as all forcings verify $\beta m_{1,2}\omega^2 A^2 << 1$). According with expression (11), the forcing on heavier particles is going to induce a larger effecive diffusion coefficient. We want to emphasize that the diffusion coefficients $D_{1,2}$ also recover their classical value $D^0_{1,2}$ in absence of forcing.

4. Pattern formation

In order to understand the mechanism of Turing pattern formation, we perform a standard stability analysis [21] of the reaction-diffusion equations (4, 5). Thus, we test a general solution for chemical waves (concentration waves):

$$\bar{c} = \begin{pmatrix} c_1 \\ c_2 \end{pmatrix} = \bar{C}\exp\left(\lambda t + i\frac{\bar{k}\cdot\bar{r}}{\sqrt{n}}\right) \tag{13}$$

where λ is the eigenvalue, $\boldsymbol{k}$ is the wavenumber and n is the number of space dimensions. Through a linear stability analysis we obtain the following expression for the Turing eigenvalue:

$$\lambda_{1,2} = \frac{1}{2}\left[tr(J) - k^2 tr(D) \pm \sqrt{\left[k^2 tr(D) - tr(J)\right]^2 - 4p(k^2)}\right] \tag{14}$$

with,

$$p(k^2) = k^4|D| - k^2 tr(D^*J) + |J|, \tag{15}$$

$$D = \begin{pmatrix} D_1 & 0 \\ 0 & D_2 \end{pmatrix}; \quad D^* = \det(D)D^{-1} \tag{16}$$

here J is the Jacobian matrix associated with the linearization. Turing patterns occur when the eigenvalue is positive, for some wavenumber. Then, only one wavenumber can generate the pattern with a characteristic wavelength. It is important to note that the final wavelength of the pattern comes from the wavenumber which provides the most unstable value for the eigenvalue, k_0. Thus, by maximizing the eigenvalue we obtain:

$$\left.\frac{dp(k^2)}{dk}\right|_{k_0^2} = 0 \Rightarrow k_0^2 = \frac{tr(D^*J)}{2|D|} \geq 0 \tag{17}$$

Figure 2 shows the set of wavenumbers[1] that may produce Turing instability and, thus, Turing structures and how this region decreases as the forcing amplitude is increased. For larger values of the forcing ($\omega^2 A^2 > 6.22\ m^2 s^{-2}$) there are not wavenumbers that produce unstable eigenvalues. Moreover, the most unstable wavenumber, k_0, is shifted to lower values when the forcing is applied. These results suggest that the Turing wavelength ($\Lambda = 2\pi/k_0$) is increasing with $\omega^2 A^2$. Thus, we establish that the forcing modifies the configuration of the patterns (given by their wavelength) and it can even inhibit their formation.

[1] i.e. k^2 such that $Re\ [\lambda(k^2)] > 0$.

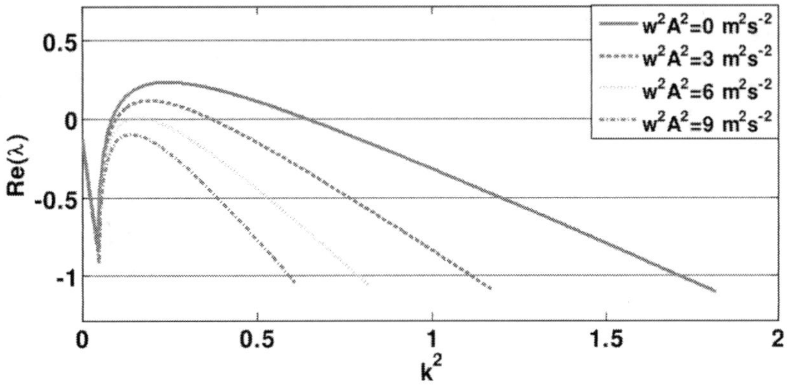

Figure 2.. Real part of the dispersion curves, $Re(\lambda)$, where λ is he eigenvalue of the linear stability analysis. Parameters $f=1.85$, $q=0.004$, $m=10$, $\varepsilon=0.4$ and $\varepsilon_1=0.02$; D_1 and D_2 vary with the applied forcing. Increasing ω^2A^2 the most probable wavenumber shifts to lower values due to dispersion relation changes.

5. Numeric Simulations

In order to verify the predictions of the previous theoretical model, we present numeric simulations of the reaction-diffusion equations (2) and (3). First of all, the effect of the forcing amplitude on the parameter space was analyzed. This phase diagram is plotted in figure 3 where the regions were Turing structures appear are plotted for the different values of the parameters f and m. The main effect of increasing the forcing amplitude is that the Turing region shrinks until it completely disappears. This diagram reflects the tendency observed in figure 2; large values of the forcing makes the Turing region diminishes. The largest forcing amplitude considered in this figure ($A^2\omega^2 = 15\ m^2s^{-2}$) is marked in the figure with empty circles (denoted by region T15 in the figure) and covers the smallest area in the diagram. On the other hand, when no forcing is considered ($A^2\omega^2 = 0\ m^2s^{-2}$, region T0 in figure 3 and marked with dark spots) the range of parameters that yield to Turing patterns is maximum. Please note that as the forcing amplitude is increased, the Turing region is included into the previously calculated, i.e., the region for $A^2\omega^2 = 0\ m^2s^{-2}$ contains all other plotted regions (T3 to T15); the region for $A^2\omega^2 = 3\ m^2s^{-2}$ (called T3 in figure 3) contains all other Turing regions calculated for higher forcing amplitudes (T6 to T15); and so on. Also note that the symbol marking each point in the f-m phase diagram shows the maximum forcing the system can experience still exhibiting Turing structures[2]; if the symbol corresponds with T0 region then any forcing inhibits the appearance of structures. If the point in the phase diagram is marked with T3, then any forcing above $A^2\omega^2 = 3\ m^2s^{-2}$ inhibits Turing structures.

We considered a set of parameter values that we considered as standard in the following simulations. This is marked in figure 3 with an arrow and correspond with the following values, $f=1.85$, $m=10$, $q=0.004$, $\varepsilon_1=0.02$ and $\varepsilon=0$. In this case, Turing structures are observed for forcing amplitudes up to $A^2\omega^2 = 6.22\ m^2s^{-2}$. Above such threshold, i.e. $A^2\omega^2 > 6.22\ m^2s^{-2}$, the forcing changes the characteristic eigenvalue from unstable ($Re(\lambda)>0$) to stable ($Re(\lambda)<0$) and the system is not able to exhibit the Turing instability.

[2] For example, for the point f=1.9 and m=60 (inside T0 region) in the phase diagram in fig 3, the maximum forcing that still exhibits Turing structures is 0 m^2s^{-2} so this region is very sensitivity to the forcing, however for the point f=1.7 and m=39 (inside T12 region), the maximum forcing supported is 12 m^2s^{-2}.

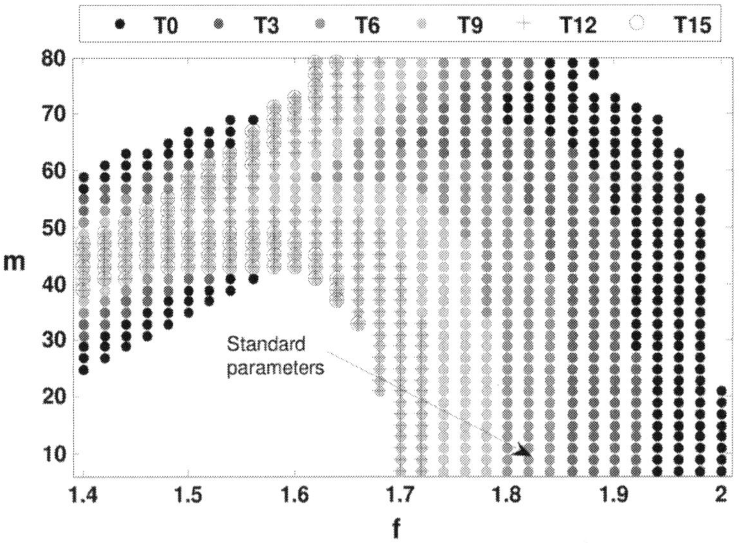

Figure 3. Phase Diagram in the *f-m* space of the different Turing regions (T) corresponding with different forcing amplitudes ($A^2\omega^2$). Other model parameters are kept constant and equal to $q=0.004$, $\varepsilon_I=0.02$ and $\varepsilon=0.4$. T0, T3, T6, T9, T12 and T15 stands for the Turing regions for forcing amplitudes equal to 0 m^2s^{-2}, 3 m^2s^{-2}, 6 m^2s^{-2}, 9 m^2s^{-2}, 12 m^2s^{-2} and 15 m^2s^{-2} respectively. Note that the different Turing regions (corresponding with different values of the forcing amplitude) are overlapped, i.e., T0 region contains all T3 to T15 regions, T3 region contains regions T6 to T15, and so on (T0⊃T3⊃T6⊃T9⊃T12⊃T15).

Numerical simulations of the equations (2) and (3) were performed by using a three level explicit algorithm with a centered in space discretization [22]. The computational domain consists in a 300x300 grid elements with space and time steps values of 0.001 t.u. and 0.5 s.u., respectively. For initial conditions we set both variables with random initial concentration around the 5 percent of the steady-state concentrations (i.e. the fixed points of the equations (2) and (3)). We present in figure 4 a sequence of pictures of the 2D Turing pattern for different forcing amplitudes.

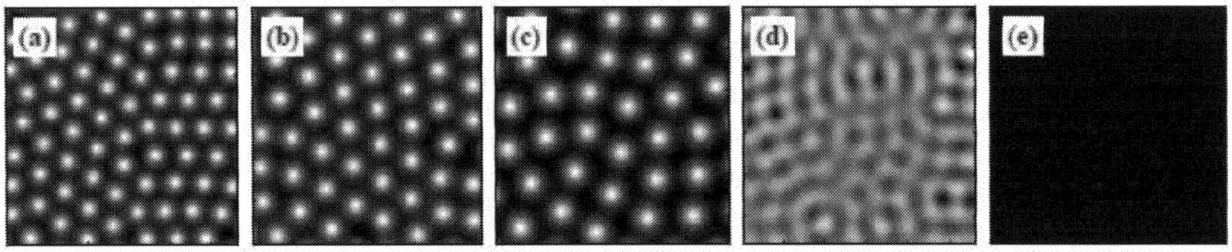

Figure 4. Two-dimensional Turing structures obtained with the model (4),(5) and (11). Parameters: $f=1.85$, $q=0.004$, $m=10$, $\varepsilon=0.4$ and $\varepsilon_I=0.02$. The value of the diffusion coefficients (D_u and D_v) depends on the applied forcing following the equation (11). Figure (a) shows the pattern at rest, i.e. $\omega^2A^2=0\ m^2s^{-2}$. In figures (b) and (c) the system was forced with a magnitude of $\omega^2A^2=3.00\ m^2s^{-2}$ and $\omega^2A^2=6.00\ m^2s^{-2}$, respectively. In figure (d) the system is very close to the transition region (*see figure* 2), the forcing is $\omega^2A^2=6.22\ m^2s^{-2}$. In figure (e) the forcing is $\omega^2A^2=9.00\ m^2s^{-2}$ and the system is clearly out of the Turing region and patterns disappear totally.

Note that the different Turing patterns exhibit a different wavelength depending on the applied forcing (figures 4a-c). Thus, we obtain numerically an increase on the Turing wavelength, in a good agreement with the theoretical predictions. When the forcing is $\omega^2A^2=6.22\ m^2s^{-2}$ (figure 4d) the maximum of the eigenvalue is zero, the onset of Turing instability. For this forcing, the system evolves to a diffuse structure with a wavelength clearly not defined. Finally, with even larger forcings ($\omega^2A^2>6.22\ m^2s^{-2}$), the maximum of the eigenvalue is negative and the system, which is in a stable configuration, evolves to a uniform value of the concentration (figure 4e) determined by the fixed points.

6. Discussion and conclusions

In this work, we studied Turing pattern formation under external harmonic vibrations in the BZ-AOT active media. The characteristics of the system are such that convective flows are not allowed, and so, only diffusion processes can take place. In this case, the energy of the forcing is spent on increasing the internal degrees of freedom of the system, and, therefore, the molecular diffusion is modified producing macroscopic effects. Pattern structure is highly sensitive to the diffusion process. We determine the relationship between the modified diffusion coefficients (due to the presence of forcing) and the resulting pattern.

We developed a mesoscopic model, based on statistical mechanics, to determine the changes in the diffusion coefficients with the forcing. By coupling this model to the reaction-diffusion equations we calculate the threshold of Turing instability. Using a linear stability analysis we find a shortening in the set of wavenumbers that are able to produce Turing instability as the forcing increases. In addition, an increase in the forcing induces a shift of the most probable eigenvalue to lower values, and therefore the pattern wavelength increases. Moreover, large values of the forcing amplitude (up to $\omega^2A^2=6.22\ m^2s^{-2}$) produce stable values for the corresponding eigenvalues, and consequently, the system moves from the Turing regime into a uniform state of concentration.

Numerical simulations of the reaction-diffusion model, demonstrate a good agreement with the predictions. In figure 4 we show a Turing pattern with hexagonal arrangement and how the associated wavelength increases with the forcing until the Turing structure disappears.

The mesoscopic approach, that we present in this work, can be applied to other situations. We believe it may result helpful in understanding other systems where an external forcing modifies the transport mechanism through diffusion (in particular when the system is highly influenced by boundary layer effects).

References

[1] Painter J K 2000, *Mathematical Models for Biological Pattern Formation, Springer-Verlag, Nederlands.*
[2] Kondo S and Asai R 1995 *Nature* **376** 765-768.
[3] Turing A M 1952, *Philos. Trans. R. Soc. London Ser. B* **237** 37.
[4] Belousov B P 1959, *Collected Abstracts on Radiation Medicine. edited by A. V. Lebedinskii (Medgiz, Moscow).*
[5] Zaikin A N and Zhabotinskii A M 1970, *Nature* **225**. 535.
[6] Vanag V K and Epstein I R 2001, *Science* **294** 835.
[7] Vanag V K 2004, *Uspekhi Fizicheskikh Nauk* **174** 991.
[8] Gyorgye L, Turanyi T and Field R J 1997, *J. Phys. Chem.* **94** 7162.
[9] Vásquez D A and Lengacher C 1998, *J. Phys. Chem.* **58** 6865.
[10] Navy I P and Pojman J A 1993, *J. Phys. Chem.* **97** 3443. Carballido-Landeira J, Vanag V K and Epstein I R 2010, *Phys. Chem. Chem. Phys.* **12** 3656.
[11] Rüdiger S, Míguez D G, Muñuzuri A P, Sagués F and Casademunt J 2003, *Phys. Rev. Lett.* **90** 12.
[12] Pérez-Villar V, Porteiro J L F and Muñuzuri A P 2006, *Phys. Rev. E* **74** 046203.
[13] Fernández-García G, Roncaglia D I, Pérez-Villar V, Muñuzuri A P and Pérez Muñuzuri V.

2008, *Phys. Rev. E* **77**, 026204.
[14] von Kameke A, Huhn F, Fernández-García G, Muñuzuri A P and Pérez-Muñuzuri V 2010, *Phys. Rev. E* **81** 066211.
[15] Guiu-Souto J, Carballido-Landeira J, Muñuzuri A P and Pérez-Villar V 2010, *Phys. Rev. E* **82** 066209.
[16] Kaminaga A, Vanag V K and Epstein I R 2005, *J. Chem. Phys.* **122** 174706.
[17] Kaminaga A, Vanag V K and Epstein I R 2006, *Angew. Chem. Int.* **118** 3159.
[18] Cohen I M and Kundu P K 1990, *Fluid Mechanics, Academic Press, San Diego.*
[19] Vekstein G E 1991, *Physics of Continuous Media (Hardcover).*
[20] Huang K 1987, *Statistical Mechanics. Massachusetts: John Wiley & Sons.*
[21] Murray J D 2002, *Mathematical Biology, Springer, Berlin.*
[22] Strikwerda J C 1989, Finite *Difference Schemes and Partial Differential Equations. Philadelphia: SIAM.*

FLUIDOS 2010: XI Meeting on Recent Advances in the Physics of Fluids and their Applications IOP Publishing
Journal of Physics: Conference Series **296** (2011) 012017 doi:10.1088/1742-6596/296/1/012017

Topography evolution in convergent flows of two liquid layers

Carlos Alberto Perazzo[1] **and Julio Gratton**[2]

[1] Universidad Favaloro and CONICET, Solís 453, 1078 Buenos Aires, Argentina.
[2] INFIP–CONICET, Dpto. de Física, FCEyN, UBA, Ciudad Universitaria, Pab. I, 1428 Buenos Aires, Argentina.

E-mail: `jgratton@tinfip.lfp.uba.ar, perazzo@favaloro.edu.ar`

Abstract. We have recently developed a two–layer model that considers the convergent motion of two initially uniform liquid layer with different densities and viscosities and assumes that the flow is due to the basal traction that acts at the bottom of the lower layer. We have used this model to describe successfully the evolution of mountains belts (Perazzo & Gratton, Phys. Fluids **22**, 056603, 2010). In this work we discuss how to modify our model to also describe the formation of plateaus. To this end we assume that below of a given level the viscosity of the upper layer drops abruptly, and in consequence the flow of this layer becomes decoupled of the motion of the lower region of the system.

1. Introduction

The structure, properties and motions of the outer layers of the Earth determine the topography of its surface, that changes in processes that take place very slowly: the timescale T_o of the evolution of mountain ranges is of the order of $10^6 - 10^7$ years. Rocks subject to stresses for such long times deform plastically, with a rate that increases rapidly with the temperature, that in turn increases with the depth. Let us now summarize some concepts relevant to our purpose in order to introduce the geophysical terminology.

The *lithosphere* (figure 1) is the outer solid layer of the Earth and comprises the *crust* and the *lithospheric mantle*, that are separated by the Mohorovičić discontinuity (also called *Moho*), where the composition and the density of the rocks changes abruptly. The lithosphere rests on the *asthenosphere* that is the remaining portion of the upper mantle and that in the timescale in which we are interested behaves as a liquid. The lithosphere–asthenosphere boundary (*LAB* in brief) corresponds to the 1500 °C isotherm, where the viscosity drops by 2–3 orders of magnitude. The upper crustal rocks are rigid and fracture if the stresses overcome their mechanical strength, but deform plastically at greater depths. The level of this transition depends on the strain rate and on the temperature and is closer to the surface for slow deformations.

The term *isostasy* denotes the state of gravitational equilibrium of the lithosphere in which the pressure at the LAB is uniform. It implies that underneath any visible relief there is a anti–relief of the Moho, called *root* (see figure 1 for the case of a continent). If ρ_c (≈ 2.7 g/cm^3) is the density of the crust, R the topographic elevation, $\rho_m = \rho_c + \rho'$ (≈ 3.2 g/cm^3) the density of the lithospheric mantle ($\rho' \approx 0.5$ g/cm^3) and R' the depth of the root, isostasy requires that

$$\rho' R' = \rho_c R, \tag{1}$$

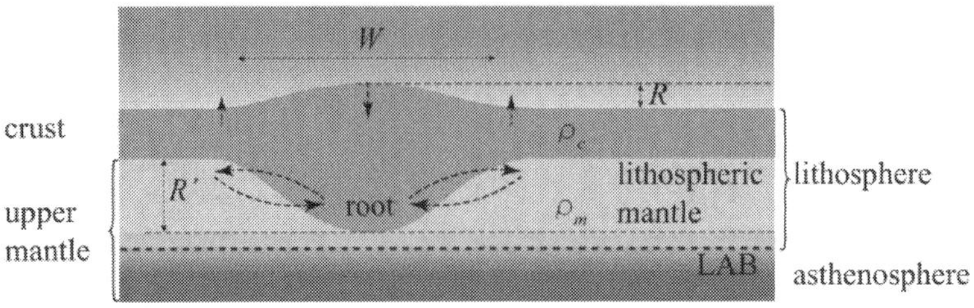

Figure 1. The lithosphere and the crust in isostatic balance.

that means that $R' \approx 5.4R$. Any departure from the condition (1) (as happens if the relief flattens due to erosion) produces an isostatic recovery that takes place on a timescale $T_i < T_o$, which means that mountain ranges are isostatically compensated. This compensation occurs on regional scale and not locally because the rigidity of the crust distributes stresses over a larger area.

The lithosphere is divided into a number of *plates* that are in relative motion. The formation of reliefs such as mountain ranges (the *orogeny*) is due to the shortening and thickening of the crust that occurs when two continental plates collide (like the India–Asia collision that produced the Himalaya and the Tibet) or when an oceanic plate is subducted beneath a continent (as happens in the case of the Andes).

Various physical models have been used to describe the build–up and the evolution of mountain belts. All are extensions and variations of the thin viscous sheet model of England and McKenzie [1; 2]. A discussion and a classification of these models has been made by Medvedev and Podladchikov [3; 4], to which the reader is referred for more details. Many numerical calculations based on these models have been performed to describe specific orogenies as realistically as possible, but little effort has been aimed to gain a deeper physical insight of the process.

Several years ago Gratton [5] derived scaling laws for the evolution of cordilleras, based on simple assumptions about the viscous flow due to the shortening of the crust. To achieve a better understanding about these kind of flows we recently investigated a simple model that consists of an initially uniform layer of a Newtonian fluid that rests over an horizontal substrate [6; 7]. This substrate is divided in two parts, that for $T > 0$ are pushed one against the other with constant velocity. This convergent motion drags the liquid and produces a ridge, and we showed that there are two self–similar regimes that occur in different space–time domains. These regimes and their corresponding scaling laws were obtained analytically. Next we extended these results for liquids with a power–law rheology [8] because it describes better the behavior of the lithosphere. However, these simple models do not take into account isostasy so that are nor adequate to describe the behavior of the lithosphere. For this reason, more recently we used a two–layer model to investigate the formation of mountain belts [9].

In this paper we discuss the hypotheses and approximations involved in the theoretical models that describe the hydrodynamics of the orogenic processes that take place in the lithosphere. In Section 2 we discuss the basic assumptions of the two–layer model and its implementation. In Section 3 we comment its application to the evolution of peaked ridges. In Section 4 we deal with the changes that must be introduced to describe flat–topped plateaus. Finally in Section 5 we summarize the main conclusions.

2. The lubrication approximation and the two–layer model

The simplest way to take into account isostasy and the convergent motion that leads to orogeny is to model the lithospheric plates as two–layer systems, to allow for the different properties of the crust and the lithospheric mantle (see figure 2). For simplicity we shall consider here that the problem depends on a single horizontal coordinate X (not all orogens are linear, but the generalization to two horizontal dimensions X, Y is straightforward). The crust has a viscosity μ_c and a thickness $H_c(X,T)$. The lithospheric mantle has a viscosity μ_m and its thickness is $H_m(X,T)$. In general $\mu_c \neq \mu_m$ and for the moment we shall assume that both are uniform, but later on we shall discuss what happens if μ_c depends on the depth. We shall also assume that initially both layers are uniform and there is no relief, so that $H_c(X,0) = C$ and $H_m(X,0) = M$, and that isostasy is maintained at any time, so that for $0 \le Z \le H_m$ the pressure does not depend on X. The visible topography is then given by

$$R = \frac{\rho'}{\rho_m}(H_c - C) \tag{2}$$

and H_m can be expressed in terms of H_c as

$$H_m = M + \frac{\rho_c}{\rho_m}(C - H_c). \tag{3}$$

To model the basal traction that drives the convergent motion of the plates we assume that at $T = 0$ the bottom of the lithosphere ($Z = 0$) starts moving with a prescribed velocity $U_b(X) = -U_0 \operatorname{sign}(X)$. The two–layer model conserves the mass of the crust, but not the mass of the lithospheric mantle. The flow Q (per unit transversal length) of lithospheric mantle material across the LAB is

$$Q = 2U_0 M\left(1 + \frac{\rho_c C}{\rho_m M}\right). \tag{4}$$

Notice that Q is larger than the inflow of lithospheric mantle due to the convergence of the plates ($2U_0M$).

The vertical, horizontal and temporal scales of the problem are, respectively:

$$C, \quad X_0 = \rho' \frac{\rho_c g M C^2}{\rho_m \mu_m U_0}, \quad T_0 = X_0/U_0. \tag{5}$$

A characteristic of all mountain ranges is that on the regional scale the slopes are gentle ($X_0 \gg C$) so that the vertical component of the velocity can be neglected and the pressure is hydrostatic. The theory (see [9]) is then based on the lubrication approximation [10–12], for details see [13]. Briefly, a single equation can be obtained for H_c in terms of the vertically averaged horizontal velocity in the crust:

$$V_c = \frac{1}{H_c}\int_{H_m}^{H_m+C} U dZ. \tag{6}$$

Introducing the dimensionless quantities

$$h = H_c/C, \quad v = V_c/U_0, \quad x = X/X_0, \quad t = T/T_0, \tag{7}$$

the equations of the model are

$$v = \mp 1 - \left(1 + \frac{\rho_c C}{\rho_m M}\right) h \frac{\partial h}{\partial x} - \frac{C}{M}\left(\frac{\mu_m}{3\mu_c} - \frac{\rho_c}{\rho_m}\right) h^2 \frac{\partial h}{\partial x} \tag{8}$$

(-1 for $x > 0$ and $+1$ for $x < 0$) and the continuity equation

$$\frac{\partial h}{\partial t} + \frac{\partial (vh)}{\partial x} = 0. \tag{9}$$

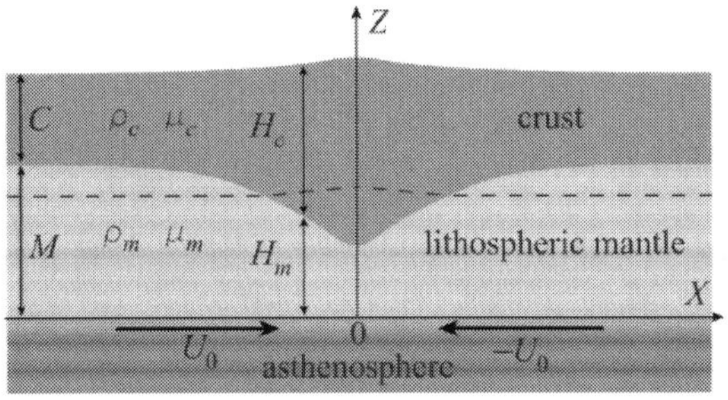

Figure 2. Geometry of the two–layer model. The dashed line is an isobar.

3. The two layer model and the evolution of cordilleras

Omitting details (see [9]) it can be shown that there is a self–similar asymptotics of the phenomenon in the limit $R \ll C$, that describes with good approximation the growth of ranges whose height does not exceed 5 km. According to this self–similar solution the height and the width of the range vary as $t^{1/2}$, and the aspect ratio (width/height) is given by X_0/C. This value is of the correct order of magnitude for actual mountain ranges, within the uncertainties of the parameters.

The scaling law $t^{1/2}$ can be justified [14] by means of a dimensional argument based on isostasy, conservation of the crustal mass, and a balance between gravitational and viscous stresses, as was done by Gratton [5]. However in that paper different scaling laws were obtained because the viscous stresses were not estimated correctly (it was assumed that the relevant velocity gradients occurred near the root since it was not realized that the whole lithospheric mantle participated in the dynamics).

In [9] we compared the results of the theory with the average profiles of 10 approximately straight segments of various mountain belts and we obtained a very good agreement.

We conclude that the two–layer model describes reasonably well the formation of peaked ranges, and that with good approximation their evolution is self–similar. Although the properties of the lithosphere involve many parameters, orogeny depends only on U_0, C and X_0.

These results have been obtained assuming a Newtonian rheology for both layers, but our previous results for a convergent flow of a non–Newtonian liquid on moving solid substrates [8] indicates that the $t^{1/2}$ scaling does not depend on the rheology, although the profile of the ridge depends on it.

4. The two layer model and the evolution of plateaus

The two–layer model with uniform viscosity in the crust used in [9] does not describe the formation of almost flat plateaus like the Tibet and the Altiplano. It is generally believed (see for example the review of Hodges [15]) that a flat top appears when the upper crust becomes decoupled from the basal traction. The formation of plateaus has been investigated by several authors [3; 4; 16; 17] by means of the *channel flow* hypothesis. In these papers it is assumed that there is a low viscosity layer at the bottom of the crust, just above the Moho. It is also assumed that the basal traction acts on the Moho, so that the lithospheric mantle is taken into account only for isostasy, but does not participate in the dynamics of the process. For example, Royden [16] assumes that the crust consists of an upper part in which the viscosity is uniform, and a lower part where the viscosity falls exponentially with depth. Obviously, as long as the root is shallow the viscosity of the crust is uniform and convergence produces a peaked ridge. However, as the thickness of the crust increases, the root eventually reaches the depth where μ_c becomes very small and then the basal traction decouples from the crust. In these places

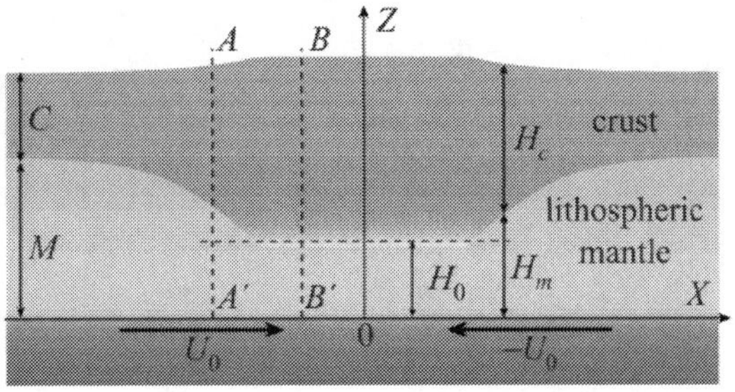

Figure 3. A plateau develops when the root arrives at the level H_0 where μ_c falls abruptly and the crust becomes decoupled.

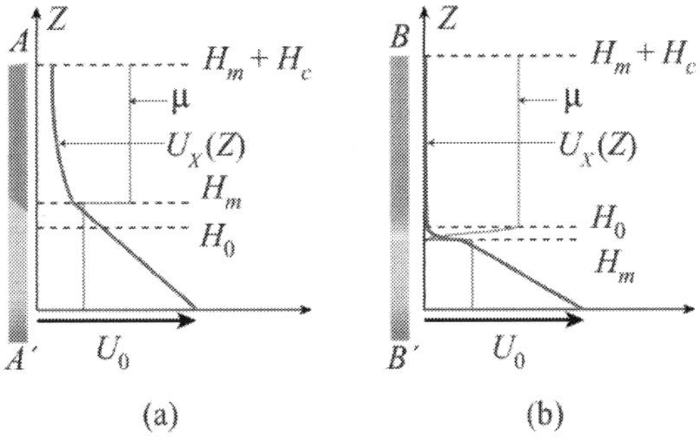

Figure 4. Formation of a plateau: (a) in the sides μ_c is uniform and the profile can be calculated by means of the model of Ref. [9]; (b) when the root arrives at H_0 the crust decouples and the relief cannot grow. The figures show the dependence on Z of μ_c and μ_m (thin, red) and of U (thick, blue) in both cases.

the height of the relief ceases to grow and a flat plateau develops, whose width increases as the convergent motion accumulates cortical mass at its sides (figure 3).

According to the two–layer model of Ref. [9] a decoupling that might produce a plateau happens only if the root touches the LAB, but the resulting plateau would be absurdly high (about 20 km).

Clearly, to reproduce real plateaus it is necessary to modify the two–layer model and assume that μ_c depends on Z and that falls abruptly when the root arrives at a given height H_0 above the LAB (figure 4). The height R_0 of the plateau depends on H_0:

$$R_0 = \frac{\rho'}{\rho_c}(M - H_0). \tag{10}$$

5. Conclusiones

The two–layer model based on the lubrication approximation allows to describe the essential aspects of the hydrodynamics of orogeny, provided it is assumed that the viscosity of the crust may fall abruptly at a certain depth in order to produce the decoupling that is needed to generate a plateau. We are presently investigating these processes.

Acknowledgments

We acknowledge support from Universidad Favaloro for participation in Fluidos 2010.

References

[1] England P and McKenzie D 1982 *Geophys. J. Int.* **70** 295

[2] England P and McKenzie D 1983 *Geophys. J. Int.* **73** 523

[3] Medvedev S E and Podladchikov Y Y 1999 *Geophys. J. Int.* **136** 567

[4] Medvedev S E and Podladchikov Y Y 1999 *Geophys. J. Int.* **136** 586

[5] Gratton J 1989 *J. Geophys. Res.* **94** 15627

[6] Perazzo C A and Gratton J 2008 *Phys. Fluids* **20** 043103

[7] Perazzo C A and Gratton J 2009 *J. Phys.: Conf. Ser.* **166** 012012

[8] Gratton J and Perazzo C A 2009 *J. Phys.: Conf. Ser.* **166** 012011

[9] Perazzo C A and Gratton J 2010 *Phys. Fluids* **22** 056603

[10] Buckmaster J D 1977 *J. Fluid Mech.* **81** 735

[11] Huppert H E 1982 *J. Fluid Mech.* **121** 43

[12] Oron A, Davis S and Bankoff S 1997 *Rev. Mod. Phys.* **69** 931

[13] Gratton J and Perazzo C A *J. Phys.: Conf. Ser.* this volume

[14] Perazzo C A and Gratton J *Anales AFA* in press

[15] Hodges K V 2006 *Channel Flow, Ductile Extrusion and Exhumation in Continental Collision Zones* ed Law R D *et al* (Geol. Soc., London, Special Pub. vol. 268) p 71

[16] Royden L 1996 *J. Geophys. Res.* **101** 17679

[17] Medvedev S E 2002 *J. Geophys. Res.* **107** 2123

Effects of gravity on the stability of the steady propagation of a liquid plug in a small conduit

D M Campana, [1,2] **S Ubal,** [1,2] **M D Giavedoni,** [1] **F A Saita** [1]

[1] Instituto de Desarrollo Tecnológico para la Industria Química (CONICET – UNL), Güemes 3450, 3000 Santa Fe, Argentina

[2] Facultad de Ingeniería, UNER, Ruta Provincial 11, km 10, 3101, Oro Verde, Argentina

madelia@santafe-conicet.gov.ar

Abstract. In this work we numerically study the stability of the steady state displacement of a liquid plug in a capillary tube when gravity, inertia and surface forces are important. The methodology employed is based on the analysis of steady state solutions and has been presented in previous publications. Gravity is assumed to act only along the axis of the tube.

1. Introduction

The propagation of liquid plugs in small conduits has been studied for many applications including the extraction of oil from a porous rock and medical treatments requiring a liquid plug to be instilled in the pulmonary airway such as in mechanical ventilation and surfactant replacement therapy. Also, due to their small size distal airways are prone to closure in various respiratory disorders creating liquid plugs which obstruct the airflow.

A simplified model of those problems is the motion of a liquid slug inside a capillary whose wall is coated with a thin film of the same fluid. Therefore, the liquid plug propagates on that preexisting film, the precursor film, whose thickness is H_∞^F and leaves behind a trailing film with thickness H_∞^T which, for a given liquid and tube radius, depends on the speed of plug propagation. If the displacement of the plug is steady, mass conservation requires $H_\infty^T = H_\infty^F = H_\infty$ since otherwise liquid would accumulate in ($H_\infty^F > H_\infty^T$) or evacuate from ($H_\infty^F < H_\infty^T$) the plug. This quantity (H_∞) as well as the length of the slug, L_P, will depend on the relative strength of capillary, inertia, viscous and gravity forces. When the length of the plug exceeds a certain value (L_P^{max}) which depends on the magnitude of the forces acting on the system, the phenomena is similar to the propagation of the rear and front menisci of two consecutive semi-infinite bubbles separated by a distance L_P^{max}. The propagation of semi-infinite bubbles has received considerable attention since the pioneering work by Bretherton [9]; in fact it has been studied analytically ([1], [2]), experimentally ([3]-[5]), and numerically ([6]-[8]) by a number of authors.

When the plug is short, the flow is affected by the interaction between the menisci and the velocity and pressure fields must be simultaneously solved in the whole domain. This case has received considerable less attention in the literature, and most works have been published by the team led by Professor Grotberg; they are motivated by the displacement of liquid plugs in the pulmonary airways. In particular, Fujioka and Grotberg [10] presented a numerical study on the steady propagation of a liquid plug in the gap formed by two horizontal parallel plates separated by a small distance. The governing equations and their boundary conditions were solved using the finite volume technique; the authors could not obtain solutions for capillary numbers larger than 0.4 and plugs shorter than half the gap of the channel. Based on this issue, they conjecture that the lack of convergence of the numerical algorithm employed might be related to the non existence of stable steady states, meaning that if the steady state solution is perturbed, the distance between the menisci will either continuously increase or decrease until the collapse of the plug occurs.

To determine if the above speculation was correct, we numerically investigated the stability of the displacement of a liquid plug in a capillary tube under the effects of surface, inertia, and viscous forces. The following two cases were considered in previous papers: i) an instant perturbation to the plug length occurs [11], and b) the pressure drop between the rear and front gas phases suffers a perturbation [12]. The main outcome of those works is that the stability of a liquid plug may be derived from the shape of the curves illustrating the steady state film thickness as a function of the plug length. In none of those papers the effect of gravity was considered. From a practical point of view, the more attractive case is that in which the disturbance affects the driving pressure drop; therefore, in this work we apply the methodology presented in [12] to study the effect of gravity on the stability of a liquid plug moving inside a capillary tube when the constant pressure difference applied between the front and rear gas phases is slightly perturbed. We consider the case in which the tube radius is small enough so that the flow can be regarded as symmetric around the centre of the tube. The Bond number that measures the ratio between gravity and surface tension forces may be positive or negative depending on whether the plug propagates against or with the direction of gravity, respectively.

The paper is organized as follows. In the next section we present the governing equations with their boundary conditions and briefly discuss the numerical methodology used to solve them. In Section 3, we present selected numerical results that show how gravity affects the stability of the plug. Finally, Section 4 concludes the paper.

2. Mathematical formulation and numerical solution

2.1 Governing equations and boundary conditions

We investigate the stability of the displacement of an incompressible and Newtonian liquid plug following the methodology presented in [12]. To that end, we consider the steady-state propagation of a plug of length L_P inside a capillary tube of radius R coated with a film of the same liquid whose thickness is uniform and equal to H_∞^F. Viscosity (μ), density (ρ), and surface tension (σ) are constant, and the gas phase is regarded as inviscid. The pressure drop, $\Delta P = P_{BT} - P_{BF}$, between the rear and front air phases drives the motion of the liquid plug with a constant velocity U. The pressure of the leading bubble, P_{BF} is taken as reference and it is arbitrarily set equal to zero. As the plug displaces inside the tube, it leaves behind a liquid film, the trailing film, with thickness H_∞^T. Since the displacement is steady, $H_\infty^T = H_\infty^F = H_\infty$; i.e., the trailing and precursor films have the same thickness. In this formulation, the frame of reference moves with the displacement velocity of the liquid plug, U. Then, the flow is described by the Navier-Stokes and continuity equations, that in dimensionless form read as follows,

$$\lambda Ca\, \mathbf{v} \cdot \nabla \mathbf{v} = -\frac{1}{Ca} \nabla p + \nabla^2 \mathbf{v} + \frac{Bo}{Ca} \mathbf{e}_z, \qquad (1)$$

$$\nabla \cdot \mathbf{v} = 0, \qquad (2)$$

where $\mathbf{v}(v_r, v_z)$ is the velocity vector scaled with U, the pressure is scaled with σ/R, and the coordinates (r, z) are scaled with R; $\lambda = Re/Ca = \rho\sigma R/\mu^2$ is the Laplace number, $Ca = \mu U/\sigma$ is the capillary number, and $Bo = \rho g R^2/\sigma$ is the Bond number and measures the ratio between gravitational and surface tension forces. It is worth to note that a positive value of the Bond number represents a plug moving in the opposite direction of the gravity force; on the contrary, a negative value of that parameter corresponds to a plug which moves in the direction of gravity (see Fig. 1).

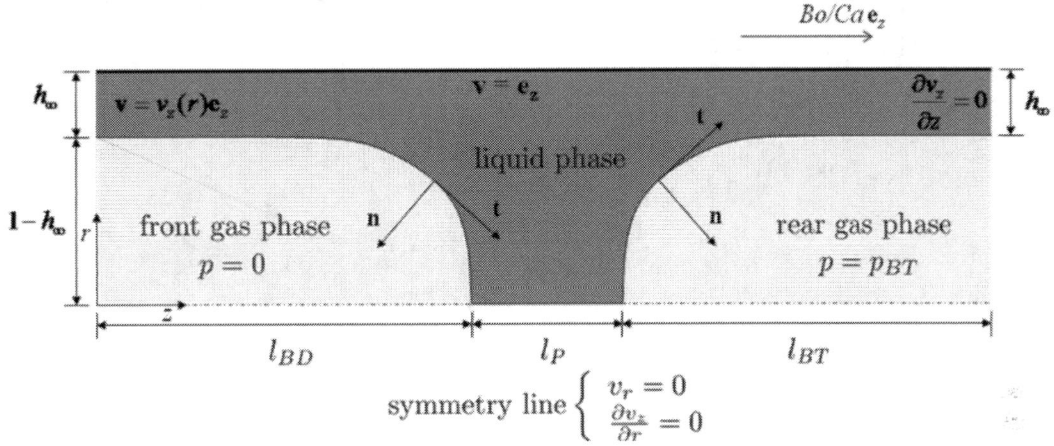

Figure 1. Sketch of the flow domain and coordinate system adopted.

The radius of the tube is small enough so that surface tension tends to dominate over other forces and the flow can be regarded as symmetric around the tube centerline; at the tube wall the no-slip condition is imposed,

$$v_r = \frac{\partial v_z}{\partial r} = 0, \text{ at } r=0, \qquad (3)$$

$$v_r = 0, v_z = 1, \text{ at } r=1. \qquad (4)$$

At both ends of the domain, the interfaces are parallel to the z-axe and the flow is fully developed; therefore, the following conditions must be satisfied at the ends of the domain (see Fig. 1):

$$v_r = 0, v_z = -\frac{Bo}{4Ca}\left[r^2 - 1 - 2(1-h_\infty)^2 \ln r\right] + 1, \text{ at } z=0, (1\text{-}h_\infty) \le r \le 1; \qquad (5)$$

$$\frac{\partial v_z}{\partial z} = 0, \text{ at } z=l_{BD}+l_P+l_{BT}, (1\text{-}h_\infty) \le r \le 1. \qquad (6)$$

In (6) l_P is the dimensionless length of the plug. The gas/liquid interfaces are material surfaces; therefore, the kinematic condition applies at both free surfaces

$$\mathbf{v} \cdot \mathbf{n} = 0. \qquad (7)$$

The tangential component of the stress balance vanishes at the interfaces and the normal component adopts the following expression,

$$-p + Ca\, S_{nn} = \begin{cases} 2H - 0 \\ 2H - p_{BT} \end{cases}, \qquad (8)$$

where $S_{nn} = \mathbf{n} \cdot \mathbf{S} \cdot \mathbf{n}$, $\mathbf{S}$ being the rate of strain tensor and $H = -1/2\nabla_s \cdot \mathbf{n}$ the interface curvature. The governing equations (1) and (2) with boundary conditions (3) – (8) are numerically solved as described in the next paragraphs; their solution provide pressure and velocity fields as well as the location of the free surface. Since under steady state propagation the length of the plug is fixed, the position of the tips of the front and rear gas bubbles is fixed. The symmetry of the interface at those points supplies the two additional equations needed to obtain the film thickness, h_∞, and the dimensionless pressure of the back gas phase, p_{BT}, to complete the solution.

2.1. Numerical method.

The numerical technique employed uses the Galerkin/finite element method with the parameterization of the free surface by means of spines. The physical domain is tessellated into quadrilateral elements; mixed interpolation is used to approximate the velocity and pressure fields, and the coefficients that locate the free surface are interpolated by the one-dimensional specialization of the bi-quadratic basis functions that interpolate the velocity.

Weighted residuals are built in the usual form and the resulting set of algebraic equations is simultaneously solved through a Newton loop. The iterative process is stopped when the norm of the difference between two consecutive iterations is smaller than or equal to 10^{-6}. Further details are provided in refs. [11] and [12].

Similar tests to those described in ref. [12] were carried out to select an appropriate finite element mesh; in particular we seek for the invariance of the film thickness and the pressure drop with the mesh and elements size. As a result of those trials, the meshes used to compute the solutions presented in this work have between 10016 and 12608 elements depending on l_P.

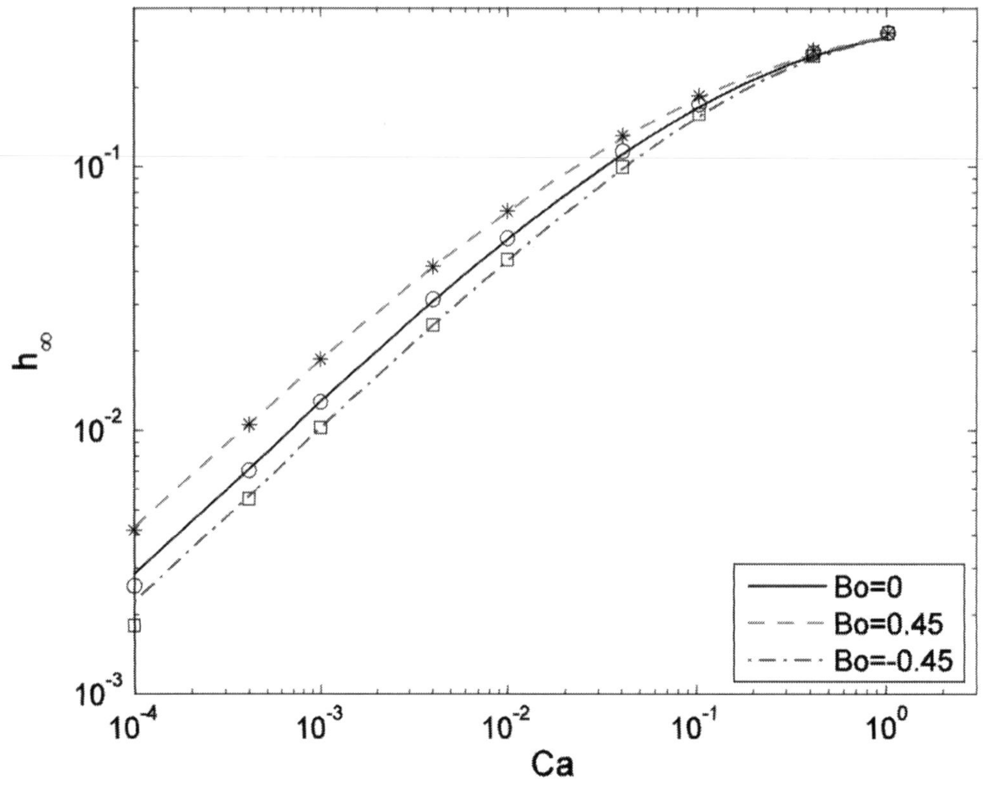

Figure 2. Comparison of the predictions of the film thickness of a large plug (lines) with those for a semi-infinite bubble reported by Hazel and Heil (symbols).

When the plug is sufficiently large, the front and rear menisci do not interact and the problem can be split into two smaller ones. Moreover, under these circumstances the film thickness is determined by the propagation of a semi-infinite bubble for the same value of λ, *Ca*, and *Bo*. Hazel and Heil [13] investigated the propagation of an air finger into a tube when gravity acts parallel to the tube axis. Even when that paper considers flexible tubes of elliptical or rectangular sections, the authors report numerical solutions for a circular tube within a wide range of capillary number, when inertia forces are negligible and the Bond number is equal to 0.45 and -0.45. In order to validate our numerical code, we carried out computations for a large plug and the same values of *Ca* and *Bo* used by Hazel and Heil; in Figure 2 we compare our prediction for the film thickness with those reported in ref. [13].

3. Results

As we have mentioned in the Introduction to establish whether or not the steady displacement of a liquid plug is stable, it suffices to analyze the steady state results conveniently organized. To that end the curves of h_∞ for a fixed value of both the capillary number and the driving pressure drop are drawn as a function of l_P for selected values of the Bond and Laplace numbers. The analysis presented in ref. [12] for *Bo*=0, shows that the slope of those curves can be related to the stability of a given plug driven by a constant pressure drop (Δp) as follows:

- ✓ A liquid plug corresponding to a point where $(\partial h_\infty / \partial l_p)_{\Delta p} > 0$, is stable to small disturbances of Δp.
- ✓ Any steady plug located at a point where $(\partial h_\infty / \partial l_p)_{\Delta p} < 0$ is unstable.

Briefly, this result can be understood from the following analysis: when the driving pressure is perturbed, the speed of plug propagation is also affected, leading to a corresponding variation in the thickness of the trailing film (the precursor film is not affected by the perturbation). The resulting inflow/outflow imbalance modifies the volume —i.e. length— of the liquid plug. Once the perturbation disappears, whatever the change underwent by l_P, h_∞ will suffer a variation of the same sign if $(\partial h_\infty / \partial l_p)_{\Delta p} > 0$, leading to a new inflow/outflow imbalance that tends to recover the original plug volume. At this point the inflow/outflow ratio becomes one. The opposite happens when $(\partial h_\infty / \partial l_p)_{\Delta p} < 0$.

The selection of the range of values of the dimensionless parameters that characterize the problem depends on the particular application. In this work we consider a liquid with properties ρ=1 g/cm^3, μ=6 cP, and σ=50 dyn/cm; therefore for liquid plugs propagating with velocity between 0.08 m/s and 0.8 m/s in a conduit of radius approximately equal to 0.7mm, the resulting set of dimensionless parameters is λ=1000, *Bo*=0.1 and $0.01 \leq Ca \leq 0.1$. This set may represent a liquid slug moving in the lung airways [14]. It is important to remember that in the coordinate system adopted, a positive Bond number represents a plug propagating against the direction of gravity while a negative Bond number accounts for the opposite situation (see Fig. 1). Also, the radius of the conduit chosen in this work is consistent with the assumption that gravity force does not affect the axial symmetry of the flow.

We carried out computations for plug lengths between 0.05 and 4.3 and *Bo* equal to ±0.1 and ±0.2. The predictions of h_∞ and Δp for each set (Ca, l_P) were used to build the stability charts corresponding to a selected value of the Laplace and Bond numbers. Each of the charts shown in Figs. 3-6 demanded about 400 individual simulations; curves for a fixed value of *Ca* (thin dashed blue lines) were plotted directly, while those for a fixed value of Δp (black solid lines) were obtained by interpolation. In Figures 3-6, we illustrate the maps for *Bo* equal to 0.1, 0 and -0.1. In those figures, the stable region is approximately enclosed by the thick-red-dashed line.

When the effect of gravity is negligible, i.e. when the axis of the tube is normal to gravity, the stable region encompasses short plugs (l_P<0.6) moving with a velocity corresponding to capillary numbers between 0.0562 and 0.112, approximately (see fig.3).

The map illustrated in Figure 4 represents plugs moving against gravity. If these results are compared with those depicted in Figure 3 for *Bo*=0, we see that the stable region is now smaller. The length of the largest stable plugs is now equal to or smaller than 0.5 and the range of displacing velocities in which the steady motion is not affected by a perturbation of Δp is slightly narrower.

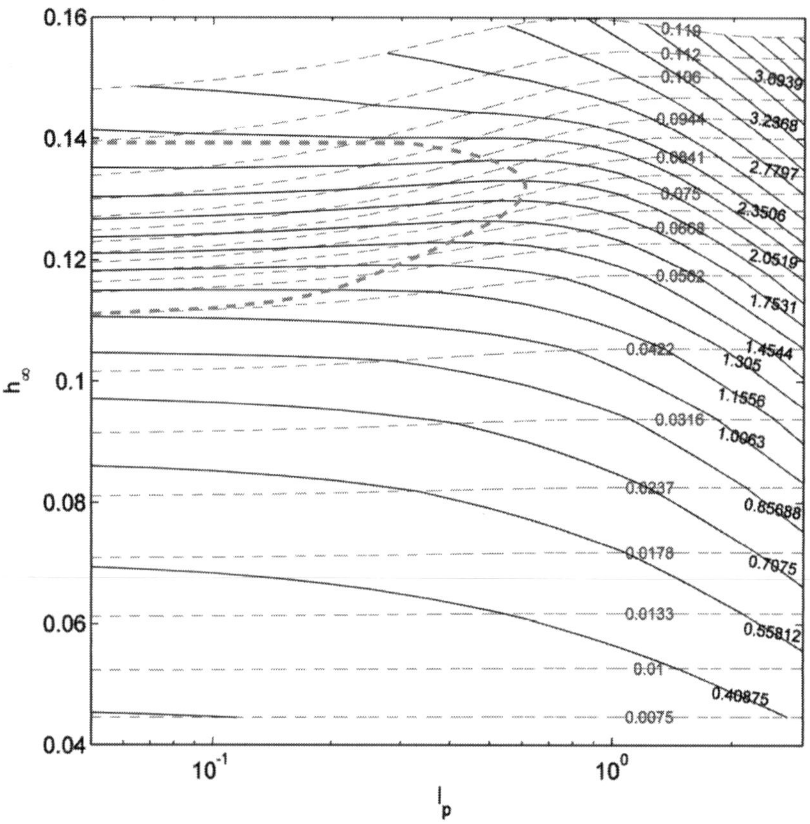

Figure 3. Steady state film thickness as a function of the plug length for λ=1000, *Bo*=0, and selected values of the capillary number (thin dashed blue lines). Black solid lines are curves of constant Δp. Steady state solutions within the region delimited by the thick dashed red line —where $\left(\partial h_\infty / \partial l_p\right)_{\Delta p} > 0$— are stable.

Solutions presented in Figure 5 for *Bo*= -0.1, i.e. for plugs moving in the same direction as gravity, show two regions of stable displacements. The upper one expands the stable zone delimited in Figure 3 for *Bo*=0: it comprises larger plugs which can propagate within a wider range of velocities. The lower one appears for *Ca* smaller than approximately 0.017 where it can be easily verified that the numerical predictions of the film thickness for a given value of the driving pressure drop monotonically increases with the length of the plug.

The above results point out that depending on the direction of the displacement of the slug with respect to gravity, the steady state propagation of a given plug may be stable or unstable. Also, for the three *Bo* considered the upper limit of the stable region seems to be a line where h_∞ is approximately constant for a certain Δp. This limit represents marginally stable plugs which move with decreasing velocity as the length of the plugs increases. Also, the propagation of a long liquid plug may be stable depending on the values of the capillary and the Bond number when λ is fixed. For instance, for λ=1000 and *Bo*= -0.1, the above statement is true for *Ca*< 0.013335, approximately.

Numerical solutions computed for other values of the Bond number (see Figure 6) show that the size of the stable region continuously increases (or decreases) as that parameter becomes bigger for liquid plugs moving in the same (opposite) direction as gravity force. Moreover, the two stable regions present for *Bo*= -0.1 have merged into a larger one when that parameter is equal to -0.2.

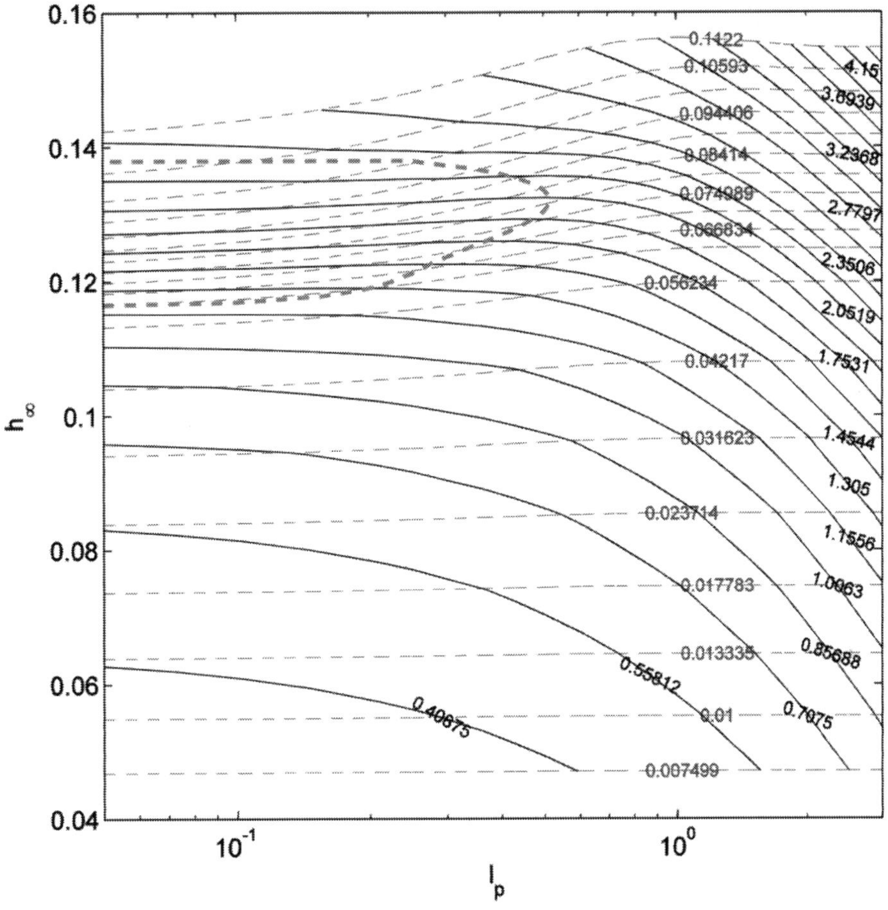

Figure 4. Steady state film thickness as a function of the plug length for λ=1000, *Bo*=0.1, and selected values of the capillary number (thin dashed blue lines). Black solid lines are curves of constant Δp. Steady state solutions within the region delimited by the thick dashed red line —where $\left(\partial h_\infty / \partial l_p\right)_{\Delta p} > 0$ — are stable.

4. Conclusion

In this paper we have applied the methodology presented in a previous work to analyze the effects of gravity on the stability of the steady state propagation of a liquid plug in a capillary tube lined with a uniform thin liquid film.

The stability maps show that the size of the stable region for a given Laplace number shrinks as the Bond number is augmented when the plug moves against gravity. On the other hand, if the plug propagates in the direction of gravity, one or two stable regions are detected depending on the value of *Bo*. Moreover, in this case the steady displacement of large plugs can be stable, a situation which is not observed when gravity effects are negligible.

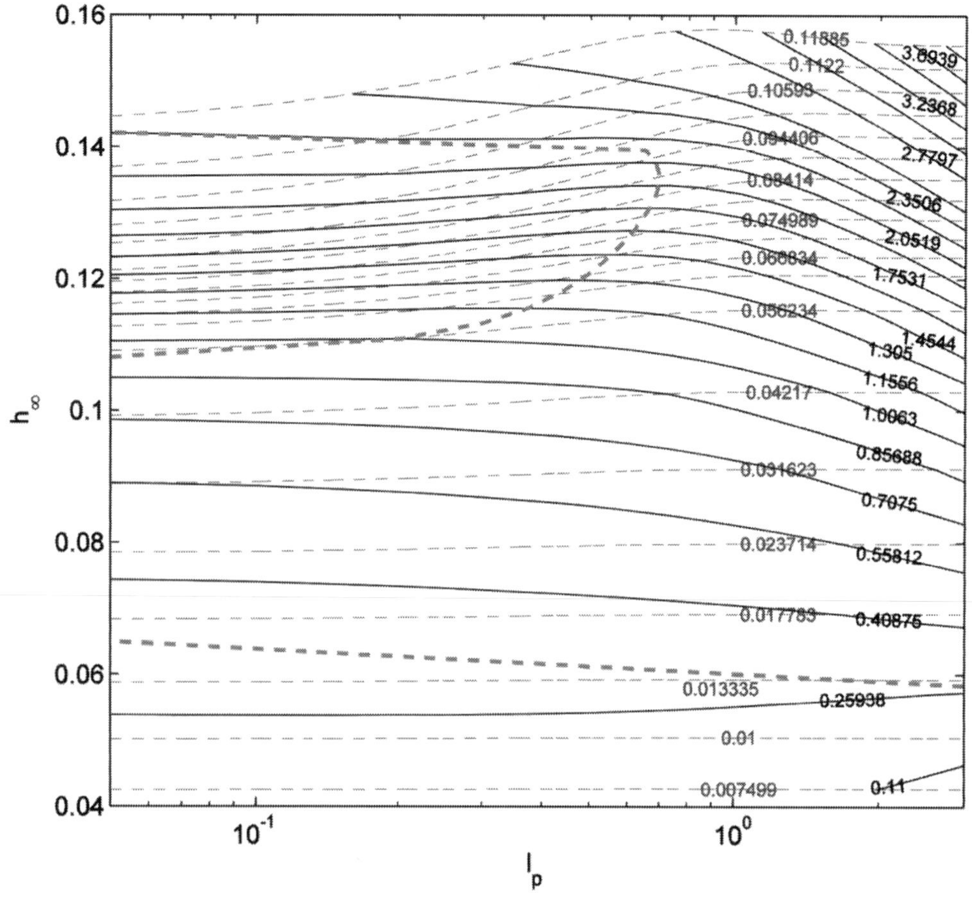

Figure 5. Steady state film thickness as a function of the plug length for λ=1000, *Bo*=-0.1, and selected values of the capillary number (thin dashed blue lines). Black solid lines are curves of constant Δp. Steady state solutions within the region delimited by the thick dashed red line —where $\left(\partial h_\infty / \partial l_p\right)_{\Delta p} > 0$ — are stable.

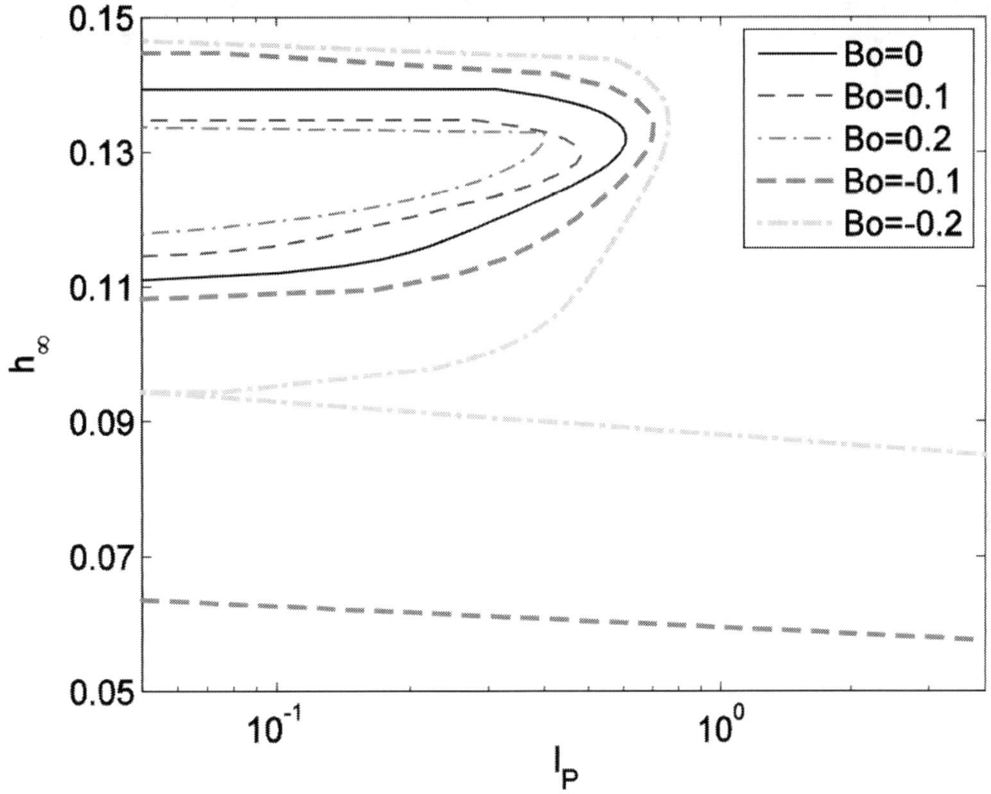

Figure 6. Marginally stable plugs for λ=1000 and selected Bond.

References

[1]Park, C. -W., Homsy, G. M. (1983). Two phase displacement in Hele-Shaw cells. Theory. *J. Fluid Mech.* **139**, 291.

[2]Ratulowski, J., Chang, H. -C., (1989). Transport of gas bubbles in capillaries. *Phys. Fluids A* **1**, 1642.

[3]Taylor, G.I., (1961). Deposition of a viscous fluid on the wall of a tube. *J. Fluid Mech.* **10**, 161.

[4]Hen, J. D., (1986). Measuring the film thickness surrounding a bubble inside a capillary. *J. Colloid Interface Sci.* **109**, 341.

[5]Schwartz, L. W., Princen, H. M., Kiss, A. D., (1986). On the motion of bubbles in capillary tubes. *J. Fluid Mech.* **172**, 259.

[6]Reinelt, D. A., Saffman, P. G., (1985). The penetration of a viscous fluid in a channel and a tube. *SIAM J. Sci. Stat. Comput.* **6**, 542.

[7]Shen, E. I., Udell, K. S., (1985). A finite element study of low Reynolds number two-phase flow in a cylindrical tube. *Trnas. ASME J. Appl. Mech.* **52**, 253.

[8]Giavedoni, M. D., Saita, F. A., (1997). The axisymmetric and plane cases of a gas phase steadily displacing a Newtonian liquid – A simultaneous solution of the governing equations. *Phys. Fluids* **9**, 2420.

[9]Bretherton, F. P, (19610). The motion of long bubbles in tubes. *J. Fluid Mech.* **10**, 166.

[10]Fujioka, H, Grotberg, J. B., (2004). Steady propagation of a liquid plug in a two-dimensional channel. *ASME J. Biomech. Eng.* **126**, 527.

[11]Campana, D. M., Ubal, S., Giavedoni, M. D., Saita, F. A., (2007). Stability of the steady motion of

a liquid plug in a capillary tube. *Ind. Eng. Chem. Res.* **46**, 1803.

[12]Ubal, S., Campana, D. M., Giavedoni, M. D., Saita, F. A., (2008). Stability of the steady-state displacement of a liquid plug driven by a constant pressure difference along a prewetted capillary tube. *Ind. Eng. Chem. Res.* **47**, 6307.

[13]Hazel, A. L, Heil, M., (2002). The steady propagation of a semi-infinite bubble into a tube of elliptical or rectangular cross section. *J. Fluid Mech.* **462**, 79.

[14]Zheng, Y., Fujioka, H., Grotberg, J. B., (2007). Effects of gravity, inertia, and surfactant on steady plug propagation in a two-dimensional channel. *Phys. Fluids* **19**, 082107-1.

Journal of Physics: Conference Series **296** (2011) 012019 doi:10.1088/1742-6596/296/1/012019

Spray formation with complex fluids

S Lustig, M Rosen

Grupo de Medios Porosos, Facultad de Ingeniería, Universidad de Buenos Aires, LIA (Laboratoire International Associe).Argentina.

mrosen@fi.uba.ar

Abstract. Droplet formation through Faraday excitation has been tested in the low driving frequency limit. Kerosene was used to model liquid fuel with the addition of PIB in different proportions. All fluids were characterized in detail. The mechanisms of ejection were investigated to identify the relative influence of viscosity and surface tension. It was also possible to characterize the type of instability leading to the emission drop process.

1. Introduction

The formation of spray from a liquid film on a vibrating surface is used by ultrasonic atomizers for applications ranging from humidification to metal- powder manufacturing.

However, this subject of industrial interest requires detailed knowledge of the preconditions for the formation of drops, because it is necessary to control size and conditions of the ejection.

In Faraday's experiment, a layer of fluid in a container is excited vertically by periodic oscillation. On the flat surface, at threshold frequency and applied acceleration, instability appears in the form of parametric waves. When these control parameters (acceleration and frequency) are increased, the initial alignment of these standing-wave patterns is lost due to the appearance of secondary instabilities.

With further increased excitation, a sharp transition is observed, to a state with spikes on the surface with droplets being ejected from the tip [1]. The characteristics of the instability produced during the ejection process are analyzed here, taking into account the influence of rheology and surface tension of the fluids used to interpret the different behaviors of the system. We also test some models to check the influence of each of the variables which influence ejection.

The chosen fluids are similar to those present in fuels, with different amounts of a viscoelastic component to control the properties of the drops produced.

2. Experimental Details

The experimental system has already been described in previous publications [1]. The different components are illustrated in Fig. 1: the signal from a digital generator is amplified to excite an electromechanical transducer, which is rigidly connected to the sample cell with an aluminium bar, monitored by a speedometer with an oscilloscope. The system is completed by suitable illumination and in line computer processing.

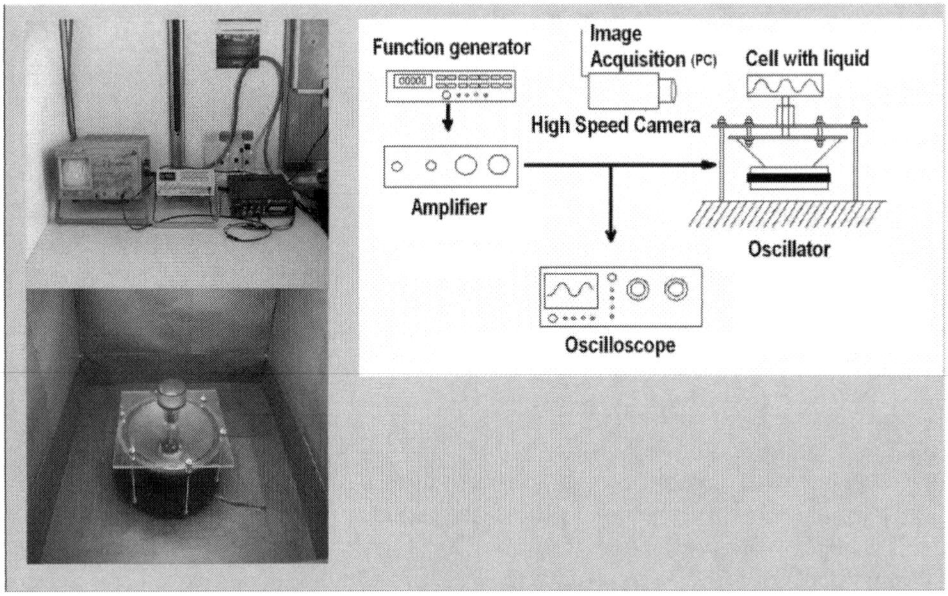

Figure 1. Experimental set-up

3. Characteristics of selected fluids

Fuel feed systems for certain types of engines are investigated for exhaust gas control (CO, NO_X), to improve fuel efficiency in thermodynamic terms, because a non-homogeneous mixture of compounds with different vapour pressures is required.

When a viscoelastic polymer of high molecular weight is added, the physical form of hydrocarbon is altered and the size distribution of fuel droplets is narrowed [2].]. Interfacial tension also increases promoting drop encapsulation and favoring simultaneous component ignition.

The rheology of the chosen components (kerosene and polyisobutilene (PIB) are represented in Fig. 2.

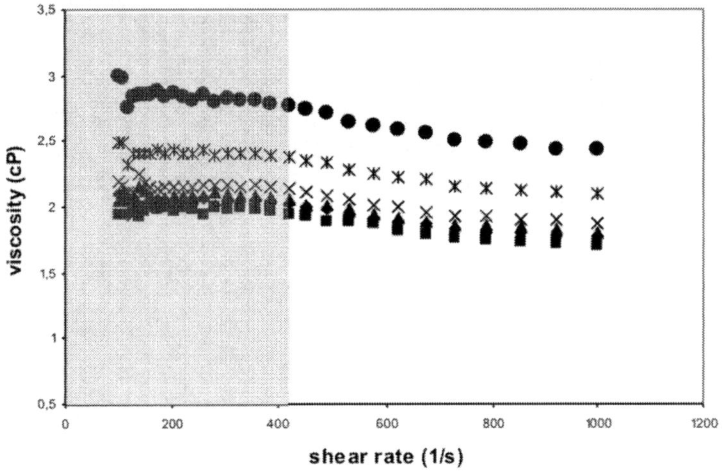

Figure 2. Rheology of the Kerosene + PIB solutions● Kerosene+2% PIB, * Kerosene + 1% PIB, x Kerosene + 0.5% PIB, ▲Kerosene+0.3% PIB, ■ Kerosene+0.2% PIB. . The shaded area corresponds to quasi-Newtonian behavior.

The rheology of the solutions employed shows "shear thinning" behaviour, preceded by a quasi-Newtonian region which corresponds to the area of interest. Also the viscosity increases with PIB concentration.

The relevant feature of this system lies in interfacial tension behavior, which grows when increasing the PIB added (cfr. Fig.3).

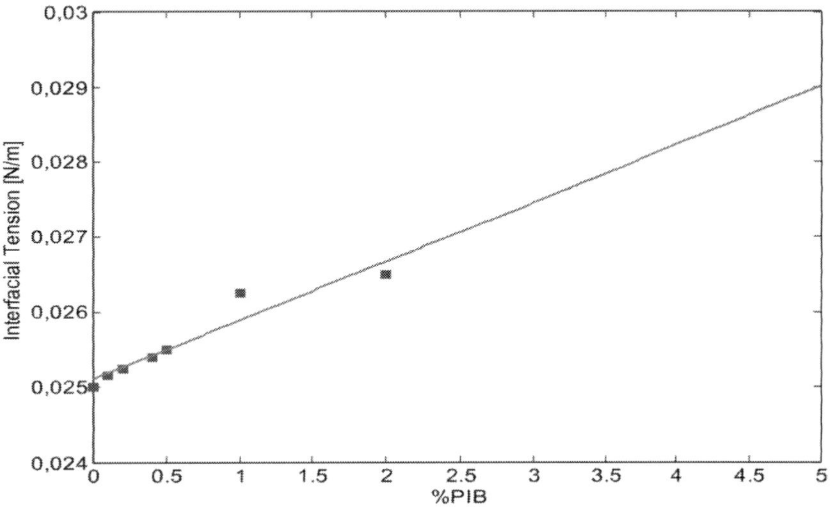

Figure 3. Interfacial tension as a function of PIB %.

4. Threshold of ejection

To identify the critical acceleration at which ejection starts acceleration was gradually increased up to the ejection of one or two droplets detected in a time span of ten seconds.

Goodriges [3] proposed an expression which combines density and interfacial tension to yield a critical acceleration ac

$$a_c \approx \omega_0^{4/3}\left(\frac{\sigma}{\rho}\right)^{\frac{1}{3}} \tag{1}$$

where σ stands for the interfacial tension, ρ density and ω frequency. This scaling law was obtained from the dispersion relation for capillary-gravity waves and it is valid for fluids of constant viscosity.

The critical acceleration as a function of frequency is represented in Fig. 4.

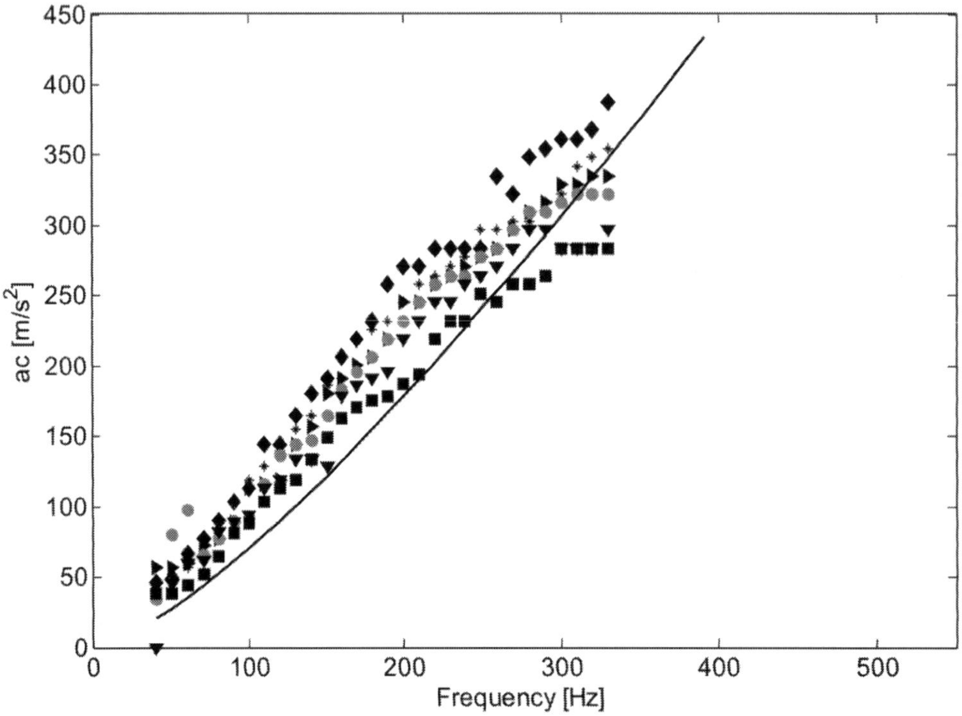

Figure 4. Ejection acceleration threshold in terms of frequency. ♦Kerosene +2%PIB, ∗ Kerosene + 1%PIB, ▸ Kerosene+ 0.5%PIB, ● Kerosene +0.3%PIB, ▾ Kerosene +0.1%PIB, ■ Kerosene, ---- Theoretical expression (eq. 1).

Goodrige's model (Eq.1) is represented with dotted lines in Fig. 4, jointly with our experimental results. The best fit corresponds to a Newtonian fluid (constant viscosity) and experimental a_c values increase and depart from this model with increased PIB concentration.

5. Dimensionless Parameters (Newtonian fluids)

Goodriges [4] proposes dimensionless expressions for acceleration and frequency which depend on the surface tension and viscosity, which define the boundary for dominant dependence:

$$a^* = \frac{a\upsilon^4}{\left(\sigma/\rho\right)^3} \qquad \omega^* = \frac{\omega\upsilon^3}{\left(\sigma/\rho\right)^2} \tag{2}$$

Dimensionless acceleration Dimensionless frequency

Depending on the values of w * is possible to identify the dominant influence for ejection.

viscosity dominates if $\omega^* > 10^{-5}$

surface tension dominates if $\omega^* < 10^{-5}$

These results are represented in Fig. 5

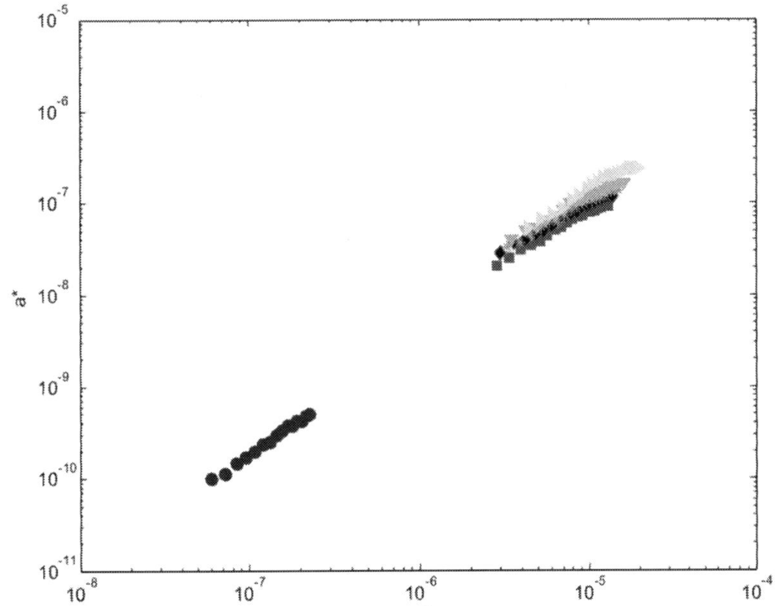

Figure 5. Dimensionless acceleration as a function of dimensionless frequency for water and different concentrations of polymer in Kerosene. ■ Kerosene, ♦Kerosene +0.1%PIB, *Kerosene+0.2%PIB, ▾ Kerosene+0.3%PIB, ▸ Kerosene +0.5%PIB, ● Water.

Journal of Physics: Conference Series **296** (2011) 012019 doi:10.1088/1742-6596/296/1/012019

Our results match the alignment of a Newtonian liquid such as water, validating the choice of dimensionless parameters, as well as the importance of surface tension in the process.

6. Rayleigh-Plateau Instability (drops formation) and ejection mechanisms

A liquid column of capillary dimensions may be formed on the surface of the fluid where irregularities appear as control parameters are increased [5]. These irregularities mediate the formation of necks which mark the onset of the Rayleigh Plateau instability.
The two possible mechanisms are illustrated in Fig. 6.

Kerosene + 0.3% PIB Kerosene + 5% PIB

Figure 6. Formation of ejection tubes is more noticeable with increased PIB concentration.

The height of the capillary columns increases with PIB concentration (Fig.6), illustrating the effect on the elastic properties of the solution, because greater polymer content increases this property.
To characterize the ejection problem [6] the dimensionless Ohnesorge number Oh is used:

The number of Weber (We) $\frac{\rho d v^2}{\sigma}$ (3)

Reynolds (Re) number $\frac{\rho d v}{\mu}$ (4)

and the Ohnesorge (Oh) number $Oh = \frac{\mu}{\sqrt{\rho \sigma d}} = \frac{\sqrt{We}}{\mathrm{Re}}$ (5)

where d is the characteristic length scale (typically drop diameter). Ohnesorge number reflects the effect of viscosity compared with surface tension influence.

7. The column formation

The type of dripping produced directly from the surface of the liquid: with or without column formation before ejection (jetting) is represented in Fig. 7, for the Weber number as a function of Oh number.

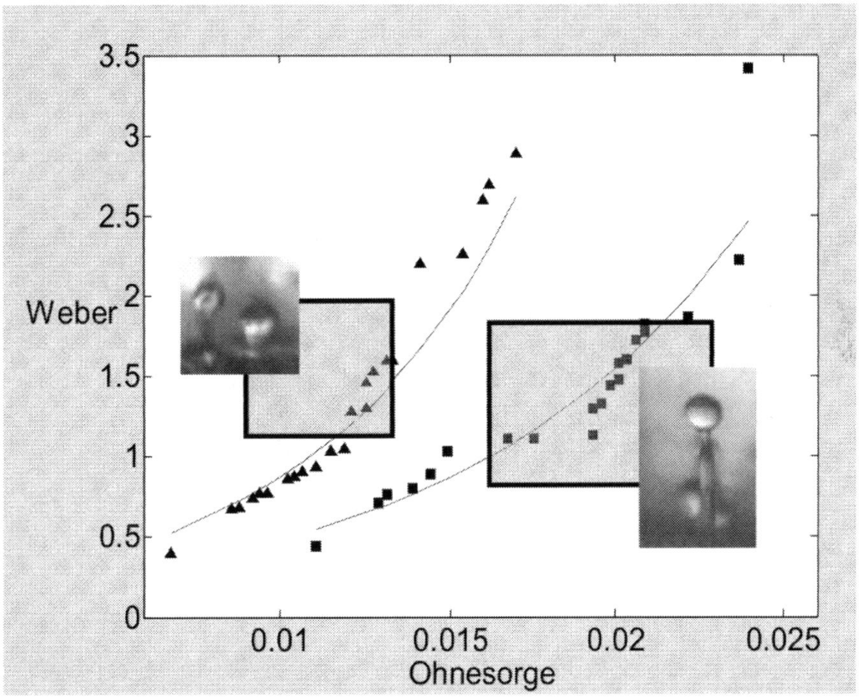

Figure 7. Weber number as a function of Ohnesorge number for 19 g and 70 Hz in accordance with the model of Ambravaneswaran [7] ■ Kerosene+ 5%PIB, ▲ Kerosene +0.3%PIB.

The type of ejection can be defined as simple dripping, complex dripping or jetting depending on the Weber and Ohnesorge [7, 8] dimensionless numbers. As the We value increases, the behaviour of the fluid changes from a simple dripping to a complex dripping and attains jetting when Oh is about 10-1.

In our case, the increased of the surface tension promotes the formation of a liquid column in which necks at a certain height will induce subsequent formation of droplet. The height of the capillary columns increases with a larger increased PIB content (5%).

Working at low frequencies, have access to study ejection conditions for each type of fluid and explorer different behaviors that can occur.

Critical acceleration (threshold) follows a law proportional to applied frequency to 4/3 according with Goodriges model.

As the We value increases, the behaviour of the fluid changes from a simple dripping to a complex dripping until it reaches jetting

8. Instability: Convective or Absolute

We attempt to establish-whether the observed instability can be classified as convective or absolute [9].

Starting from the neck, a droplet is ejected which may be generated above (convective instability) or above and below the neck (absolute instability). Both cases are illustrated in Fig. 8 corresponding to Kerosene + 5% PIB, 60 Hz, 19 g

Figure 8 Two kinds of ejection produced in the same experiment.

We have also observed that as the frequency increases, the Reynolds number decreases and the ejection departs from the absolute instability.

9. Conclusions

Working at low frequencies, it is possible to investigate ejection conditions for each type of fluid and explore various possible alternative responses. Our chosen solution increases its interfacial tension with PIB concentration.

Critical acceleration (threshold) is proportional to the 4/3 power of applied frequency, in accordance with Goodriges model. With increased We value, the behaviour of the fluid changes from simple dripping to complex dripping and finally induces jetting.

Both convective and absolute instabilities could be observed in this system.

References

[1] Cabeza C, Rosen M, Ferreyra G and Bongiovanni G, 2006 Dynamical behavior of digitations state in Faraday waves with a viscoelastic fluid , *Physica A* : **371(1)**, 54-58

[2] Waters P, 2000 Global Warming Reduction by Polymers in Automotive Fuels, *Polymer:* **41(2)**, 1807-1808

[3] Goodridge L C , Tao Shi W and Lathrop D P , 1996 Threshold Dynamics of Singular Gravity-Capillary Waves. *Physical Review Letters* .**Vol 76(11)**,1824-1827

[4] Goodridge L C , Tao Shi W, Hentschel H G E and Lathrop D P, 1997 Viscous effects in droplet-ejecting capillary Waves *Physical Review E.* **Vol 56(1)**,472-475

[5] Vukasinovic B,Marc K S and Glezer A , 2007.Mechanisms of free-surface breakup in vibration-induced liquid atomization *Physics of Fluids* **19**

[6] Tuffillaro N B , Ramshankar R and Gollub J P,1989 Order-Disorder Transition in Capillary Ripples. *Physical Review Letters* **62** , 422 -425

[7] Ambravaneswaran B , Subramani H J , Phillips S D and Basaran O A, 2004 Dripping-Jetting Transitions in a Dripping Faucet *Physical Review Letters* **93(3)**

[8] Longuet M S and Higgins J, 1983. Bubbles, breaking waves and hyperbolic jets at a free surface. *J. Fluid Mech.* **127**,103

[9] Leib S J and Goldstein M E , 1986 Convective and absolute instability of a viscous liquid jet. *Physics of Fluids* **29** , 952-954

Reverse flow and vortex breakdown in a shear-thinning fluid

C. Cabeza, G. Sarasua, N. Barrere, A.C. Martí

Instituto de Física, Facultad de Ciencias, UDELAR, Uruguay

e-mail: cecilia@fisica.edu.uy

Abstract. The effect of polymer concentration on the development of reverse secondary flow and vortex breakdown was studied using a viscoelastic solution of polyacrlylamide in water. The fluid was contained in cylindrical containers of two different radii, the top end wall of which rotated at a varying speed, thus, imparting a circulating motion to the fluid. Whereas using a newtonian fluid, streamlines will occupy the entire container, the flow of a shear-thinning fluid may divide into two cells of opposite circulating motion. The curve of critical Reynolds and elasticity numbers (*Re, E*) values corresponding to the development of reverse flow was obtained over a wide range of *Re* values. Vortex breakdown was found to occur at extremely low *Re* values.

1. Introduction

The instabilities and structural changes undergone by confined rotating flows have been a matter of interest from the scientific as well as the technological standpoint.

One largely studied system consists of a newtonian or viscoelastic fluid confined in a cylindrical container fitted with a rotating end wall. Using newtonian fluids, an increased angular velocity eventually results in a dominant centrifugal force that induces secondary flow in the positive radial direction in the proximity of the rotating end wall and in the opposite direction in the vicinity of the fixed end [1,2]. However, using viscoelastic fluids, the resulting flow may divide into two cells, in one of which the direction of secondary rotation is opposite to that found in newtonian fluids [3,4,5,6]. In addition, under certain conditions, viscoelastic fluids may flow in more varied patterns than those found in newtonian fluids.

Both newtonian and non-newtonian fluids show structural changes referred to as vortex breakdown (VB), characterized by the appearance of a point of stagnation and the onset of reverse flow development. Stokes et al [4,5] studied the development of VB in polymer solutions in a water-glycerine mixture, using a large proportion of the latter to ensure that the overall viscosity was only slightly dependent on the shear rate. The authors reported that VB occurred at an increasing value of the Reynolds number (Re) with increasing polymer concentration, and that no VB was observed above a critical concentration value, the two polymer types used by the authors (polyacrylamide and xantham gum) leading to qualitatively similar behaviour.

Confined viscoelastic fluids flowing at low *Re* values have been reported to develop elasticity-driven turbulence similar to that developed by newtonian fluids at high *Re* values [7,8], in addition to certain instabilities similar to those found in laminar flow [4].

Despite the multitude of parameters that may be used to describe flow patterns, the overall behaviour of viscoelastic fluids is generally described by *Re* and the Deborah number (*De*). The former is defined as that used for newtonian fluids, $Re = U\,L/v$, where the viscosity v depends on the shear rate; and the latter is defined as $=\tau/\gamma$, where τ is the polymer relaxation time and γ is the shear rate. Whereas other dimensionless parameters may also influence the behaviour of viscoelastic fluids, their influence has been assumed to be low and has generally been neglected (see above references).

In this work, an aqueous solution of polyacrylamide at a concentration varying from 0.03% to 1.0% was used to study the structural changes undergone by shear-thinning fluids. The conditions that define the development of reverse flow were studied in detail, and the capacity of (*Re,De*) plots to adequately describe the critical conditions resulting in such flow revised. The relative relevance of additional parameters was considered. The elasticity number ($E = De / Re$, a measure of the relative importance of elastic and inertial effects on the flow dynamics) and the *Re* number were used for dimensionless characterization of reverse flow. The *Re* value required for the occurrence of VB was determined under the assayed conditions and compared with the critical value leading to VB in newtonian fluids.

2. Experimental set-up

Two cylindrical containers of internal diameter R1=40.0 mm and R2=67.5 mm, respectively, were constructed of acrylic plastic. The top end wall of the containers was fixed to the shaft of a DC motor rotating at an electronically-controlled speed varying between ω=0.06 rad/s and ω=42.77 rad/s. The shaft-motor system was so designed that no shaft precession was observed. The height of the container end wall was adjustable to within an accuracy of 1.0%. An aspect ratio H/R=2 was used. In order to optimize the optical system and minimize the effects of light refraction, the cylindrical container was placed inside an outer container, square in cross-section, of side L=150.0 mm.

An aqueous polyacrlylamide solution was used as study fluid. A water-soluble, non-ionic polyacrylamide powder (Sigma nº 92560) was dissolved to different concentrations (0.03%, 0.1%, 0.5% and 1.0% by weight). Sample rheology was analysed on a Physica MCR 300 rheometer[1].

Digital particle image visualization (DPIV) was used to obtain velocity fields within the fluid. A 2-mm-thick vertical laser sheet (500mW green laser) crossing the container along its axis was used as illumination means. The fluid was seeded with 50-μm neutral polyacrylamide particles. Images were recorded on a Cmos camera (PL 741 Pixelink) at 90 fps. To obtain velocity profiles, standard cross-correlation algorithms were used between consecutive images with 64x64-pixel interrogation windows and 16-pixel overlapping.

The *Re* number, defined as $Re=\rho\omega R^2/\eta$ (where ρ is the fluid density, ω is the angular velocity, R the container radius and η is the dynamic viscosity), varied between 0.009 and 1800, as determined with an accuracy lower than 2%. The fluid temperature was kept at 20ºC at all times during all runs. The rotation speed was varied very slowly, allowing the necessary time (between 3 and 5 min, depending on the polymer concentration) between increments for flow patterns to become steady before the visualization of structures developing within the fluid. At a concentration of 0.5% or 1.0%, the fluid flowed in transient state for around 4 min. Moreover, at these concentrations, the system also

[1] Samples were characterized by Grupo de medios Porosos, Facultad de Ingeniería, Universisdad de Buenos Aires, Argentina

showed hysteresis –*i.e.*, critical *Re* values at which structures developed differed according to the increasing or decreasing nature of speed variation.

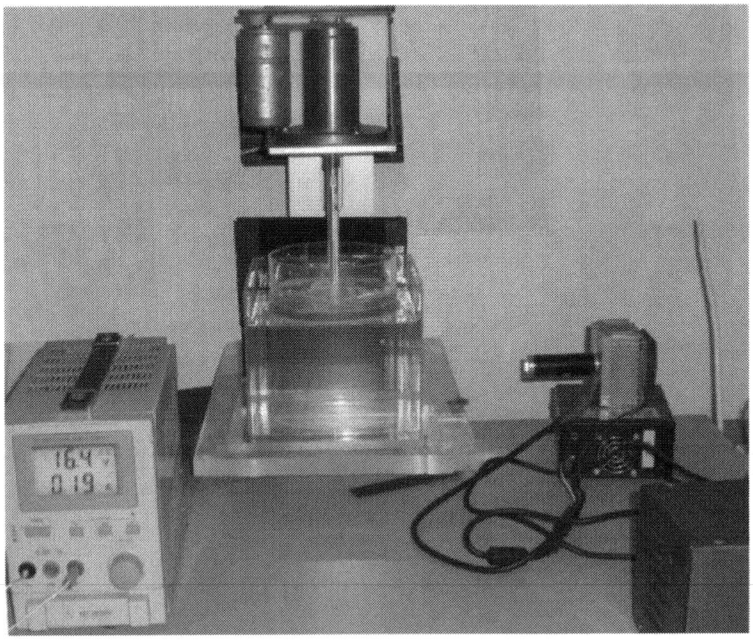

Figure 1: Experimental set-up

3. Results

Figure 2 shows typical reverse flow patterns, as well as velocity profiles on the z and radial axes. The reverse flow zone is situated between the fixed end wall of the cylinder and a plane crossing its axis perpendicularly at a distance z*, defined as the point of stagnation –*i.e.*, where the vertical velocity v_Z along the cylinder axis changes sign. As the rotation speed increases, the reverse flow zone decreases in size, eventually disappearing above a critical value.

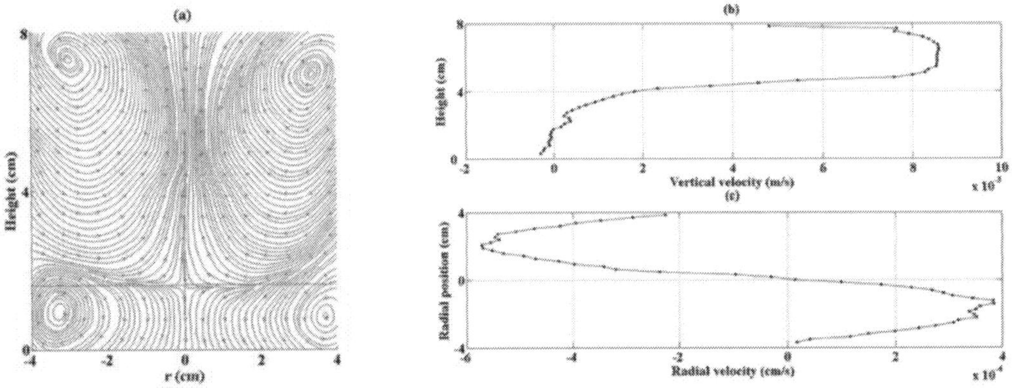

Figure 2: a) Streamlines of an aqueous solution of 0.5% w/w PAA; container radius: R1; Re=9x10^{-3}, De= 0.45. b) Vertical velocity profile as a function of height along the cylinder axis, represented in a) by a vertical red line. c) Radial velocity as a function of radius on the plane represented in a) by a horizontal red line

Figure 3 shows the experimental flow pattern and v_Z values within an interval around the point of stagnation.

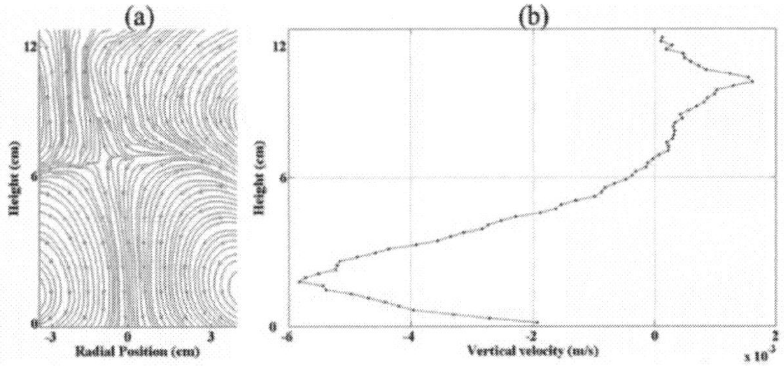

Figure 3: a) Streamlines of an aqueous solution of 0.5% w/w PAA; container radius: R2; Re= 0.22, De=3.8 b) Vertical velocity profile as a function of height along the cylinder axis within a zoomed-in plot area around its centre.

Table 1 shows critical *Re* and De values leading to the disappearance of reverse flow, using fluids of different polymer concentration in the two containers. As both the fluid viscosity and relaxation time were found to depend on the angular velocity of the rotating end wall, *Re* and *De* were calculated accordingly, using the η and τ values determined at the angular frequency being used. Table 1 also shows those critical values leading to bubble formation, associated with VB. It did not occur at high polymer concentrations in the fluid, consistently with previously reported results [4,5]. However, these results show that VB developed at fairly low *Re* values (*Re* = 100), as opposed to the reported data (*Re* = 1000).

	Flow reversal	Flow reversal	Vortex breakdown	Vortex breakdown
Concentration	Radius=4.00cm	Radius=6.75 cm	Radius=4.00cm	Radius=6.75 cm
0.03%	No flow reversal	No flow reversal		
0.1%	Re = 17.85 De = 3.36	No flow reversal	Re = 58.85 De = 8.35	Re = 57.52 De = 3.78
0.5%	Re = 1.51 De = 13.20	Re = 4.81 De =11.10	Re=128.00 De = 101.10	Re = 45.40 De = 10.90
1%	Re = 0.71 De = 17.4	Re = 0.22 De = 7.50	No bubble formation	No bubble formation

Table 1: Critical values of the Reynolds and Deborah numbers according to polymer concentration. Values show the disappearance of reverse flow and the formation of recirculation bubbles (vortex breakdown)

The experimental data appear to follow a *Re* vs *E* curve. Figure 4 describes the development of reverse flow as a function of *Re* and *E* on logarithmic axes. As shown in Figure 4, the data appear to fit a smoothly varying curve, suggesting that *Re* and *E* may be used as suitable descriptors of the reverse flow zone, with reverse flow developing at higher (*Re, E*) values than those shown by the curve.

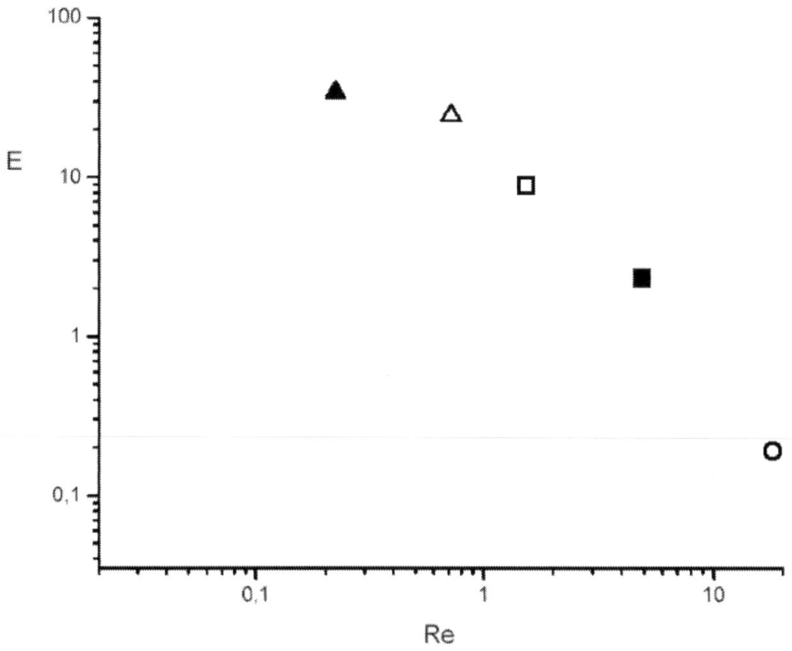

Figure 4: Critical *E* and *Re* values leading to the disappearance of reverse flow. Full symbols correspond to the cylinder of radius R2 (67.5 mm); open symbols correspond to the cylinder of radius R1 (40.0 mm); triangles, squares and circles correspond to PAA concentrations of 1.0%, 0.5% and 0.1%, respectively.

As $E = \lambda\eta/\rho R^2$, and η decreases with increasing rotation speed (ω) , *E* increases with decreasing ω, which should be expected to favour the stabilization of reverse flow. As shown in Table 1, reverse flow did not develop at low concentrations, even when very low ω values were used. There appears to be a finite critical concentration value below which reverse flow does not develop. The fact that *E* is a measure of the relative importance of elastic and inertial forces is consistent with the existence of a minimal *E* value at which the influence of elastic forces is sufficiently strong for reverse flow to develop.

4. Conclusions

Using different configurations of cylinder radius and rotation speed, these results showed that the critical *Re* and *E* values leading to flow reversal follow a smooth curve such that reverse flow occurred at (*Re,E*) values above this curve, suggesting the descriptive capacity of *Re* and *E* for dimensionless characterization of systems similar to the one being described.

Under the studied conditions, VB occurred at extremely low *Re* values. The effect of polymer concentration on VB was found to be similar to that on turbulence, suggesting a possible connection between the two phenomena. Unlike the case with the development of reverse flow, VB did not follow a simple (*Re,E*) curve, suggesting the need of a third parameter to adequately describe the VB phenomenon.

A preliminary study of the dimensionless parameters resulting from the Oldroyd-B model [9] suggests that a dimensionless variable accounting for the characteristic polymer length may play an important role in the overall flow characterization. The predictive relevance of such a parameter should be the focus of further research.

Acknowledgements

We thank CSIC and PEDECIBA of Uruguay for financial support, and Dr. Marta Rosen, of Grupo de Medios Porosos, Universidad de Buenos Aires, Argentina, for the characterization of samples.

References

1. Vogel, H. U. Experimentelle Ergebnisse über die laminare Strömung in einen zylindrischen Gehäuse mit darin rotierender Scheibe, Max-Planck-Inst. Bericht 6. (1968).
2. Escudier, M. P. Obervations of the flow produced in a cylindrical container by a rotating endwall. Exps Fluids **2**, 189 (1984).
3. Hill, C. T. Nearly visometric flow of viscoelastic fluids in the disk and cylinder system II: Experimental. Trans. Soc. Rheol. **16**, 213-245, (1972).
4. J. R. Stokes, L. J. W. Graham, N. J. Lawson, and D. V. Boger, "Swirling flow of viscoelastic fluids. Part 1. Interaction between inertia and elasticity," J. Fluid Mech. **429**, 67 (2001).
5. J. R. Stokes, L. J. W. Graham, N. J. Lawson, and D. V. Boger, "Swirling flow of viscoelastic fluids. Part 2. Elastic effects," J. Fluid Mech. **429**, 117 (2001).
6. S. Tamano, M. Itoh, Y. Ide, and K. Yokota, "Vortex shedding in confined swirling flow of polymer solutions" Phys. Fluids **19**, 023103 (2007).
7. A. Groisman and V. Steinberg, "Elastic turbulence in a polymer solution flow," Nature London **405**, 53 (2000)
8. A. Groisman and V. Steinberg, "Elastic turbulence in curvilinear flows of polymer solutions," New J. Phys. **6**, 29 (2004).
9. R. B. Bird, R. C. Armstrong, and O. Hassager, *Dynamics of Polymeric Liquids. Volume 1: Fluid Mechanics*, 2nd ed. Wiley Interscience, New York, 1987, pp. 353–369.

FLUIDOS 2010: XI Meeting on Recent Advances in the Physics of Fluids and their Applications IOP Publishing
Journal of Physics: Conference Series **296** (2011) 012021 doi:10.1088/1742-6596/296/1/012021

Modeling of CO_2 storage in aquifers

Gabriela B. Savioli[1] **and Juan E. Santos**[2]

[1] Laboratorio de Ingeniería de Reservorios, IGPUBA and Departamento de Ingeniería Química, Facultad de Ingeniería, Universidad de Buenos Aires, Av. Las Heras 2214 Piso 3 C1127AAR Buenos Aires, Argentina
[2] CONICET, Facultad de Cs Astron. y Geofísicas, Universidad Nac. de La Plata, Paseo del Bosque S/N, La Plata, B1900FWA, Argentina, and
Department of Mathematics, Purdue University, 150 N. University St., West Lafayette, IN 47907, U.S.A

E-mail: gsavioli@di.fcen.uba.ar, santos@math.purdue.edu

Abstract. Storage of CO_2 in geological formations is a means of mitigating the greenhouse effect. Saline aquifers are a good alternative as storage sites due to their large volume and their common occurrence in nature. The first commercial CO_2 injection project is that of the Sleipner field in the Utsira Sand aquifer (North Sea). Nevertheless, very little was known about the effectiveness of CO_2 sequestration over very long periods of time. In this way, numerical modeling of CO_2 injection and seismic monitoring is an important tool to understand the behavior of CO_2 after injection and to make long term predictions in order to prevent CO_2 leaks from the storage into the atmosphere. The description of CO_2 injection into subsurface formations requires an accurate fluid-flow model. To simulate the simultaneous flow of brine and CO_2 we apply the Black-Oil formulation for two phase flow in porous media, which uses the PVT data as a simplified thermodynamic model. Seismic monitoring is modeled using Biot's equations of motion describing wave propagation in fluid-saturated poroviscoelastic solids. Numerical examples of CO_2 injection and time-lapse seismics using data of the Utsira formation show the capability of this methodology to monitor the migration and dispersal of CO_2 after injection.

1. Introduction

Fossil-fuel combustion generates carbon dioxide (CO_2), which is mainly discharged into the atmosphere, increasing its temperature (greenhouse effect). To minimize climate change impacts, geological sequestration of CO_2 is an immediate option [1]. Geologic sequestration involves injecting CO_2 into a target geologic formation at depths typically greater than 1000 m where pressure and temperature are above the critical point for CO_2 (31.6C, 7.38 MPa).

The CO_2 injection operation at the Sleipner gas field in the North Sea, operated by Statoil and the Sleipner partners, is the world first industrial scale CO_2 injection project designed specifically as a greenhouse gas mitigation measure [1]-[2]. CO_2 separated from natural gas produced at Sleipner is currently being injected into the Utsira Sand, a saline aquifer some 26000 km^2 in area. Injection started in 1996 and is planned to continue for about twenty years, at a rate of about one million tonnes per year.

Time-lapse seismic surveys aim to demonstrate storage integrity, provide early warning should any leakage occur and monitor the migration and dispersal of the CO_2 plume. Recent papers [3]-[4] successfully apply seismic modeling for monitoring the spatio-temporal distribution of CO_2 using synthetic generated CO_2 saturation fields. Instead, in this work we employ numerical simulations of CO_2 injection; therefore saturation fields are obtained as a result of the simultaneous flow of CO_2 and brine in porous media.

The final objective is to test that underground storage is a safe and verifiable technology in the long term.

2. The Black-Oil formulation of two-phase flow in porous media

The simultaneous flow of brine and CO_2 is described by the well-known Black-Oil formulation applied to two-phase, two component fluid flow [5]. In this model, CO_2 may dissolve in the brine but the brine is not allowed to vaporize into the CO_2 phase. This formulation uses, as a simplified thermodynamic model, the following PVT data, determined using the Hassanzadeh's correlations [6]: R_s: CO_2 solubility in brine; B_{CO2}: CO_2 formation volume factor, and B_b: brine formation volume factor. The nonlinear system of partial differential equation is,

$$\nabla\cdot(\underline{k}(\frac{k_{rCO2}}{B_{CO2}\mu_{CO2}}(\nabla p_{CO2}-\rho_{CO2}g\nabla D)+\frac{R_s k_{rb}}{B_b\mu_b}(\nabla p_b-\rho_b g\nabla D)))+q_{CO2} = \frac{\partial\left[\phi\left(\frac{S_{CO2}}{B_{CO2}}+\frac{R_s S_b}{B_b}\right)\right]}{\partial t}, \tag{1}$$

$$\nabla\cdot(\underline{k}(\frac{k_{rb}}{B_b\mu_b}(\nabla p_b-\rho_b g\nabla D))+q_b = \frac{\partial\left[\phi(\frac{S_b}{B_b})\right]}{\partial t}. \tag{2}$$

The unknowns are the fluid pressures p_{CO2}, p_b and saturations S_{CO2}, S_b for the CO_2 and brine phases. The parameters k and ϕ are the absolute permeability and porosity respectively. Also, for $\beta = CO2, b$, the funcions $k_{r\beta}$, μ_β and ρ_β are the relative permeability, viscosity, and density of the β-phase, respectively.

Two algebraic equations relating the saturations and pressures, complete the system:

$$S_b + S_{CO2} = 1, \qquad p_{CO2} - p_b = P_C(S_b), \tag{3}$$

where P_C is the capillary pressure.

The solution of the Black-Oil fluid-flow model was obtained employing the public domain software BOAST [7], which solves the differential equations using IMPES, a semi-implicit finite difference technique [8].

3. Biot's Equations of Motion

Let us consider a 2D isotropic fluid-saturated porous material Ω. The oscillatory motion of Ω at the angular frequency ω subject to external sources $F^{(s)}$ and $F^{(f)}$ obeys Biot's equation of motion [3]

$$-\omega^2\rho_b u^{(s)} - \omega^2\rho_f u^{(f)} - \nabla\cdot\sigma(u) = F^{(s)} \tag{4}$$

$$-\omega^2\rho_f u^{(s)} - \omega^2 g u^{(f)} + i\omega b u^{(f)} + \nabla p_f(u) = F^{(f)}. \tag{5}$$

The unknowns are $u^{(s)}$ and $u^{(f)}$, the time Fourier transforms (FT) of the averaged displacement vectors of the solid and fluid phases, respectively. Also, ρ_f and ρ_b denote the densities of the single-phase fluid and the bulk material, g and b are mass and viscous coupling coefficients, σ_{ij}

is the FT of the stress tensor of the bulk material and p_f is the FT of the fluid pressure. See [4] for the definition of the variables involved in (4)-(5).

Biot's equations were solved with the finite element method, employing a 2D non-conforming finite element space for each component of the solid displacement vector and the vector part of the Raviart Thomas Nedelec space of zero order for the fluid displacement [9].

4. Numerical Experiments

4.1. Idealized model of the Utsira formation

To test the proposed methodology, we consider an idealized geometrical and physical domain consisting of 5 regions as shown in Figure 1. The upper 100 m is region Ω_1, a sand of permeability 60 mD and porosity 0.32. Ω_2 is a sealed shale 2 m thick, the top of the Usira formation. Regions Ω_3 and Ω_5 are the Utsira formation, of permeability 1000 mD and porosity 0.37. We assume that Ω_4 is a sealed shale layer within the Utsira sand. The medium was excited with a compressional point source located at x= 400 m, z= 710 m.

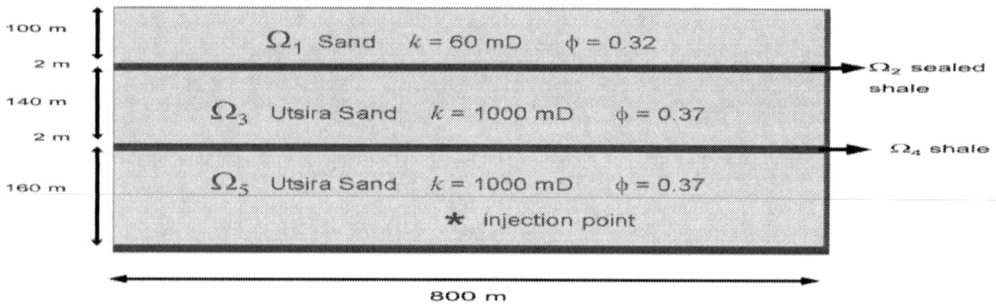

Figure 1. Idealized model of the Utsira formation. The injection point is located at x= 400 m, z= 1060 m.

The Biot model assumes a single-phase fluid, therefore effective fluid density, viscosity and bulk modulus were obtained using the properties of the CO_2 and brine weighted by the corresponding saturations computed by the fluid-flow simulator.

4.2. Injection Modeling

Figure 2 shows the CO_2 saturation distribution after 5 years of injection obtained by the BOAST simulator. A CO_2 accumulation beneath the Ω_4 seal can be clearly observed.

4.3. Seismic Monitoring

Figure 3 displays traces of the vertical component of the particle velocity of the solid phase before (black curve) and after (red curve) 5 years of CO_2 injection. The strong arrival at about 240 ms corresponds to a reflection due to the CO_2 accumulation beneath the seal.

Time histories measured near the surface before (left plot) and after 5 years of CO_2 injection (right plot) are shown in Figure 4.

The first reflection in both figures is due to the direct wave coming from the point source located at x= 400 m, z= 710 m. The second reflection in the time histories after 5 years, not observed before the injection, is generated by the CO_2 accumulations below the thin shale layer at depth z = 940 m.

Figure 5 displays the vertical component of the solid phase velocity before and after 5 years of CO_2 injection at 200 ms. At this time, the waves generated by the point source have generated

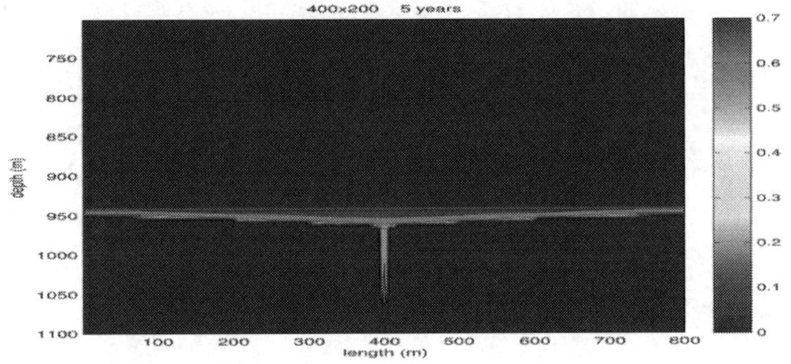

Figure 2. CO_2 saturation distribution after 5 years of injection. The injection point is located at x= 400 m, z= 1060 m.

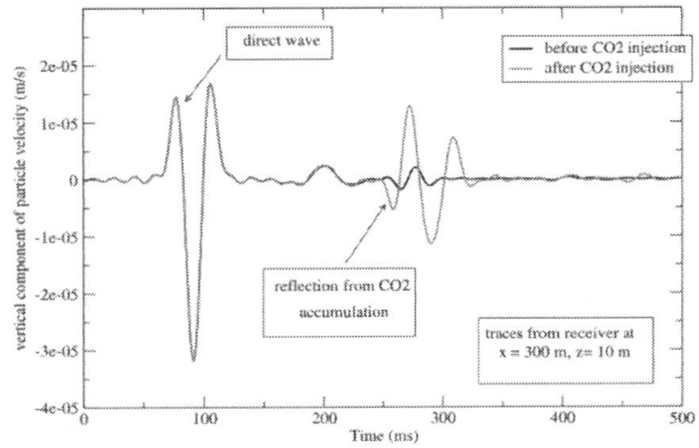

Figure 3. Traces of particle velocity of the solid phase before and after 5 years of CO_2 injection.

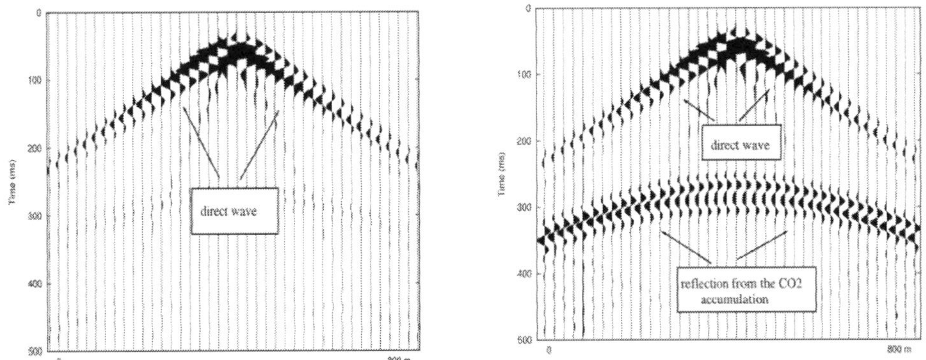

Figure 4. Time histories measured near the surface before and after 5 years of CO_2 injection.

reflected and transmitted waves due to the CO_2 accumulation below the thin shale layer at z = 940 m.

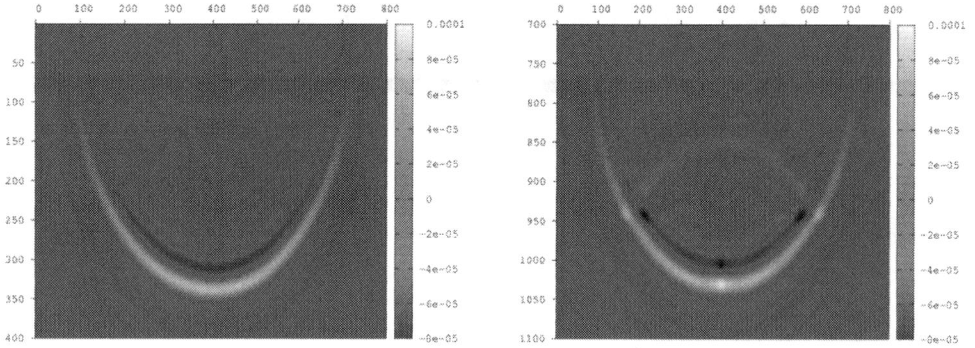

Figure 5. Vertical component of the solid phase velocity at 200 ms before and after 5 years of CO_2 injection.

5. Conclusions

In this work we introduced a methodology to model and monitor CO_2 sequestration using numerical simulations. For that purpose we integrated numerical simulators of CO_2-brine flow and seismic wave propagation. This metodology was tested with a numerical example showing the capability of seismic monitoring to identify the horizontal and vertical accumulations of CO2. Therefore, combining fluid-flow simulations with seismic methods constitute an important tool to analyze storage integrity, provide early warning should any leakage occur, and monitor the migration and dispersal of the CO2 plume.

6. References

[1] Arts R, Chadwick A, Eiken O, Thibeau S and Nooner S, 2008 Ten years of experience of monitoring CO_2 injection in the Utsira Sand at Sleipner, offshore Norway *First break* **26** pp 65-72

[2] Chadwick A, Arts R and Eiken O 2005 4D seismic quantification of a growing CO_2 plume at Sleipner, North Sea. In: Dore A G and Vincent B (Eds) *Petroleum Geology: North West Europe and Global Perspectives - Proc. 6th Petroleum Geology Conference* (Geological Society, London) pp 1385-99

[3] Santos J, Rubino J and Ravazzoli C 2008 Modeling mesoscopic attenuation in a highly heterogeneous Biot's medium employing an equivalent viscoelastic model *Proc. 78th Annual International Meeting SEG (Las Vegas)* pp 2212-15

[4] Carcione J, Picotti S, Gei D, and Rossi G 2006 Physics and Seismic Modeling for Monitoring CO_2 storage *Pure and Applied Geophysics* **163** pp 175-207

[5] Aziz K and Settari A 1985 *Petroleum Reservoir Simulation* (Great Britain: Elsevier Applied Science Publishers)

[6] Hassanzadeh H, Pooladi-Darvish M, Elsharkawy A, Keith D and Leonenko Y 2008 Predicting PVT data for CO_2-brine mixtures for black-oil simulation of CO_2 geological storage *International Journal of Greenhouse Gas Control* **2** pp 65-77

[7] Fanchi J 1997 *Principles of Applied Reservoir Simulation* (Houston, Texas: Gulf Professional Publishing Company)

[8] Savioli G and Bidner M S 2005 Simulation of the oil and gas flow toward a well - A stability analysis *Journal of Petroleum Science and Engineering* **48** pp 53-69

[9] Santos J E and Sheen D 2007 Finite element methods for the simulation of waves in composite saturated poroviscoelastic materials *SIAM J Numer Anal* **45** pp 389-420

Convection generated by a small heat source in a box with a cooled upper contour at constant temperature

L P Thomas[1,2], B M Marino[2], R Tovar[3], J A Castillo[3]

[2] Facultad de Ciencias Exactas, Universidad Nacional del Centro de la Provincia de Buenos Aires, Pinto 399, B 7000 GHG Tandil, Argentina
[3] Centro de Investigación en Energía, Universidad Nacional Autónoma de México Priv. Xochicalco s/n, Col. Centro, Temixco, Morelos, México

E-mail: lthomas@exa.unicen.edu.ar

Abstract. The flow generated by a linear heat source inside a thermally insulated box with the upper boundary maintained at constant temperature is analyzed by means of a series of experiments. The attention is focused on the steady state during which the heat provided by the source is absorbed by the upper boundary giving place to a particular convective process in two well defined zones. One of them occupies most part of the box from the lower boundary and is characterized by a turbulent convective flow; the other is the thermal boundary layer developed below the cooled top contour where strong fluctuations are detected. Special interest is found in the analogies with the results obtained employing a configuration similar to that used to research the Rayleigh-Benard convection with lower and upper boundaries at constant temperatures but without internal sources. This study has particular relevance to understanding and predicting the use of the cooled-ceiling systems in buildings as a passive solution to enhance the comfort of its inhabitants during hot days with the consequential energy savings.

1. Introduction

The free convection of a fluid that is heated from below and cooled from above is a classic fluid-dynamics problem that has received much attention in the last years. It is important in theoretical, experimental and numerical studies of the Rayleigh-Bénard instabilities associated with the formation of the patterns of convective cells, the turbulence developed when spatial coherence is lost, and the boundary layer structures. Thermally driven flows are of utmost importance in industrial applications and in natural phenomena. Examples of convective processes are found in the atmosphere and oceans, in geophysics and astrophysics, and in technological applications such as solar energy collectors, cooling of nuclear reactors and electronic equipment, ventilation of rooms, and crystal growth in materials processing [1]. Other applications include natural waterbodies (e.g. lakes, ponds, reservoirs and estuaries), shallow and salt gradient stratified ponds, artificial lakes for seasonal energy storage, sensible energy storage systems, etc. (see [2] and references therein). In natural waterbodies the flow and thermal structures have important effects on chemical and physical properties, dissolved oxygen,

[1] To whom any correspondence should be addressed.

biological and chemical oxygen demand, aquatic life, chemical and thermal pollution, ecology and mixing processes in water.

Particularly, in the recent years, building developers, architects and engineers are increasingly deciding to use a radiant "cooled-ceiling" instead of air-conditioned spaces with high cooling loads, as the experiences with cooled-ceilings are persuasive. This kind of systems consists of a thermally insulated room with a cooled upper contour. The heat is generated by means of one or more internal sources as persons, domestic appliances, computers, etc. which are within the room and have smaller sizes than the overall space. The comparison between the flows that takes place in systems like this and in the traditional one to study the Rayleigh-Benard convection presents some aspects that deserve to be studied more deeply.

To our knowledge, much work was done to study the thermally-driven flow under the conditions imposed in the well-known Rayleigh-Benard experiments [3], or when there are one or more plumes in a insulated room [4-8], but the combination of these two problems have not been studied yet. In order to analyze the thermal structure and heat transfer in a cooled-ceiling system with an internal source, we performed a series of specific laboratory experiments using a scale model and dynamic similarity criteria as explained by Linden [9]. The experimental setup consists of a closed thermally insulated box filled with water instead of air, an electrical heat source on the bottom and a top contour maintained at constant temperature equal to the fluid´s initial one. The heat flux supplied by the source is constant throughout each experiment, but varies from one experiment to another thus covering a good range of situations. An array of thermocouples is used to measure the temperature of the fluid at several heights and of the cold top contour, while synthetic schlieren technique is applied to visualize the flow.

At the beginning of the experiment, a plume of smaller density than the environment one is generated by the source. Once the lighter fluid reaches the top contour, it spreads horizontally until it gets to the lateral walls; the continuous arrival of warmer fluid caused the formation of a well-defined interface that descends gradually until it covers the whole receptacle. During the steady state the fluid fills the room transmitting heat from the source to the upper boundary by means of an efficient particular process that is described and discussed in the following. The emphasis is particularly focused on the results of the classical Rayleigh-Benard experiments to understand the heat transport process, the developed flow and the fluctuations in the boundary layer formed below the upper contour.

2. Rayleigh-Benard convection: Useful remarks

Rayleigh–Benard (RB) convection is a type of natural convection occurring in a layer of fluid confined between two horizontal planes that is heated from below and cooled from above as sketched in figure 1. Two heat conductive plates are maintained at temperatures T_1 and T_2, with $\Delta T = T_1 - T_2 > 0$, giving place to an unstable configuration of the fluid that fills the space between them. The flow is characterized by the Rayleigh number:

$$Ra = g\,\beta\,\Delta T\,L^3 / \nu\alpha\,, \tag{1}$$

the fluid transport properties by means of the Prandtl number

$$Pr = \nu / \alpha\,, \tag{2}$$

and additional parameters related to the shape of the contours and other boundary conditions, named *bc* from now on. Here g is the gravity acceleration, L is the separation between the plates, and α, β and ν are the diffusivity, the thermal expansion coefficient and the kinematic viscosity of the fluid, respectively.

The response of the system is often described as a function of the Nusselt number, or the relationship between the total heat flow and the heat flow by conduction,

$$Nu = \frac{QL}{\lambda S \Delta T}, \tag{3}$$

and the Reynolds number

$$Re = LU/\nu = \sqrt{\frac{Ra}{Pr}}, \tag{4}$$

where the velocity scale for the natural convection is given by

$$U = \sqrt{g\,\beta\,\Delta T\,L}\,, \tag{5}$$

and λ is the thermal conductivity of the fluid, Q is the heat flux and S is the surface area of the plates. In addition, the velocity scale of the natural convection introduces the characteristic frequency

$$f_C = \frac{U}{L} = \sqrt{\frac{g\,\beta\,\Delta T}{L}}\,. \tag{6}$$

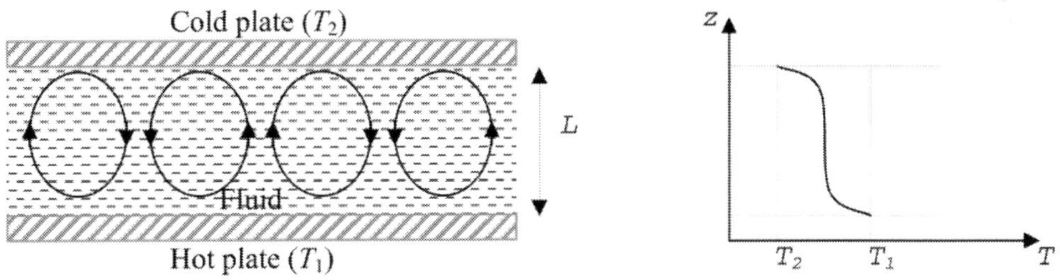

Figure 1. (a) Physical configuration to study the Rayleigh-Benard convection.
(b) Temperature profile of the fluid in the steady state for $Ra >> 1$.

It is well known that, when Ra is smaller than a given critical value Ra_c, the fluid can remain at rest and the heat between plates is driven by conduction through the liquid ($Nu = 1$). For $Ra > Ra_c \approx 1700$ the instabilities increase in the fluid near both plates causing small random plumes to form. Eventually the plumes close to the upper plate move in opposite directions to those located next to the lower plate generating a flow of scale L [10-12]. Such a flow contains convective cells of different shapes according to the values of Ra, Pr and bc, as for example those shown in figure 1(a), by means of which the fluid transports heat from the lower plate to the upper one. A typical mean temperature profile during the steady state is represented in figure 1(b).

The heat transfer driven by free convection is given by the relationship $Nu = Nu(Ra,Pr,bc)$ in the form

$$Nu = aRa^r Pr^p\,, \tag{7}$$

where the values of the coefficient a and exponents r and p vary according to the regime developed in the bulk of the fluid and in the boundary layers of the plates [1]. Here we adopt the values $a = 0.14$, $r = 0.29$, $p = 0$ as a first order approximation coming from different experimental and numerical results obtained using different fluids and shown in figure 2. Naturally, for a limited range of Ra numbers, the values of the parameters in (7) may be better fitted to the range of interest.

As observed in figure 2, as Ra increases (for example, by increasing the temperature difference between the plates) the heat transfer also increases due to the increase of fluid velocity U and the change of the flow regime. For $Ra >> 1$ we have $Nu >> 1$, which indicates that the heat transfer by conduction is negligible according to (3).

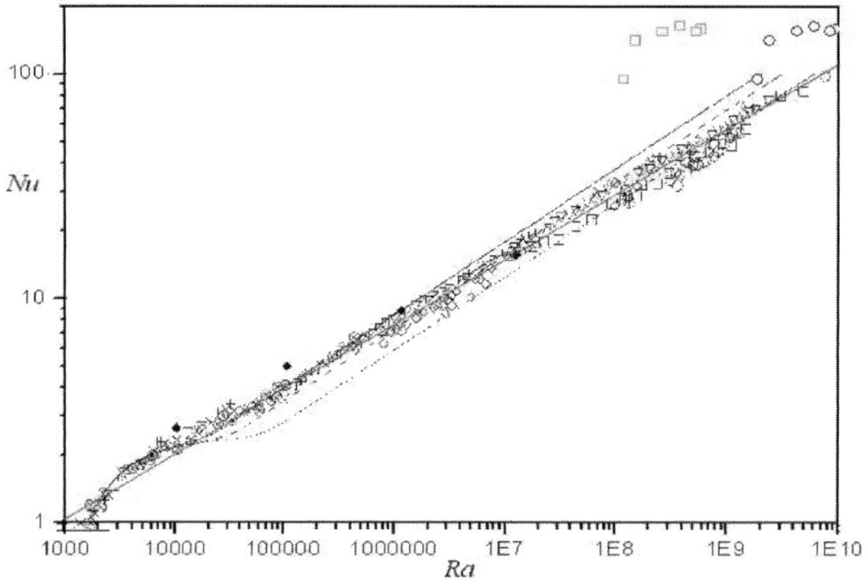

Figure 2. Relationship between Nusselt and Rayleigh numbers found in RB experiments. Black points and best fit lines represent values and trends, respectively, obtained by other authors. This figure is adapted from that reported in [13]. The solid red line is the approximation adopted in this work. The green squares and blue circles result from adopting the *Nu* and *Ra* values obtained in our experiments according to that discussed in Section 5.

A variety of regimes (laminar, transient, oscillatory, steady, etc.) has been detected for different ranges of *Ra*, *Pr* and *bc*. The oscillatory regimes are usually associated with structures of the flow (big vortexes, for example) changing the sense or the position periodically with a frequency that may be related to (6). In this case the characteristic frequency and its multiples appear when the Fourier transforms of velocity and temperature are calculated; in addition, the variations of these magnitudes in different points of the cell are correlated.

When *Ra* is greater than a given characteristic value Ra^* ($\approx 10^6$ for water, for example), the convection becomes turbulent and the positions of the fluid elements fluctuate in a scale that is smaller than *L*. In such a case, the Reynolds number increases for increasing ΔT according to (4) and (5). Turbulence is then added to the flow transporting fluid (phenomenon known as *wind of turbulence*) with velocity *U* between different parts of the room, by means of structures of scale *L*. The related structures and circulation are usually described as of *large scale*, in contrast to the *small scale* motions associated with turbulence.

The completely developed turbulence in the bulk of an isothermal fluid involves energy transfer from structures of large scales to smaller ones with an energy dissipation spectrum proportional to $k^{-5/3}$ that is known as the "Kolmogorov spectrum", where *k* is the wave number or the frequency in the cases of spatial or time dependence, respectively.

Boundary layers with kinetic and thermal features (that is boundary layers related to velocity *U* and energy transfer, respectively) form near the contours. According to Grossmann and Lohse [14], the kinetic and thermal boundary layers´ thicknesses may be estimated as

$$\delta_u \sim \frac{L}{\sqrt{Re}} \tag{8}$$

and

$$\delta_\theta \sim \frac{L}{2Nu}, \tag{9}$$

Journal of Physics: Conference Series **296** (2011) 012022 doi:10.1088/1742-6596/296/1/012022

respectively. Hence, for *Re* >> 1 and *Nu* >> 1, the boundary layers have a characteristic thickness that is smaller than *L* and, inside them, zones with different flow patterns than those formed in the fluid bulk can take place.

Grossmann and Lohse [14] obtained scale laws for temperature and velocity variations associated with the Rayleigh-Benard convection and found two components of the variations originated in the background turbulence and the plumes emerging from the contours. Thus, the amplitude *T'* of the temperature fluctuations is obtained as a *Ra*- and *Pr*- power law, consistently with experimental results. Using (7) it follows that

$$T' \sim Ra^{q} \sim Nu^{q/r} \quad . \tag{10}$$

For a particular fluid (Pr = constant), it is found that the T'- power law has an exponent $-0.11 < q < -0.16$ according to the relative importance of the background turbulence and the emerging plumes. Castaing *et al.* [15] and Hayakawa and Tsuji [16] found a similar value of this exponent ($q \sim -0.14$) in the fluid bulk. The boundary layers adjacent to the horizontal conductive plates and those next to the total or partial insulating vertical walls have different scale relationships compared to the ones characterizing the bulk. Grossmann and Lohse [17] argued that this explains the dependencies of geometry, aspect ratio and boundary conditions of the particular setup employed to experimentally study the RB convection.

The large scale convection connects the upper and lower thermal boundary layers, and may influence the thermal fluctuations and the temperature profile in the proximity of such boundary layers. However, when the large scale convective flow is suppressed or modified no significant variations of heat transport through the room is observed, that is to say *Nu* values variations are unimportant [12]. In other words, the heat transport is determined mainly by the efficiency of the mechanisms involved in the boundary layers where a greater temperature gradient exists (figure 1b).

All the concepts recalled in this Section will give support to the comprehension of the flow generated under the conditions of our interest described in the following and discussed in Sections 4 and 5.

3. Experimental setup description

Thermally-driven flows have been experimentally studied mainly using photographs [19], holographic interferometry [20], laser Doppler velocimetry [21] and PIV [22]. Here, synthetic schlieren technique is applied to visualize and quantify the temperature gradient variations, while an array of thermocouples is used to measure temperatures at different height levels.

Figure 3 illustrates the experimental arrangement in the laboratory. The physical model consists of a 0.60 m long, 0.20 m wide and 0.25 m deep tank (the room) with transparent acrylic walls, filled with water initially at uniform ambient temperature. On the center of the floor there is a cylindrical electric heater (0.10 m long, 0.014 m diameter) with a nominal power of 100 W. The input power is regulated with a *Variac*, and maintained constant during each experiment with a 2-4% variation. This energy is transferred to the water and absorbed eventually by the upper contour that consists of a cooper plate of thickness 0.005 m. An array with two series of five cooper straight tubes of 0.01 m diameter, connected with their entrances and exits at opposite positions, is soldered to the top plate. Water circulates in the tubes from a *NESLAB* refrigerated bath that maintains the cooper plate temperature under control. A neoprene layer of thickness 0.02 m covers the tubes and the top plate to thermally insulate the cooper plate.

The temperatures at five points of the cooper plate, in the heater, within the fluid and ambient are measured at intervals of 5 s with a T-type thermocouples net connected to an *Agilent* 34970A device. Ten thermocouples separated 0.023 m from each other were mounted on a rod inside the box. A video-camera *Basler* A102k with a 1392×1040 pixels resolution was employed to register the experiments taking one image every 5 s initially and every 10 min after the first hour during 12 hours. The images were processed mainly using the software *Digi*Flow [23] while the synthetic schlieren technique allows visualizing the flow generated in the fluid [24].

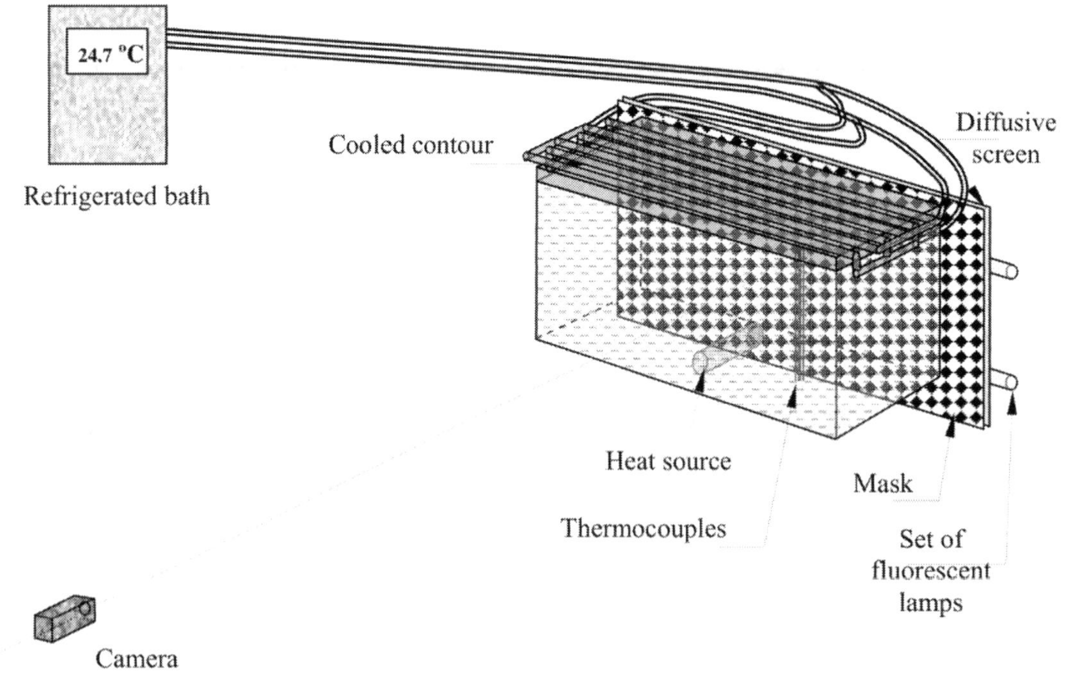

Figure 3. Experimental setup.

4. Results

Figure 4 shows the measurements of the fluid temperature registered by the thermocouples inside the room. Except for the thermocouple located nearest the bottom that measures an approximately constant temperature during the first 800 s, all register a progressive increase of the temperature during the transient regime and a constant difference between the lowest and highest heights temperatures. The heat supplied by the source is gradually extracted by the upper plate, and after around 15000 s a constant temperature of 24°C (that is, 2°C greater than the initial one) is reached and maintained constant thus obtaining the steady state regime in which all the heat supplied by the source is absorbed by the ceiling.

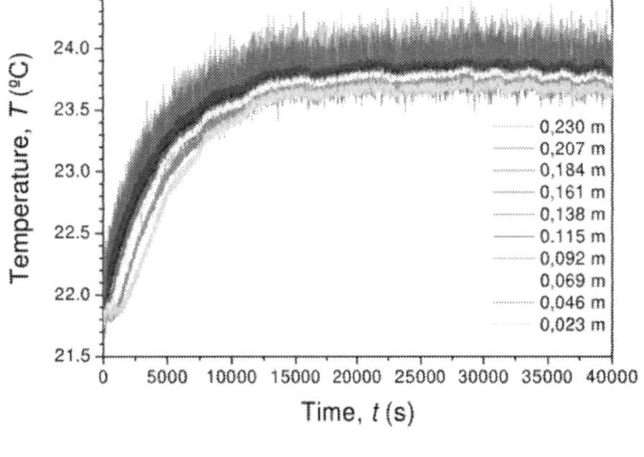

Figure 4. Fluid temperature evolution for $P = 40$ W.

Figure 5 shows the difference ΔT between the time average of the temperatures, $<T>$, at different heights inside the fluid and the top contour temperature, T_{top}, during the steady state regime. For each experiment, corresponding to a different input power, the points are connected by solid lines. These height profiles show a quite uniform temperature difference with a maximum near the upper plate and a minimum near the bottom. On the other hand, the greater the power supplied by the source is, the greater the temperature difference obtained. Figure 6 illustrates the relationship between ΔT in the bulk and P; the predicted analytical line, on which we will come back later, is also shown.

Figure 4 shows temperature fluctuations of the order of 0.5°C in the measurements of the thermocouples located near the cooled contour. In order to quantify these fluctuations, the standard deviations σ of the readings for different values of the power supplied are displayed in figure 7. It is clear that σ corresponding to the thermocouples located at 0.207 and 0.230 m from the bottom have the greatest fluctuations and those registered at 0.207 m are a bit greater than at 0.230 m. It can be also observed that the fluctuations are uniform for $h \leq 0.184$ m. Note that the amplitude of the fluctuations decreases as the power supplied by the heater increases, both in the fluid bulk and near the top.

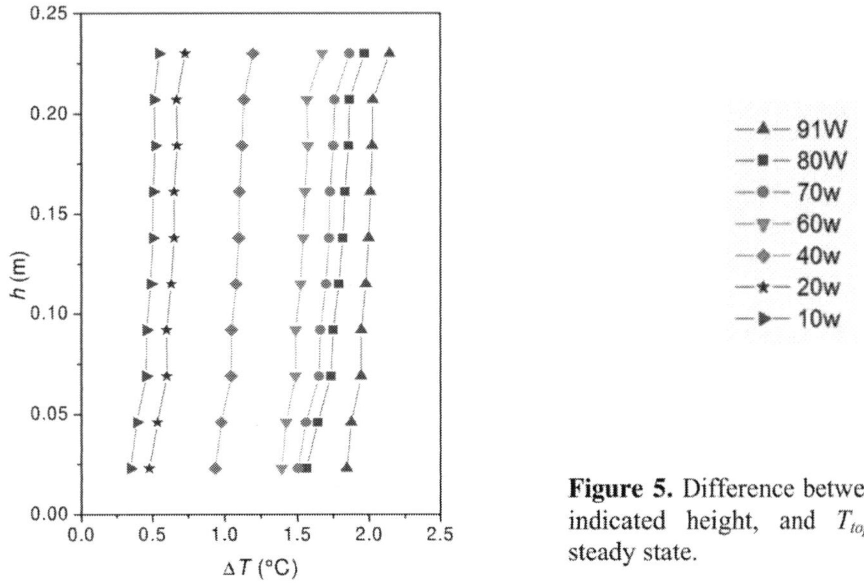

Figure 5. Difference between $<T>$ at the indicated height, and T_{top} during the steady state.

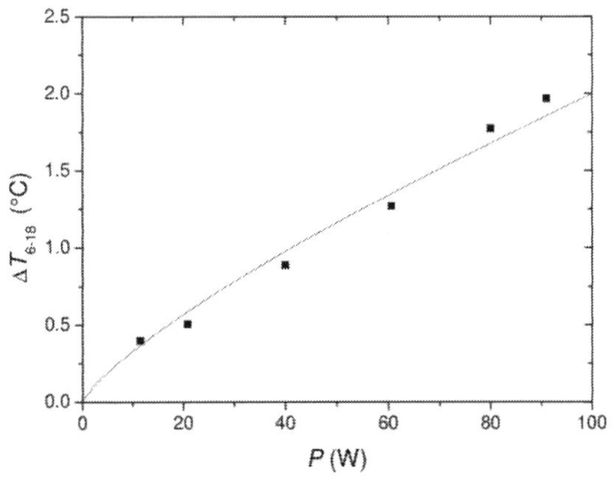

Figure 6. Difference between the average of the temperatures measured with the thermocouples located between 0.069 and 0.184 m within the fluid, and T_{top} during the steady state regime. The line represents equation (11).

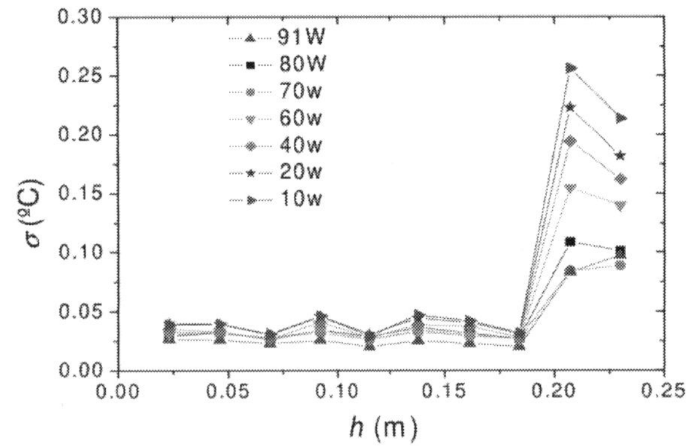

Figure 7. Standard deviation σ of the time average of the temperature values shown in figure 4 as function of the position of the thermocouples.

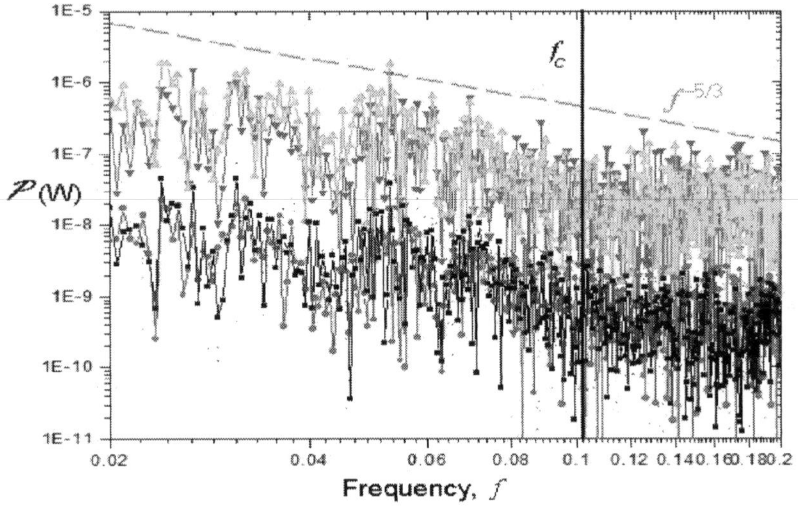

Figure 8. Power of the Fourier transform of temperature in the fluid, P , into components of different time frequencies for measurements registered at $h = 0.230$ m (olive triangles and line), 0.207 m (green triangle and line), 0.184 m (red circle and line) and 0.023 m (black squares and line), after the steady state regime is reached and for $P = 40$ W.

The response in terms of frequency of the temperature measurements is another significant feature observed. Figure 8 shows a typical result obtained when the Fourier transform of temperature is determined. The algorithm of the fast Fourier transform was used to estimate the spectrum of frequencies that are multiples of the minimum frequency $(t_{end}\text{-}t_{initial})^{-1}$ up to the maximum value $(2dt)^{-1}$, where dt is the interval between measurements, and $t_{end}\text{-}t_{initial}$ is the processing interval. Similar graphs are obtained for other P values. Consistently with figure 7, the temperature fluctuations detected near the cooled top plate (green and olive lines and points) have greater amplitudes than those in the fluid bulk (black and red lines and points). The frequency spectrum is similar in both cases and seems to follow a power law with exponent –5/3 even if one decade of frequency resolution is hardly enough data to confirm that the flow is described by such a law. Figure 8 does not show any frequency for which the amplitude is greater than the others, thus suggesting that there is not a characteristic frequency in the measurements range.

In addition, the heights where the major fluctuations are detected coincide with the places where strong variations of the temperature gradient occur as shown by the synthetic schlieren image in figure 9. Dark zones corresponding to negative horizontal gradients appear alternating with cyan/green zones of greater gradients below the upper boundary. The dark zones resemble small plumes that separate from the top plate and they are consequently composed of colder fluid. It is found that the locations of these plumes change randomly with time. Therefore, a thermocouple registers an abrupt decrease of temperature when it is reached by one of these plumes and an increase when the plume moves to other position, consistently with the fluctuations observed in figure 4. Figure 9 shows that these plumes of cold fluid reach a given height before mixing completely, thus suggesting the position of the lower boundary of the layer (located between 0.184 and 0.207 m), above which the greater variations of temperature are registered. Significant spatial gradients are not observed in the fluid below this layer, in agreement with the results shown by figures 5 and 7.

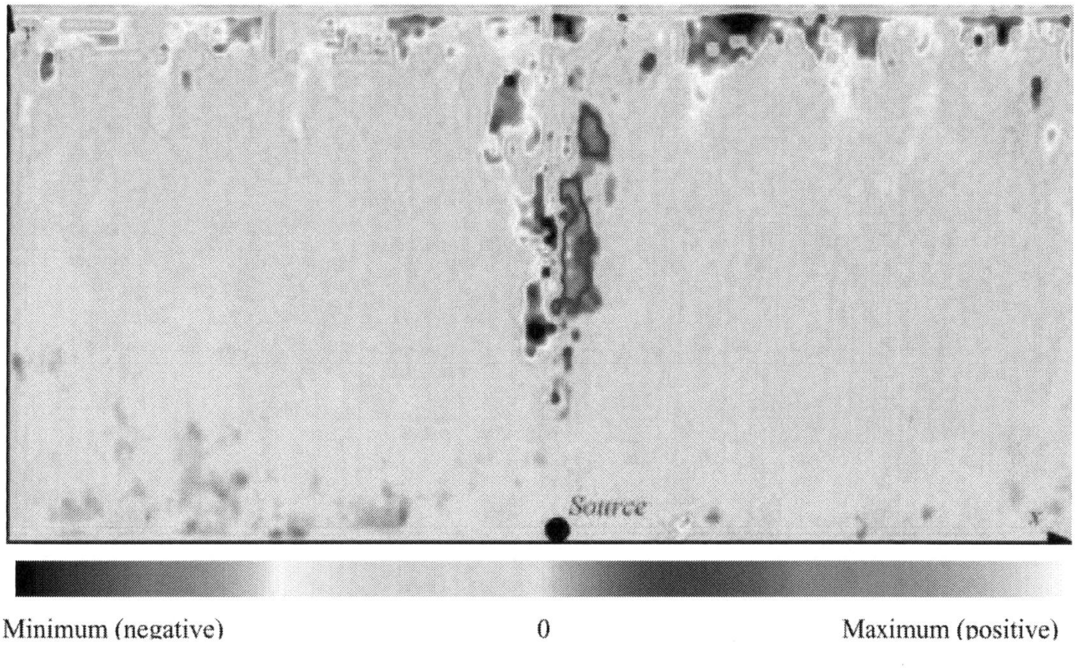

Figure 9. Distribution of the horizontal component of the temperature gradient during the steady state regime obtained by applying synthetic schlieren. The reference color scale is shown.

5. Discussion

The second column of table 1 shows the main parameters involved in the problem estimated with equations (1)-(6). These experimental values suggest that a turbulent regime is established. The green symbols (squares) in figure 2 represent $Nu = Nu(Ra)$ that drop far away from the general trend followed by the results obtained in the RB experiments.

To explain this apparent disagreement we recall that a distinctive feature of the studied system is that the bottom plate is not at constant temperature as in RB experiments but has a heat source supplying a constant heat flux. Therefore, a useful comparison may be obtained if our experimental setup is considered as an RB experiment with plates separated a distance $2L$ and maintained with a temperature difference $2\Delta T$ as illustrated in figure 10. In this case the "equivalent" values shown in the third column of table 1 are determined and $Nu = Nu(Ra)$ is represented with blue circles in figure 2. As seen, they are closer to the general trend, suggesting that the equivalent values are better to compare our results with those coming from RB experiments.

Table 1. Main experimental parameters

	Experimental values	RB equivalent values
L	0.25 m	0.50 m
S	0.12 m^2	0.12 m^2
Pr	6.4	6.4
ΔT	0.5 - 2 °C	1 - 4 °C
Ra	(1 - 6) $\times 10^8$	(2 - 10) $\times 10^8$
U	0.015 - 0.035 m/s	0.03 - 0.07 m/s
Re	(0.4-0.9) $\times 10^4$	(1.5 - 3.5) $\times 10^4$
Nu	95 – 160	95 - 160

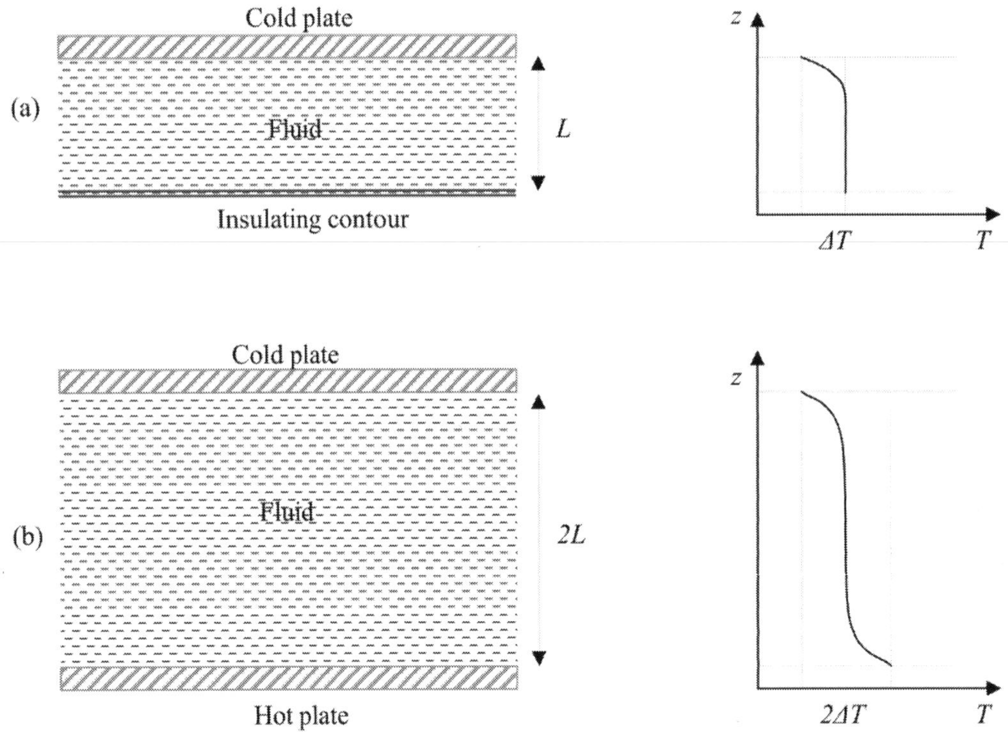

Figure 10. Sketches of (a) the experimental setup used here with a source inside (not shown), and (b) that equivalent of Rayleigh-Benard's. The corresponding average temperature profiles are represented on the right side.

The power P supplied by the source is equal to the heat flux Q in the steady-state regime. In addition, since the energy entering the fluid is known, the parameters characterizing the final state are Nu, Pr and bc while the system reaches the necessary ΔT and Ra values for the fluid to transport the energy supplied by the source. Using equations (1)-(3) and (7), it results

$$\Delta T = \left[\left(\frac{h}{\lambda S a} \right) \left(\frac{\nu \alpha}{g\, \beta\, L^3} \right)^r \left(\frac{\alpha}{\nu} \right)^p \right]^{\frac{1}{1+r}} P^{\frac{1}{1+r}} . \quad (11)$$

This relationship suggests that ΔT quasi-linearly depends on the source power because of the small values of r, reasonably fitting the experimental points as shown in figure 7.

According to the remarks made on Section 2, the fluctuations below the cooled contour seem to be related to the fluctuations in the boundary layer and not to the small scale turbulence in the fluid bulk. The eventual periodic oscillations of the free convection flow have characteristic frequencies of the order of the value indicated in equation (6) ($f_c \approx 0.1\ s^{-1}$ for the represented case). The Fourier transform (figure 8) does not show a special frequency with amplitude notoriously greater than the others, while the energy spectrum seems to follow a $k^{-5/3}$ law. The latter is the theoretical dependency of the inertial range of the spectrum when a homogeneous isotropic Kolmogorov turbulence is verified. This suggests that the turbulence may be fully developed both in the fluid bulk and in the measurement points within the boundary layer and the chaotic motion of the fluid suppresses the formation of structures that might periodically vary in time.

6. Conclusions

The steady-state convection generated by a heat source in an insulated room with a cooled top contour at constant temperature is experimentally analyzed. After a short initial regime, the temperature in the fluid bulk becomes uniform increasing with time up to reach a value a bit greater than that of the upper contour, achieving a steady state during which the heat supplied by the source is absorbed by the upper plate. Time fluctuations of the temperature and spatial gradient variations suggest the presence of a boundary layer with a thickness of about 0.05 m below the ceiling, where a flow different from that in the fluid bulk is developed. This layer is unstable because the temperature of the ceiling is smaller than that of the fluid bulk, which facilitates the formation of small plumes and vortexes that allow a more efficient heat transfer to the upper plate.

The findings derived from the Rayleigh-Benard classic convection are used to conveniently explain the results found considering RB equivalent experiments with a separation between the lower and upper plates and a temperature difference twice the values of the configuration used in this work. However, the correspondence between the two cases compared is not perfect and some differences deserve to be considered further. Figure 2 shows that the Nu values obtained here are a bit greater than those of the equivalent RB experiments suggesting a more efficient energy transfer from the source to the fluid. On the other hand, the location of the source in the cooled-ceiling system is fixed and symmetrical, which may originate some special symmetrical flow. Moreover, there is not a lower boundary layer and the characteristic distance L enters in the definition of the main parameters of the physical problem in a non-linear way ($Ra \sim L^3$, $Re \sim L^{3/2}$, $fc \sim L^{-1/2}$). These features might produce different flow patterns, upper boundary layer structure and Nusselt number from those corresponding situations in the RB experiments. According to the results obtained, these differences seem to be minimal in the turbulent regime studied, although they might be more important for lower values of Rayleigh and Reynolds numbers, with small or negligible turbulence. These aspects are being analyzed and will be reported in a future work.

ACKNOWLEDGEMENTS

This work was performed under the International Cooperation Project CONICET-CONACYT, the Project CONICET 054/2010 (Argentina) and the Project CONACYT 25116 (México).

7. References

[1] Ahlers G, Grossmann S and Lohse D 2009 Heat transfer and large scale dynamics in turbulent Rayleigh-Bénard convection *Rev. Mod. Phys.* **81** 503-37

[2] Viskanta R and Behnia M 1982 Experimental and analytical study of heat transfer and mixing in thermally stratified buoyant flows *Int. J. Heat Mass Transfer* 25 847-61

[3] Getling A V 1998 *Rayleigh-Bénard convection: Structures and Dynamics (Advanced Series in Nonlinear Dynamics, Vol. 11)* (Singapore: World Scientific Publishing).

[4] Morton B R, Taylor G I and Turner J S 1956 Turbulent gravitational convection from maintained and instantaneous sources *Proc. Roy. Soc. A.* **234** 1-23

[5] Baines W D and Turner J S 1969 Turbulent buoyant convection from a source in a confined region *J. Fluid Mech.* **37** 51-80

[6] Hunt GR, Cooper P and Linden PF 2001 Thermal stratification produced by plumes and jets in enclosed spaces. *Build. Environ.* **36** 871-82

[7] Kaye N B and Hunt G R 2004 Time-dependent flows in an emptying filling box. *J. Fluid Mech.* **520** 135-56

[8] Fitzgerald S D and Woods A W 2010 Transient natural ventilation of a space with localised heating. *Build. Environ.* **45** 2778-89

[9] Linden P F 1999 The fluid mechanics of natural ventilation *Annu. Rev. Fluid Mech.* **31** 201–38

[10] Krishnamurti R and Howard L N 1981 Large-scale flow generation in turbulent convection (instability/transition/turbulence/large-scale order) *Proc. Nat. Acad. Sci. USA, Applied Physical and Mathematical Sciences* **78** 1981-85

[11] Xi H-D, Lam S and Xia K-Q 2004 From laminar plumes to organized flows: the onset of large-scale circulation in turbulent thermal convection *J. Fluid Mech.* **503** 47-56.

[12] Ciliberto S, Cioni S and Laroche C 1996 Large-scale flow properties of turbulent thermal convection.*Phys. Rev. E* **54** R5901-04

[13] Itamoto I, Ishida H, Higashiyama M, Miki D and Kawahara G 2009 Average and extremal properties of heat transfer and shear stress on a wall surface in Rayleigh–Bénard convection *Heat Mass Transfer* **46** 153-65

[14] Grossmann S and Lohse D 2004 Fluctuations in turbulent Rayleigh–Bénard convection: The role of plumes *Phys. Fluids* **16** 4462-4472

[15] Castaing B, Gunaratne G, Heslot F, Kadanoff L, Libchaber A, Thomae S, Wu X-Z, Zaleski S and Zanetti G 1989 Scaling of Hard Thermal Turbulence in Rayleigh-Benard Convection. *J. Fluid Mech.* **204** 1-30

[16] Hayakawa T and Tsuji Y 2010 Mean wind: Its velocity and temperature fluctuation in low-Prandtl-number thermal convection. *Physica D* **239** 1353-8

[17] Grossmann S and Lohse D 2003 On geometry effects in Rayleigh-Benard convection *J. Fluid Mech.* **486** 105–14

[18] Bernaz L, Bonnet J M and Seiler J M 2001 Investigation of natural convection heat transfer to the cooled top boundary of a heated pool. *Nucl. Eng. Des.* **204** 413-27.

[19] Krishnamurti R 1995 Low frequency oscillations in turbulent Rayleigh-Benard convection: laboratory experiments. *Fluid Dyn. Res.* **16** 87-108

[20] Shim D S, Kang B S and Park C K 2002 An interferometric investigation of natural convection in a partially opened enclosure with a discrete heat source. *Exp. Heat Transfer* **15** 121–35

[21] Qiu X-L and Tong P 2001 Large-scale velocity structures in turbulent thermal convection *Phys. Rev. E* **64** 036304

[22] Mamun M A H, Johnson D A, Hollands K G T and Leong W H 2008 PIV measurements of the flow field inside an enclosed cubical cavity in natural convection *Exp Fluids* **44** 647–59

Petracci A, Delfos R and Westerweel J 2006 Combined PIV/LIF measurements in a Rayleigh-Bénard convection cell. *13th Int Symp on Applications of Laser Techniques to Fluid Mechanics*, Lisbon (Portugal), Paper 1221

[23] Dalziel S B 2006 *Digi*Flow *User Guide*. http://www.damtp.cam.ac.uk/lab/digiflow/.

[24] Thomas L P, Marino B M and Dalziel S B 2009 Synthetic Schlieren: Determination of the density gradient generated by internal waves propagating in a stratified fluid *J Phys: Conf Series* **166** 012007

FLUIDOS 2010: XI Meeting on Recent Advances in the Physics of Fluids and their Applications IOP Publishing
Journal of Physics: Conference Series **296** (2011) 012023 doi:10.1088/1742-6596/296/1/012023

Study of a pseudo-empirical model approach to characterize plasma actuators

M Marziali Bermudez[1], R Sosa[2,4], D Grondona[3], A Márquez[3], H Kelly[3] and G Artana[2,4]

[1]Departamento de Física, Facultad de Ciencias Exactas y Naturales, UBA, Ciudad Universitaria Pab. I, Buenos Aires 1428, Argentina
[2]Laboratorio de Fluidodinámica, Facultad de Ingeniería, UBA, Av. Paseo Colón 850, Buenos Aires 1063, Argentina
[3]Instituto de Física del Plasma (CONICET) - Departamento de Física, Facultad de Ciencias Exactas y Naturales, UBA, Ciudad Universitaria Pab. I, Buenos Aires 1428, Argentina
[4]CONICET

E-mail: `rsosa@fi.uba.ar`

Abstract. The use of plasma actuators is a recent technology that imposes a localized electric force that is used to control air flows. A suitable representation of actuation enables to undertake plasma actuators optimization, to design flow-control strategies, or to analyse the flow stabilization that can be attained by plasma forcing. The problem description may be clearly separated in two regions. An outer region, where the fluid is electrically neutral, in which the flow is described by the Navier-Stokes equation without any forcing term. An inner region, that forms a thin boundary layer, where the fluid is ionized and electric forces are predominant. The outer limit of the inner solution becomes the boundary condition for the outer problem. The outer problem can then be solved with a slip velocity that is issued from the inner solution. Although the solution for the inner problem is quite complex it can be contoured proposing pseudo-empirical models where the slip velocity of the outer problem is determined indirectly from experiments. This pseudo-empirical model approach has been recently tested in different cylinder flows and revealed quite adapted to describe actuated flow behaviour. In this work we determine experimentally the influence of the duty cycle on the slip velocity distribution. The velocity was measured by means of a pitot tube and flow visualizations of the starting vortex (i.e. the induced flow when actuation is activated in a quiescent air) have been done by means of the Schlieren technique. We also performed numerical experiments to simulate the outer region problem when actuation is activated in a quiescent air using a slip velocity distribution as a boundary condition. The experimental and numerical results are in good agreement showing the potential of this pseudo-empirical model approach to characterize the plasma actuation.

1. Introduction

Electro-hydrodynamic (EHD) actuators for flow control with aims to improve aerodynamic design have been receiving special attention in the last years [1, 2, 3]. Through the ionization of the flowing air close to the surface of a body, EHD devices produce a modification of the condition of the flow at the wall proximity. The charged particles present in the ionized air (plasma) experience an electric force that is transmitted through particle collision to the neutrals of the flowing air. The expression of this force on the fluid depends on the discharge characteristics, and in some cases there exists a two-way interaction created by electric forces as they elicit a mechanical response that, in turn, alters the fields. The implementation of this force in numerical flow simulations is of interest, as it could be used to optimize the design and placement of plasma actuators for different applications for flow control.

Detailed studies on how to implement the electric force coupling in validated codes of computational fluid mechanics as direct numerical simulations (DNS) have been reported for the relatively simple cases of classic corona discharges [4]. For the specific case of dielectric barrier device (DBD) actuators, which we study here, efforts have been also reported recently by different researchers [1, 2, 5].

A recent computational exploration has been conducted by Visbal et al. (2006) [6] using large-eddy simulation approaches at a Reynolds number $Re \approx 10^4$ in order to capture the effects of the local forces of the DBD device on turbulent, separated flows. Unfortunately, these researches also concluded that further work was still needed in order to construct detailed models for the spatio-temporal distribution of the plasma-induced body forces suitable for incorporation into flow simulations.

Corke et al. [5] derived a simplified expression for the body force distribution and implemented it in the numerical flow solver CFL3D [7] to estimate the aerodynamic coefficients involved in the control of leading-edge separation on wing sections. Likhanskii et al. [8] developed a more detailed physical model for DBD in air, and noted the computational difficulties of plasma numerical simulation over large temporal and spatial scales. They concluded that in general, the cell size for the rectangular grid has to be chosen according to the characteristics of plasma and voltage scales and geometrical size (on the order of 10 microns); this in turns requires an extremely high number of computational points in order to perform calculations in large scale geometries (i.e., beyond the domain which is next to the electrode configuration) or multielectrode arrangements.

The other difficulty in plasma-effects simulations is the choice of the time step. Because the time step should allow for the resolution of plasma formation and streamer propagation dynamic time (which is of the order of nanoseconds), this time scale must be several orders of magnitude smaller than the time scale over which the plasma is producing an effect on the surrounding air. Therefore, both factors make prohibitive an easy approach to detailed plasma actuation over large spatio-temporal scale using DNS.

The use of some approximated models (empirical or quasi-empirical) to evaluate the plasma actuation are in consequence attractive. The possibility to attain realistic models is however better when flow characteristic frequencies result much lower than those of cyclic excitation of the discharge (usually some kHz). Flows can then be considered steadily excited with a time averaged plasma forcing that is determined under controlled experimental conditions and then extrapolated to the case of interest.

On the other hand, the region where the momentum transfer from ions to neutrals can occur is confined to a very narrow region quite similar to streamer radius. This region is extremely thin compared to most fluid dynamics characteristic lengths of reference.

Under these circumstances the full problem description may be clearly separated in two: an outer and an inner problem. At the outer region we have fluid that is electrically neutral and is described by the Navier-Stokes equation without any forcing term. In the inner region at the discharge region we have a thin boundary layer where the electric forces are confined.

Opaits et al. [9] have proposed a new approach for non-intrusive diagnostics of plasma actuator induced flow in quiescent gas. This approach consisted in using the force calculated by the plasma model [7] as an input to a two-dimensional viscous flow solver to predict the time-dependent DBD induced flow and then to match this numerical result with Schlieren images of the induced flow. Thus this approach allowed to restore the entire two-dimensional unsteady plasma induced flow pattern as well as characteristics of the plasma induced force.

In the present work we propose to solve the outer problem imposing a slip velocity as a boundary condition. A similar approach has been used in other problems with similar characteristics like for instance in electrosmotic flows [10]. The solution for the inner problem is quite complex in general but can be contoured proposing pseudo-empirical models where the slip velocity of the outer problem is determined indirectly from experiments. A large number of efforts have been done by different authors that have measured for different kind of plasma actuators the velocity profiles induced by the discharge with different techniques like laser doppler velocimetry or pitot tubes. It can be observed that actuation produces in most cases a pulsed type distribution, with its width and peak value depending on electrical parameters of the actuator excitation signal (amplitude, frequency, shape type) and constructive characteristics (spacing, dielectric thickness, dielectric material).

This pseudo-empirical model approach has been also recently tested in different cylinder flows and revealed quite adapted to describe actuated flow behavior [11, 12, 13].

The goal of this article is to determine how the actuation representation has to be adapted when considering different actuation intensities.

2. Experimental set-up

2.1. Electrodes arrangement and electric circuit

Figure 1 shows a scheme of the electrode arrangement and the electric circuit. The electrode arrangement consists of an air-exposed electrode flush mounted on a Kapton® foil (electrode (1), width 20 mm) and an electrode located on the opposite side (electrode (2), width 5 mm). Both electrodes are made of aluminum foils and they are about 50 μm thick. The Kapton® foil, with the electrodes, was stuck on the surface of a 4mm thick PMMA flat plate.

The electrode exposed to the air (electrode (1)) was connected to the output of an AC power supply (peak-to-peak voltage V_{AC} in the range 0–23 kV) and the encapsulated electrode (electrode (2)) was grounded. Thus the plasma actuator induces a mean flow as sketched on figure 1.

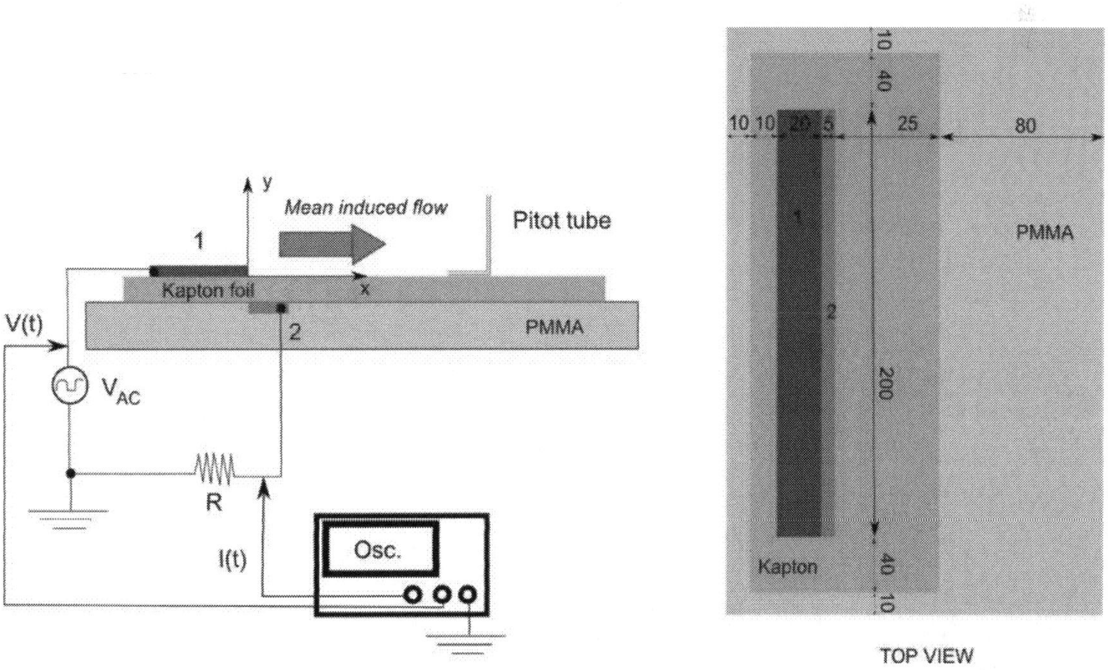

Figure 1. Electrodes arrangement and electric circuit (dimensions are in mm)

The AC power supply consisted of a function generator coupled to an audio-amplifier (of 700 W) that fed a high voltage transformer coil. In practice there was an optimal matching frequency, established by the resonance between the transformer inductance and the stray capacity of the electrode arrangement, including the wire connections. For our circuit geometry, the optimum excitation AC frequency was $f_{AC} = 5.5$ kHz ($T_{AC} = 1/f_{AC} = 180$ μs).

The DBD current ($I(t)$) measurement was performed by connecting electrode (2) through a $50\,\Omega$ resistance to the ground and registering the voltage drop on this resistance. The instantaneous voltage applied to electrode (1), $V(t)$, was measured with a HV probe (1000X, 3.0 pF, 100 MΩ). These electrical signals were registered by using a four-channel digitizing oscilloscope with an analog bandwidth of 60 MHz and 1 GS/s sampling rate.

The signal generator can operate the DBD by burst modulation of the sinusoidal high voltage signal. The effect of the duty-cycle (defined as: $100 \times T_{ON}/T_{BM}$, being T_{ON} the time during which the plasma is produced, see figure 2) was explored for values from 35% up to 100%. This parameter quantifies the fraction of time that the discharge, and consequently the forcing, is active. Thus by changing this parameter it is possible to modify the time averaged intensity of actuation.

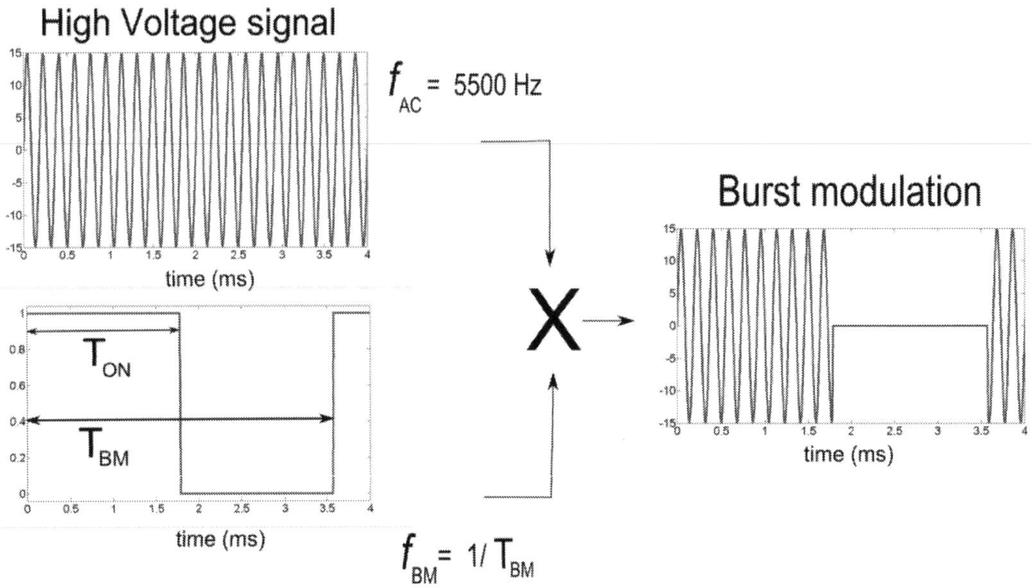

Figure 2. Burst modulation mode ($f_{BM} = 280\,$Hz, duty-cycle 50% as an example)

In our studies T_{BM} was fixed to a value of $T_{BM} = 20 \times T_{AC} = 3.6$ ms in order to assure at least seven cycles of the high voltage signal for the lower duty-cycle considered. This choice guarantees that the distribution of charges on the surface settles into an equilibrium configuration [14].

2.2. Schlieren visualization setup

To visualize the induced ionic wind a Schlieren system disposed in a Z configuration was used [15]. It comprised two spherical mirrors of 50 cm in diameter and 3.0 m in focal length. The light (a tungsten-halogen automotive lamp) was cut off with two razor blades at the meridional and sagittal focal planes. Thus 2D density gradients could be detected. The images were recorded on a monochromatic digital image camera of 12 bits that enabled fast frame acquisition (up to 200 fps). The Schlieren technique is able to detect changes in index refraction associated to density gradients in a fluid media. In general the small changes in density associated to the heating produced by the discharge enable to obtain images with sufficient contrast. In this context the tracers are the same fluid particles (the studied ones) and therefore there is not need to introduce any external particles (like smoke particles in the PIV systems).

2.3. Pitot tube measurement setup

Velocity measurements are conducted with the help of a Pitot tube. This was made of electrically insulating glass tube (internal diameter: ID = 0.70 mm, outer diameter OD = 0.97 mm) and was connected by mean of a tygon tube to a differential pressure transducer of variable reluctance type (0 to 55 Pa, accuracy of 0.25% full scale). The reference pressure value was measured far away from the electrodes. This pressure sensor may be displaced along the x and y axes (see figure 1), and the goal is to plot velocity the profiles above the wall (with the pitot on the wall, that is $y = 0.49$ mm) at different x positions.

3. Numerical simulation

3.1. General characteristics of the simulation code

The geometry has been non-dimensionalized with the electrode (2) width. The computational geometry is defined by a rectangle placed with dimensions $L = 12$ and $H = 4$. The center of coordinates is placed at the bottom surface of the box and 4 units ahead of the left vertical surface (see figure 3).

The boundary conditions can be divided in three sections: Γ_P composed of the Kapton® surface above the electrode (2) and associated to the plasma region, Γ_W composed of the rest of the solid surfaces (Kapton® and PMMA) where no plasma is produced and no slip is considered. Finally for the rest of the surfaces, Γ_N, where natural boundary conditions (NBC) have been used (see equation 4). For the Kapton® surface, above the electrode (2), Γ_P the boundary condition comes from the slip velocity imposed by the plasma actuator and is given by the equation

$$v_i(x) = \frac{A}{W\,x} \mathrm{e}^{-\frac{1}{2}\left(\frac{\log(x-x_0)}{W}\right)^2} \hat{\mathbf{e}}_x \tag{1}$$

where the log-normal function $v_i(x)$ has three parameters: A, W and x_0. W is related to the pulse width, A is related to the pulse amplitude and x_0 is related to the position of the maximum velocity. In the present study W and x_0 were fixed and only the influence of the pulse amplitude was explored (see figure 3).

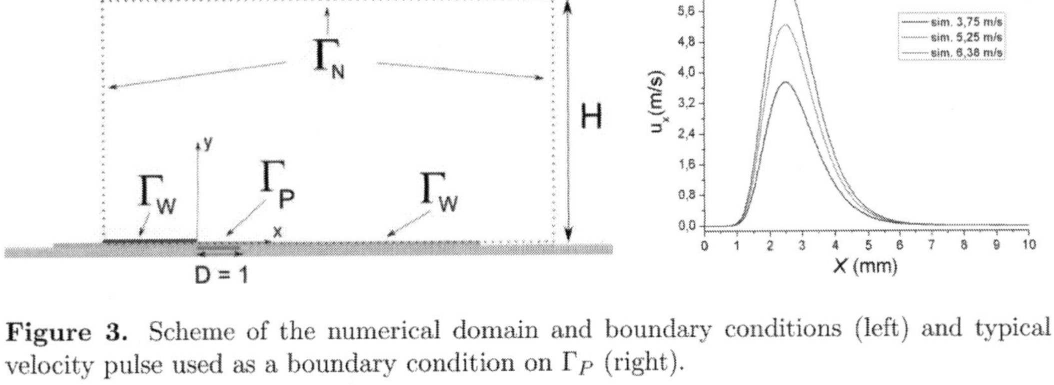

Figure 3. Scheme of the numerical domain and boundary conditions (left) and typical slip velocity pulse used as a boundary condition on Γ_P (right).

3.2. Governing equations

Mass and momentum conservation principles are represented by the Navier-Stokes equations, which have the following form for an incompressible fluid:

$$\frac{\partial u_i}{\partial x_i} = 0 \quad \text{in } \Omega \tag{2}$$

$$\frac{\partial u_i}{\partial t} + u_j \frac{\partial u_i}{\partial x_j} = -\frac{\partial p}{\partial x_i} + \frac{1}{Re}\frac{\partial^2 u_i}{\partial x_j^2} \quad \text{in } \Omega \tag{3}$$

The boundary conditions were:

$$0 = -p n_i + \frac{1}{Re}\frac{\partial u_i}{\partial n} \quad \text{on } \Gamma_N \tag{4}$$

$$u_i = v_i \quad \text{on } \Gamma_P \tag{5}$$

$$u_i = 0 \quad \text{on } \Gamma_W \tag{6}$$

Here, Ω is the computational domain, p the pressure divided by the (constant) mass density, u_i is the ith component of the velocity field and n is the outward unit-normal to the boundary.

To solve this equation we used a previously tested DNS code whose references can be found at [16].

4. Experimental and numerical results

4.1. Schlieren visualization and vortex core trajectory

The left-hand side of figure 4 shows a typical starting vortex for a high level of actuation. Because Schlieren imaging records variations in optical density, the hot air surrounding the electrode acted as a tracer. The tracer is then advected by the stream and diffuses mainly because of thermal effects. To improve these images, two laminar circular jets of helium have been added in these tests to obtain information where the thermal tracers are not present. In the right-hand side of figure 4 we show the isolines of the modulus of velocity obtained from our numerical simulations. The momentum injected in the discharge region is transported by the fluid and diffusion occurs as a consequence of viscosity. The evident physical correspondence in both cases is notorious.

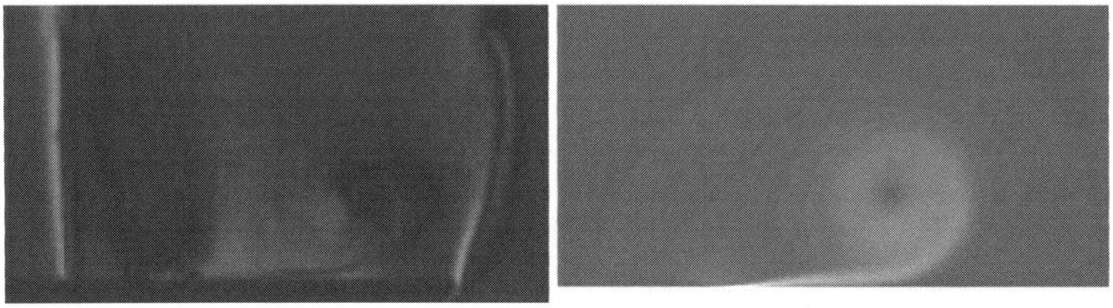

Figure 4. Results for high levels of actuation intensity: experimental imaging (left) and isolines of velocity modulus from simulation (right)

Figure 5 shows typical Schlieren images for reduced levels of actuation (no helium jets were incorporated in these experiments). The images show the starting vortex for $V_{AC} = 7$ kV, with 80% duty cycle. In order to track the starting vortex the image sequences obtained by the Schlieren technique were analyzed using the software *ImageJ*.

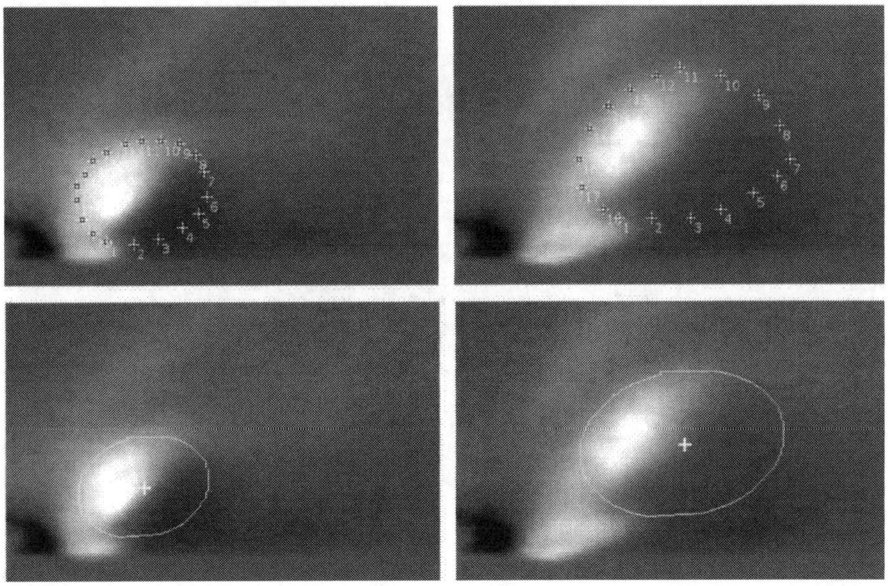

Figure 5. Schlieren images of the starting vortex for $V_{AC} = 7$ kV and 80% duty cycle. Time between images 10 ms, image size 10.5 mm x 6.5 mm.

Due to the low level of actuation intensity and due to thermal diffusion, the vortex center could be hardly determined directly. Therefore, a set of several points were laid around the vortex contour, as shown in the figure 5 (top), and an ellipse was fitted to those points (bottom). The centroid of such ellipse was considered to behave similarly enough to the actual vortex center within the scope of this study.

Figure 6 shows the x position of the centre of the vortex as a function of time (x_C vs. t, left figure) and the trajectory of the centre of the vortex ($x_C(t)$ vs. $y_C(t)$, right figure) for different duty-cycles and with $V_{AC} = 7$ kV.

As it can be observed, the vortex velocity (i.e. the slope of the curves x_C vs. t) increase with the duty-cycle. On the other hand, the starting vortex seems to have almost the same trajectories, within our experimental uncertainty, independently of the duty cycle considered.

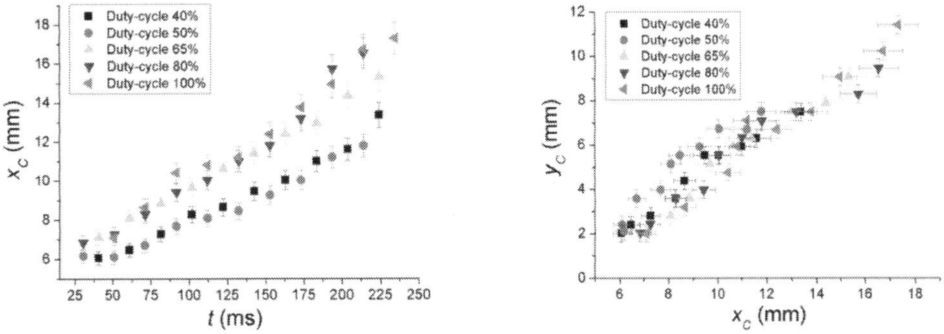

Figure 6. Left: x_C vs. t. Right: trajectory of the centre of the vortex, $x_C(t)$ vs. $y_C(t)$. For different duty-cycles and with $V_{AC} = 7$ kV.

4.2. Comparison between experimental and numerical results

4.2.1. Starting vortex trajectory Figure 7 shows a typical velocity field obtained from numerical simulation. When the absolute value of the velocity is plotted, a clear unsteady vortical structure is observed and it is quite simple to determine their centre as it is shown in figure 7. Following this procedure we can reconstruct the x position of the vortex center as a function of time and the trajectory of the centre of the vortex (for different amplitudes of the imposed slip velocity pulse) in order to compare that with the experimental results previously showed.

Figure 8 compares the curves of x_C vs. t (left) and $x_C(t)$ vs. $y_C(t)$ (right) obtained from experimental and numerical results. In the case of the experimental results we have plotted the cases of 40%, 65% and 100% of duty cycle only to simplify the graphics. On the other hand, the numerical results correspond to the cases of: 3.75 m/s, 5.25 m/s, 6.38 and 7.50 m/s of pulse amplitude.

We see that the numerical results agree quite well with the experimental ones, showing that the vortex central velocity and trajectories display the same dependence with respect to the duty-cycle value.

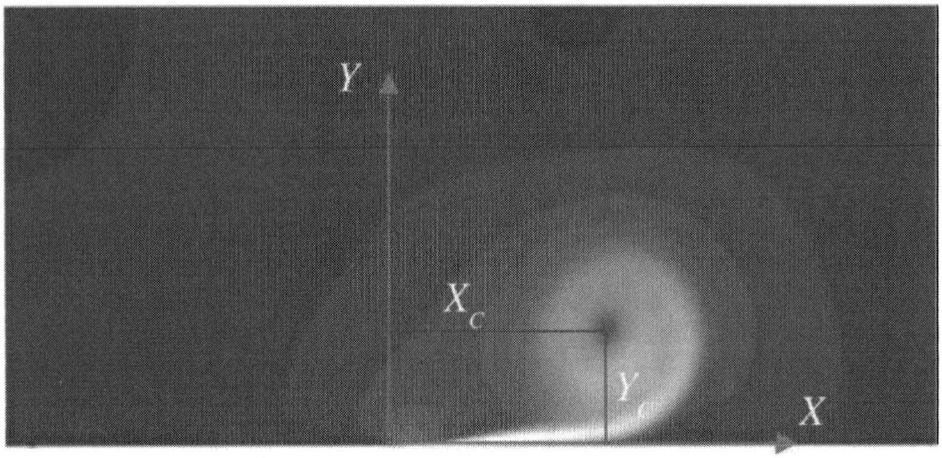

Figure 7. Typical velocity field obtained from numerical simulation.

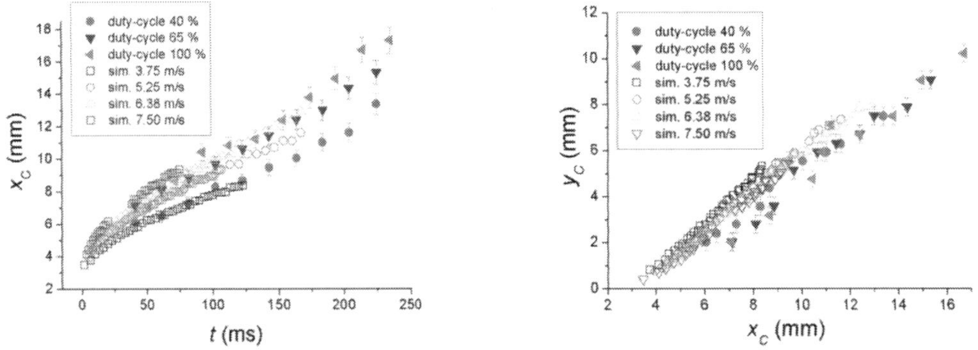

Figure 8. Plots of x_C vs. t (left) and $x_C(t)$ vs. $y_C(t)$ (right) obtained from experimental and numerical results. Experimental results for $V_{AC} = 7$ kV and duty-cycles of: 40%, 65% and 100%. Numerical results for pulse amplitudes of: 3.75 m/s, 5.25 m/s, 6.38 and 7.50 m/s.

It is important to note that the starting vortex behaviour described previously can be explained if we consider the similarity between this vortex structure and the junction vortex.

When an impulsively started moving wall slides under a stationary wall, the potential exists for a vortical structure to develop close to their junction. The formation mechanism for this vortex when the wall is set in motion, is that a Stokes layer develops over the moving surface and this layer is driven, by shear forces, past the singularity at the junction and over the stationary wall. Vorticity of opposite sign then develops over the stationary wall to enforce the no-slip condition. The flow field over the stationary wall initially resembles an unsteady wall jet. During an experiment, the circulation swept into the junction region can be defined as $\Gamma(t) = \int_0^t U_W^2(\tau)\mathrm{d}\tau$, where U_W is the wall speed. This results in a Reynolds number, or non-dimensional time, defined as: $Re_\Gamma(t) = \frac{\Gamma(t)}{\nu}$, being ν the kinematic viscosity. If the Reynolds number is large enough, then the jet separates and rolls into a vortex [17].

Allen and Naitoh [17] have demonstrated that the unsteady junction vortex flow field is universal for a given Re_Γ when the dimensions of the structure are non-dimensionalized with $\frac{U_W}{\nu}$ and the velocity is non-dimensionalized with U_W. Thus the vortex core coordinates (x_C, y_C) can be fitted with power law such as $\frac{x_C U_W}{\nu} = \kappa_x (Re_\Gamma)^{q_x}$ and $\frac{y_C U_W}{\nu} = \kappa_y (Re_\Gamma)^{q_y}$.

In our study U_W can be assumed constant (because the V_{AC} value was fixed in all the experiments) resulting the Reynolds number equal to: $Re_\Gamma(t) = \frac{\Gamma(t)}{\nu} = \frac{U_X^2 t}{\nu}\frac{T_{ON}}{T_{BM}}$. Thus, x_C and y_C are time and duty-cycle (i.e. T_{ON}/T_{BM}) dependent. This last conclusion allows to explain the observed dependence of the vortex core velocity on the duty-cycle.

On the other hand, it is easy to demonstrate mathematically that the vortex core trajectory (i.e. y_C as a function of x_C) does not depend on the duty-cycle. This can explain the fact that the experimentally observed curves $y_C(x_C)$ are almost coincident.

4.2.2. Velocity profiles In order to have another tool that allow us to validate our numerical simulations we have measured the horizontal component of the velocity (u_x) with the pitot tube resting on the surface (thus the central line of the pitot was at $y = 0.49$ mm).

For the values of pulse amplitude that recover quite well the vortex dynamics, in the numerical experiments, we have searched the value of y that best fit our pitot measures.

Figure 9 shows the measured velocity profiles considering different duty cycles and the velocity profiles from numerical simulations (for $y = 0.15$, and considering different pulse amplitudes).

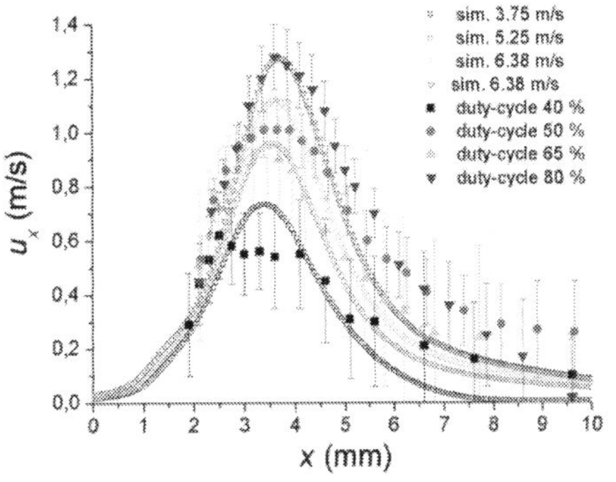

Figure 9. Velocity profiles from Pitot measurements ($y = 0.49$ mm, and duty cycles of: 40%, 50%, 65% and 80%. $V_{AC} = 7$ kV) and numerical simulations ($y = 0.15$, and considering pulse amplitudes of: 3.75 m/s, 5.25 m/s, 6.38 and 7.50 m/s.)

The measured velocity profiles behave quite similar to the numerical ones. The observed discrepancies in terms of the y coordinate can be explained considering that both the wall proximity and the intense velocity gradient close to the wall tend to produce a higher velocity measurement with the pitot tube. On the other hand, in the slip velocity model, the thickness of the inner region is neglected. If we assign to this region a thickness of 0.35 mm experimental and numerical results then are in a very good agreement.

5. Conclusions

In this work we have determined experimentally the influence of the duty cycle on the slip velocity distribution necessary to reproduce the unsteady vortex structure arising when a single DBD plasma actuator is activated in quiescent air.

By controlling the duty cycle it results easy to control the intensity of actuation when low or moderate forcing is required. This becomes of interest when air flows at low velocity and an excessive forcing may be counterproductive.

It is also a crucial parameter as its value is directly related to the power consumed by the plasma actuator. For example, in our case an actuation with a duty-cycle of 100% consumes about 20 W while the electrical consumption with a duty-cycle of η% results in (η x 20 W)/100.

The starting vortex dynamics (core trajectory and velocity) was reproduced quite well with the numerical simulation that mimics the actuation using a slip velocity distribution as a boundary condition. We also have shown that the observed behaviour of the vortex core dynamics, with respect to the duty cycle, can be satisfactorily explained using the same scaling laws that are used to analyze the junction vortex.

The experimental and numerical results are in good agreement showing the potential of this pseudo-empirical model approach to characterize the plasma actuation.

References

[1] Boeuf J P and Pitchford L C 2005 Electrohydrodynamic force and aerodynamic flow acceleration in surface dielectric barrier discharge *J. Appl. Phys.* **97** 103307

[2] Font G I and Morgan W L 2005 Plasma discharges in atmospheric pressure oxygen for boundary layer separation control *AIAA Paper* 2005-4632

[3] Moreau E 2006 Airflow control by non-thermal plasma actuators *J. Phys.* D: Appl. Phys. **39** 1

[4] Soldati A and Banerjee S 1998 Turbulence modification by large-scale organized electrohydrodynamic flows *Phys. Fluids* **10** 1742-50

[5] Corke T C and Post M L 2005 Overview of plasma flow control: concepts, optimization and applications *AIAA Paper* 2005-0563

[6] Visbal M R, Gaitonde D V and Roy S 2006 Control of transitional and turbulent flows using plasma-based actuators *AIAA Paper* 2006-3230

[7] Rumsey C L et al. 2008 *CFL3D Version 6 web page* NASA Langley Research Center (`http://cfl3d.larc.nasa.gov`)

[8] Likhanskii A V, Schneider M N, Macheret S O and Miles R B 2006 Modeling of interaction between weakly ionized near-surface plasmas and gas flow *AIAA Paper* 2006-1204

[9] Opaits D F, Neretti G, Likhanskii A V, Zaidi S, Shneider M N and Miles R B 2007 Experimental Investigation of DBD plasma actuators driven by repetitive high voltage nanosecond pulses with DC or low-frequency sinusoidal bias *AIAA Paper* 2007-4532

[10] Ghosal S 2006 Electrokinetic Flow and Dispersion in Capillary Electrophoresis *Annu. Rev. Fluid Mech.* **38** 309–38

[11] Gronskis A, Sosa R and Artana G 2009 Modelling EHD Actuation with a Slip Velocity *Joint ESA/IEEE-IAS/IEJ/SFE/IEA Conf.* Boston, MA

[12] González L M, Artana G, Gronskis A and D'Adamo J 2009 A computational moving surface analogy of an electrodynamic control mechanism in bluff bodies *4th Symposium on Global flow instability and control* Crete, Greece

[13] Peers E, Ma Z and Huang X 2010 A numerical model of plasma effects in flow control *Phys. Lett.* A **374** 1501–4

[14] Enloe C L, McLaughlin T E, Gregory J W, Medina R A and Miller W S 2008 Surface potential and electric field structure in the aerodynamic plasma actuator *AIAA Paper* 2008-1103
[15] Settles G S 2001 *Schlieren and Shadowgraph Techniques: Visualizing Phenomena in Transparent Media* (Berlin: Springer-Verlag)
[16] Rodríguez Gallego A and González Gutiérrez L M 2003 *ADFC Navier-Stokes solver* (`http://adfc.sourceforge.net`)
[17] Allen J J and Naitoh T 2007 Scaling and instability of a junction vortex *J. Fluid Mech.* **574** 1–23

The dynamics of beltramized flows and its relation with the Kelvin waves.

Rafael González[1] **and E Sergio Santini**[2]

[1]Instituto de Desarrollo Humano, Universidad Nacional de General Sarmiento
Pcia. de Buenos Aires, Argentina and Depto. de Física FCEyN, Universidad de Buenos Aires, Buenos Aires, Argentina
[2]Instituto de Ciencias, Universidad Nacional de General Sarmiento
Pcia. de Buenos Aires, Argentina and ICRA-BR, Centro Brasileiro de Pesquisas Físicas, Rio de Janeiro, Brasil

E-mail: rgonzale@ungs.edu.ar

Abstract. We define the beltramized flow as the sum of an uniform translation and an uniform rotation with a Beltrami flow. Some of their features are studied by solving the Euler equations, for different geometries, taking into account the boundary conditions, and for different symmetries. We show that the Kelvin waves are beltramized flows. Finally, we show that the variational principle found in a previous work, remains valid for the beltramized flow.

1. Introduction

The Beltrami flow defined as $\boldsymbol{\omega}_B = \nabla \times \mathbf{v}_B = \gamma \mathbf{v}_B$,. has been treated in several contexts [1], [2], [3]. Its importance lies on that it has nonlinear solutions of Euler and Navier-Stokes equations, and it has relevance for the lagrangian turbulence. And some other properties as being eigenfunctions of the curl operator and then they can be superposed linearly.

In previous works [4], [5] we have considered meanly two aspects: the formation of Kelvin waves [1] with, what we called, Beltrami structure, in an expansion, and the existence of a variational principle similar to that introduced by Woltjer in Magnetohyrodynamics concerning the force-free magnetic fields that have a *topological analogy* with the vorticity field. But in the first case [4], the whole flow is not a pure Beltrami flow but the sum of a Rankine flow plus a Beltrami flow what we will denominate *beltramized flow*. And it is related to the Kelvin waves and is also important because the *breakdown* phenomena can develops from it [8]. Nevertheless when considering Kelvin waves [9] or steady axisymmetric swirling [10] flow in an expansion the Beltrami structure of the flow is not generally considered as it is done, for example, by Dritcshel [2].

The aim of this paper is to consider the two aspects just pointed out. On the one hand we establish the dynamics of the beltramized flow and get solutions by considering different geometries and symmetries, showing that it results to be Kelvin Waves. We see that the eigenvalue of the beltramized flow is closely related to the boundary conditions. And that the steady state swirling flow is obtained as a particular case. On the other hand, we extend the variational principle introduced for pure Beltrami flows [5] to beltramized flows.

[1] Here the Kelvin waves are given in the sens introduced by Kelvin [6]

In section (2) write the dynamic equation for the beltramized flow. In section (3) we get the solutions for different geometries and symmetries and taking into account the boundary conditions. In section (4) we extend the variational principle for Beltrami flows to a beltramized ones. Finally in section (5) we write the conclusions.

2. The dynamic equation for the beltramized flow.

The Beltrami flow is defined as a field $\mathbf{v}_B$ that satisfies $\boldsymbol{\omega}_B = \nabla\times\mathbf{v}_B = \gamma\mathbf{v}_B$, with $\gamma = constant$.

On the other hand, we define a *beltramized flow* as

$$\mathbf{v} = U\mathbf{e}_z + \Omega r\mathbf{e}_\theta + \mathbf{v}_B, \tag{1}$$

being U and Ω constants, i.e a beltramized flow is the superposition of a Rankine flow with a Beltrami one. As the dynamical equations are satisfied in any inertial frame of reference, we do not consider here the uniform translation , or, in other words, we work in a frame which is translating with velocity $U\mathbf{e}_z$.

Furthermore, to take into account the uniform rotation, we will consider our analysis in a non-inertial frame rotating with the angular velocity $\boldsymbol{\Omega} = \Omega\mathbf{e}_z$, and recover the uniform translation when it corresponds trough the boundary conditions. Therefore the inertial terms, i.e. Centrifugal and Coriolis ones, are added to the RHS of Euler ' s equation (see for example [9] paragraph 14):

$$\frac{\partial\mathbf{v}}{\partial t} = \mathbf{v}\times\boldsymbol{\omega} - 2\boldsymbol{\Omega}\times\mathbf{v} - \frac{1}{\rho}\nabla H, \tag{2}$$

where $H \equiv p + \frac{v^2}{2} - \frac{1}{2}\rho(\boldsymbol{\Omega}\times\mathbf{r})^2$.

If now we set $\mathbf{v} = \mathbf{v}_B \Rightarrow \mathbf{v}_B\times\boldsymbol{\omega}_B = 0$. Then

$$\frac{\partial\mathbf{v}_B}{\partial t} = 2\mathbf{v}_B\times\boldsymbol{\Omega} - \frac{1}{\rho}\nabla H\,. \tag{3}$$

Finally, applying $\nabla\times$ to both members of the last equation and using the definition of Beltrami flow we have

$$\gamma\frac{\partial\mathbf{v}_B}{\partial t} = -2\boldsymbol{\Omega}\nabla\cdot\mathbf{v}_B + 2(\boldsymbol{\Omega}\cdot\nabla)\mathbf{v}_B\,, \tag{4}$$

which can be condensed in the following system of equations:

$$\frac{\partial\mathbf{v}_B}{\partial t} = \frac{2\Omega}{\gamma}\frac{\partial\mathbf{v}_B}{\partial z}, \tag{5}$$

$$\nabla\cdot\mathbf{v}_B = 0, \tag{6}$$

$$\nabla\times\mathbf{v}_B = \gamma\mathbf{v}_B\,, \tag{7}$$

where the eigenvalue γ will be determined together with their solutions. This system of equations in fact verifies that a Rankine flow plus a Beltrami flow gives a solution of Euler 's equations, as was demonstrated in [5].

3. Solutions of the Euler equations by considering the geometries and the symmetries

Here we consider different geometries and symmetries in order to solve the dynamical system given by (5), (6) and (7) which, from now on, will be called DS.

3.1. Rectangular geometry (plane waves).
We seek the solution of DS as a plane wave in an infinite fluid

$$\mathbf{v}_B = \mathbf{A}e^{i(\mathbf{k}\cdot\mathbf{r}-\omega t)}, \tag{8}$$

where $\mathbf{A} = constant$.

Substituting (8) in DS we have

$$\omega = -\frac{2\Omega}{\gamma}k_z, \tag{9}$$

$$\mathbf{k}\cdot\mathbf{A} = 0, \tag{10}$$

$$\frac{\mathbf{k}}{\gamma}\times\mathbf{v}_B = -i\mathbf{v}_B. \tag{11}$$

From where, according to (10) and (11) we obtain $\gamma = k$, and then Eq. (9) becomes

$$\omega = -\frac{2\Omega k_z}{k}. \tag{12}$$

which coincides with Eq. (14.8) of [9].

Now, we follow an analog approach as that given in [9] section 14. Defining $\mathbf{n} \equiv \frac{\mathbf{k}}{k}$, if we use a complex wave amplitude in the form $\mathbf{A} = \mathbf{a} + i\mathbf{b}$ with real vectors $\mathbf{a}$ and $\mathbf{b}$, it follows that $\mathbf{n}\times\mathbf{b} = -\mathbf{a}$ and $\mathbf{n}\times\mathbf{a} = \mathbf{b}$: the vectors $\mathbf{a}$ and $\mathbf{b}$ (both lying in the plane purpendicular to $\mathbf{k}$) are orthogonal and equal in magnitude. By taking their directions as the x and y axes, and separating real and imaginary parts in Eq. (8), we obtain

$$v_{Bx} = a\cos(\mathbf{k}\cdot\mathbf{r}-\omega t)\ ; v_{By} = -a\sin(\mathbf{k}\cdot\mathbf{r}-\omega t) \tag{13}$$

or

$$v_{Bx} = a\cos(kz+2\Omega t)\ ; v_{By} = -a\sin(kz+2\Omega t) \tag{14}$$

where a is a constant of finite amplitude.

3.2. Tubes with cylindrical and non-cylindrical geometries
Here we consider two possible geometries:

a) Cylindrical geometry: we consider a cylinder of radius R.

b) Non-cylindrical geometry: tubes with variable section, as for example in fig. 1. In this case the conditions of continuities between regions should be considered besides the boundary conditions, in order to find the solution for the whole tube region.

The boundary conditions for the beltramized flow are:

Case a)
given that $\mathbf{n} = \mathbf{e}_r$ it is $\mathbf{n}\cdot\mathbf{e}_z = \mathbf{n}\cdot\mathbf{e}_\theta = 0$ and then from (1)
$\mathbf{v}\cdot\mathbf{n} = 0$ and $\boldsymbol{\omega}\cdot\mathbf{n} = 0 \Rightarrow \mathbf{v}_B\cdot\mathbf{n} = 0$.

Case b)
$\mathbf{v}\cdot\mathbf{n} = 0 = U\mathbf{e}_z\cdot\mathbf{n} + \Omega r(\mathbf{n}\cdot\mathbf{e}_\theta) + \mathbf{v}_B\cdot\mathbf{n}$, and
$\boldsymbol{\omega}\cdot\mathbf{n} = 0 = 2\Omega\mathbf{e}_z\cdot\mathbf{n} + \gamma\mathbf{v}_B\cdot\mathbf{n}$, but, because of the circular geometry of the tube, $\mathbf{n}\cdot\mathbf{e}_\theta = 0$ and from last equations we obtain that $(2\Omega - \gamma U)\mathbf{e}_z\cdot\mathbf{n} = 0$. Now, for tubes with regions for which $\mathbf{e}_z\cdot\mathbf{n} \neq 0$ it must be $2\Omega - \gamma U = 0$, or

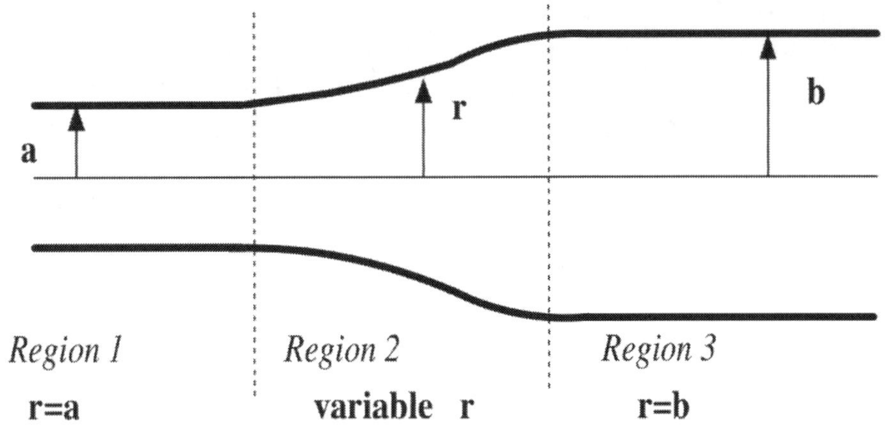

Figure 1. Non cylindrical tube geometry: it may have regions of constant and variable sections

$$\gamma = \frac{2\Omega}{U}. \tag{15}$$

From now on we consider different symmetries of the flow in a cylinder (case a) or in different regions of the non-cylindrical geometry (case b).

3.2.1. Flow with axial symmetry

For this symmetry, the general solution for the DS is:

$$\mathbf{v}_B = \mathbf{v}(r)e^{i(kz-\omega t+\Delta)}, \tag{16}$$

with k, ω and Δ constants quantities. From Eq. (5) we obtain

$$\omega = -\frac{2\Omega}{\gamma}k, \tag{17}$$

and from Eq(6) it follows

$$\mathbf{v}_B = (-\frac{1}{r}\frac{\partial \psi}{\partial z}, v_\theta, \frac{1}{r}\frac{\partial \psi}{\partial r}), \tag{18}$$

where $v_\theta = v_\theta(r, z)$. Then from (7) we have

$$\nabla \times \mathbf{v}_B = -\frac{1}{r}\frac{\partial}{\partial z}(rv_\theta)\mathbf{e}_r + \left[-\frac{1}{r}\frac{\partial^2 \psi}{\partial z^2} - \frac{\partial}{\partial r}\left(\frac{1}{r}\frac{\partial \psi}{\partial r}\right)\right]\mathbf{e}_\theta + \frac{1}{r}\frac{\partial}{\partial r}(rv_\theta)\mathbf{e}_z = \gamma \mathbf{v}_B, \tag{19}$$

and from (18) and (19) it follows that:

$$v_\theta = \gamma \frac{\psi}{r}, \tag{20}$$

$$\frac{\partial^2 \psi}{\partial z^2} + r\frac{\partial}{\partial r}\left(\frac{1}{r}\frac{\partial \psi}{\partial r}\right) + \gamma^2 \psi = 0 \tag{21}$$

from where, defining $\psi(r, z) = F(r)\sin(kz - \omega t + \Delta)$ we arrives to a solution which is well behaved at $r = 0$, i.e. it does not diverges at that point:

$$\psi = ArJ_1[(\gamma^2 - k^2)^{1/2}r]\sin(kz - \omega t + \Delta). \tag{22}$$

We consider now the boundary conditions so we have by taking into account of Eq. (17).

Case a:

$$\psi = ArJ_1[k(\frac{4\Omega^2}{\omega^2} - 1)^{1/2}r]\sin(kz - \omega t + \Delta), \tag{23}$$

with $k(\frac{4\Omega^2}{\omega^2} - 1)^{1/2}R = x_n$ as the eigenvalue equation for which $v_r(R) = 0$, where x_n are the zeros of $J_1[x]$.

Case b:

$$\psi = ArJ_1[(\frac{4\Omega^2}{U^2} - k^2)^{1/2}r]\sin[k(z + Ut) + \Delta], \tag{24}$$

from which we can get an steady state solution by moving in a reference frame with velocity $U\mathbf{e}_z$. Here A is a constant of finite amplitude.

3.2.2. Flow with cylindrical symmetry

In this situation the flow do not depend on θ nor on z so from Eq.(5) we have an steady solution, and it reads as:

case a:

$$\psi = ArJ_1[\gamma r], \tag{25}$$

case b:

$$\psi = ArJ_1[\frac{2\Omega}{U}r], \tag{26}$$

where A is a constant of finite amplitude.

3.2.3. Flow with helical symmetry

In this case defining $\phi = \theta - \kappa z$ we have $\left(\frac{\partial \mathbf{v}_B}{\partial z}\right)_{\theta,r} = -\kappa\left(\frac{\partial \mathbf{v}_B}{\partial \phi}\right)_{r,z}$, so that we can put (5) in the form:

$$\frac{\partial \mathbf{v}_B}{\partial t} = -\kappa\frac{2\Omega}{\gamma}\frac{\partial \mathbf{v}_B}{\partial \phi}. \tag{27}$$

Then we consider the solution:

$$\mathbf{v}_B = \mathbf{v}(r)e^{i(n\phi - \omega t)}, \tag{28}$$

with n integer, and then we have:

$$\omega = -\frac{2\Omega}{\gamma}\kappa n. \tag{29}$$

The Eqs. (6) and (7) are satisfied if the flow $\mathbf{v}_B$ takes the form [5]

$$v_{Br} = -C[\frac{\mu_n}{\kappa}J_n'(\mu_n r) - \frac{\gamma}{\kappa^2 r}J_n(\mu_n r)]\cos[\Lambda(n,t)], \tag{30}$$

$$v_{B\theta} = C[\frac{n}{\kappa r}J_n(\mu_n r) - \frac{\mu_n\gamma}{\kappa^2 n}J_n'(\mu_n r)]\sin[\Lambda(n,t)], \tag{31}$$

$$v_{Bz} = C[\frac{\mu_n^2}{\kappa^2 n}J_n(\mu_n r)]\sin[\Lambda(n,t)], \tag{32}$$

where C is a constant of finite amplitude, and where :

case a:

$$\mu^2 = \gamma^2 - n^2\kappa^2, \quad \Lambda(n,t) = n(\phi + \frac{2\Omega}{\gamma}\kappa t), \tag{33}$$

case b:

$$\mu^2 = n^2(\frac{4\Omega^2}{U^2} - \kappa^2), \quad \Lambda(n,t) = n(\phi + U\kappa t). \tag{34}$$

4. A variational property

In previous works [4], [5] we have considered the topological analogy between the Hydrodynamics and the Magnetohydrodynamics, and showed that the enstrophy plays the same role than the magnetic energy in Woljter's theorem [7], in the sense that the Beltrami flow equilibrium with constant eigenvalue is obtained when the enstrophy is extremized with the constraint that the helicity is conserved.

We show that the same principle is valid for beltramized flows as defined by Eq. 1. For doing that we take the flow

$$\mathbf{v} = U\mathbf{e}_z + \Omega r\mathbf{e}_\theta + \tilde{\mathbf{v}}, \tag{35}$$

$$\boldsymbol{\omega} = 2\Omega\mathbf{e}_z + \tilde{\boldsymbol{\omega}}, \tag{36}$$

and demonstrate that when $\tilde{\mathbf{v}} = \tilde{\mathbf{v}}_B$, then $\delta[\Phi - \gamma H_{\boldsymbol{\omega}}] = 0$, where $\Phi = \frac{1}{2}\int \omega^2 dV$ and $H_{\boldsymbol{\omega}} = \int \mathbf{v}.\boldsymbol{\omega}\, dV$.

In this case $\tilde{\mathbf{v}}$, and $\tilde{\boldsymbol{\omega}} = \nabla \times \tilde{\mathbf{v}}$ are the velocity and the vorticity with regards to the uniform translation and rotation.

To show that, it is necesary to prove that

$$\delta[\Phi - \gamma H_{\boldsymbol{\omega}}] = \delta[\tilde{\Phi} - \gamma H_{\tilde{\boldsymbol{\omega}}}], \tag{37}$$

where $\tilde{\Phi} = \frac{1}{2}\int \tilde{\omega}^2 dV$ and $H_{\tilde{\boldsymbol{\omega}}} = \int \tilde{\mathbf{v}}.\tilde{\boldsymbol{\omega}}\, dV$, and then to follow the steps given in [4] for the RHS of Eq.(37) assuming that the variations $\delta\tilde{\mathbf{v}} = \delta\tilde{\boldsymbol{\omega}} = 0$ on the boundary.

Really, taking into account Eqs. (35),(36) we have

$$\delta\Phi = \int \Omega\mathbf{e}_z \cdot \delta\tilde{\boldsymbol{\omega}} dV + \delta\tilde{\Phi}. \tag{38}$$

But

$$\int \Omega\mathbf{e}_z \cdot \delta\tilde{\boldsymbol{\omega}} dV = \int \Omega\mathbf{e}_z \cdot \nabla \times \delta\tilde{\mathbf{v}} dV = \int \nabla \cdot (\delta\tilde{\mathbf{v}} \times \Omega\mathbf{e}_z) dV = \int \hat{\mathbf{n}} \cdot (\delta\tilde{\mathbf{v}} \times \Omega\mathbf{e}_z) dS = 0. \tag{39}$$

In the same way

$$\delta H_{\boldsymbol{\omega}} = \int U\mathbf{e}_z \cdot \delta\tilde{\boldsymbol{\omega}} dV + \int \Omega r\mathbf{e}_\theta \cdot \delta\tilde{\boldsymbol{\omega}} dV + \int 2\Omega\mathbf{e}_z \cdot \delta\tilde{\mathbf{v}} dV + \delta H_{\tilde{\boldsymbol{\omega}}}. \tag{40}$$

And as in Eq (39), we can see that

$$\int U\mathbf{e}_z \cdot \delta\tilde{\boldsymbol{\omega}} dV = \int \Omega r\mathbf{e}_\theta \cdot \delta\tilde{\boldsymbol{\omega}} dV = \int 2\Omega\mathbf{e}_z \cdot \delta\tilde{\mathbf{v}} dV = 0, \tag{41}$$

where we have used tha $\Omega r\mathbf{e}_\theta \cdot \nabla \times \delta\tilde{\mathbf{v}} = \nabla \cdot (\delta\tilde{\mathbf{v}} \times \Omega r\mathbf{e}_\theta) + 2\Omega\mathbf{e}_z \cdot \delta\tilde{\mathbf{v}}$, and that as $\nabla \cdot \tilde{\mathbf{v}} = 0$ is $\tilde{\mathbf{v}} = \nabla \times \mathbf{A}$.

Therefore Eq (37) is accomplished, and then from $\delta[\tilde{\Phi} - \gamma H_{\tilde{\omega}}] = 0$ we get $\tilde{\omega} = \gamma\tilde{\mathbf{v}}$ that is to say that $\tilde{\mathbf{v}} = \tilde{\mathbf{v}}_B$.

5. Conclussions

We can resume our conclusions with the following points

- The beltramized flow obeys to the dynamic equations SD Eqs.(5-7).
- Their general solutions are represented by traveling waves of finite amplitude that are the so called Kelvin waves showing that these waves have a beltramized structure.
- The γ eigenvalue of the beltramized flow depends on the geometry considered by mean of the boundary conditions.
  i) In a rectangular geometry in an infinite fluid the eigenvalue is $\gamma = k$ the modulus of the vector wave.
  ii) In a cylindrical geometry of radius R, γ results from the boundary condition $v_{(r=R)} = 0$
  iii) In a non-cylindrical geometry of tubes of variable section, the eigenvalue is given by $\gamma = \frac{2\Omega}{U}$. In this case we could have different regions (see Figure 1), some of which have cylindrical geometry, but in order to accomplish the boundary conditions of the whole flow, γ must have this unique value. On the other hand, the whole solution is obtained by considering the continuity conditions among different regions, which means to determine the constant coefficients of the solutions in each region.
- The Beltramized flow is the result of extreming the enstrophy subject to the helicity conservation, extending in this way the result [4] for a pure Beltrami flow.

References

[1] Moffatt H K 1986 *J. Fluid Mech.* **166**, 359
[2] Dritschel D G 1991*J.Fluid Mech.* **222**, 525
[3] Changchun S, Yongnian H, Yaosung C 1992 *Acta Mechanica Sinica* **8**, 289
[4] González R, Sarasúa G and Costa A 2008 *Phys of Fluids* **20**, 24106
[5] González R, Costa A and Santini E Sergio 2010 *Phys of Fluids* **22**, 74102
[6] Lord Kelvin, 1880 *Phil. Mag.* **10**, 155
[7] Woltjer L 1958 *Proc. Natl Acad. Sci.* **44**, 489
[8] Benjamin B 1962 *J. Fluid Mech.* **14**, 593
[9] Landau L D and Lifshitz E 1987 *Fluid Mechanics* (Oxford, Butterwort Heinemann)
[10] Batchelor G K *An introduction to Fluids Dynamics* 1967(Cambridge, Cambridge University Press)

On the use of the Prandtl mixing length model in the cutting torch modeling

B Mancinelli[1], F O Minotti[2] and H Kelly[2]
[1] Grupo de Descargas Eléctricas, Departamento Ing. Electromecánica, Universidad Tecnológica Nacional, Regional Venado Tuerto, Laprida 651, Venado Tuerto (2600), Santa Fe, Argentina
[2] Instituto de Física del Plasma (CONICET), Departamento de Física, Facultad de Ciencias Exactas y Naturales (UBA) Ciudad Universitaria Pab. I, (1428) Buenos Aires, Argentina

E-mail: bmancinelli@arnet.com.ar

Abstract. The Prandtl mixing length model has been used to take into account the turbulent effects in a 30 A high-energy density cutting torch model. In particular, the model requires the introduction of only one adjustable coefficient c corresponding to the length of action of the turbulence. It is shown that the c value has little effect on the plasma temperature profiles outside the nozzle (the differences being less than 10 %), but severely affects the plasma velocity distribution, with differences reaching about 100 % at the middle of the nozzle-anode gap. Within the experimental uncertainties it was also found that the value $c = 0.08$ allows to reproduce both, the experimental data of velocity and temperature

1. Introduction

Plasma hydrodynamic modeling by numerical simulation in cutting torches is a common tool to predict the values of the fundamental physical quantities, namely the plasma temperature, the particles concentration and the fluid velocity. These numerical codes are employed to understand the relevant physical processes ruling the plasma behavior in order to interpret the experimental results of several plasma diagnostics, and ultimately to obtain optimized designs of such devices.

In particular, the practical use of turbulent hydrodynamic codes requires the introduction of some numerical coefficient [1] whose value has to be obtained from a comparison between the model predictions and the experiment. Unfortunately, since most of the available experimental data on cutting arcs are related to temperatures and species concentrations in the external plasma region [2], the experimental validation of the existing cutting torch models [3-6] has been restricted to the temperature distribution in the nozzle-anode gap. Only recently [7], a measurement of the plasma flow velocity was reported in cutting torches.

The purpose of this work is to validate a turbulent code based in the Prandtl mixing length model employing not only temperature but also velocity values as the experimental data to be confronted with. In order to do this, a 2-D local thermodynamic equilibrium (LTE) plasma model was developed and applied to the same 30 A oxygen cutting torch that was used in a previous velocity measurement experiment [7].

2. Mathematical model

2.1 Computational domain

The schematic of the modeled domain for the simulation is presented in figure 1. The edge of the domain EF is located at a radius of 10 R_N [6]. FG represents the anode. A mass flow rate of 0.71 g s^{-1} with a vortex injection that leads to a ratio of the azimuthal to the axial inlet velocity of $\tan(13°) \approx 0.23$ [7] was used at the torch inlet CD. More details on the torch characteristics can be found elsewhere [7].

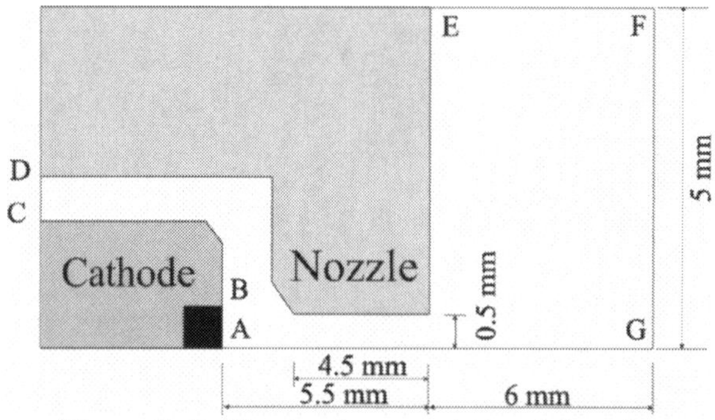

Figure 1. Cutting torch computational domain.

2.2 Model assumptions

(a) The plasma is in LTE, the fluid being characterized by a single temperature *T*.
(b) The plasma flow is two-dimensional and axisymmetric.
(c) The plasma is considered as a Newtonian fluid following the Navier-Stokes equation.
(d) The plasma gas is assumed to be pure Oxygen with thermodynamic and transports coefficients as calculated by Murphy [8].
(e) Hall currents and gravitational effects are considered negligible.
(f) In the energy equation the viscous dissipation term is considered negligible.
(g) The anode was considered as a porous free boundary characterized by its electrostatic potential.

2.3 Governing equations

The set of conservation equations describing such a flow can be expressed as follows.

Total mass conservation

$$\frac{\partial \rho}{\partial t} + \nabla \cdot (\rho \bar{u}) = 0. \quad (1)$$

Momentum conservation

$$\frac{\partial (\rho \bar{u})}{\partial t} + \nabla \cdot \left(\rho \bar{u} \otimes \bar{u} + p \bar{\bar{\delta}} - \bar{\bar{\tau}}\right) - \bar{J} \times \bar{B} = 0. \quad (2)$$

Internal energy conservation

$$\frac{\partial (\rho e)}{\partial t} + \nabla \cdot (\rho \bar{u} e + \bar{q}) - \bar{J} \cdot \bar{E} + p \nabla \cdot \bar{u} + 4\pi \varepsilon_N \frac{p}{p_{ATM}} = 0, \quad (3)$$

where ρ represents the total mass density, $\bar{u}$ the fluid velocity (having, axial $-u_x-$, radial $-u_y-$ and azimuthal $-u_z-$ components), p the pressure, $\bar{\bar{\delta}}$ the identity tensor, $\bar{\bar{\tau}}$ the stress tensor, $\bar{J}$ the current density, $\bar{B}$ the magnetic field (only the azimuthal component was considered) , e the internal energy, $\bar{q}$ the total heat flux, $\bar{E}$ the electric field and ε_N the plasma radiation net emission coefficient (NEC).

Two further equations are required to describe the electromagnetic part of the plasma model. The first is the current continuity equation

$$\nabla \cdot \bar{J} = 0, \tag{4}$$

where

$$\bar{J} = -\sigma \nabla \phi, \tag{5}$$

and the second is one of Maxwell's equations

$$\nabla \times \bar{B} = \mu_0 \bar{J}, \tag{6}$$

where σ is the electric conductivity, ϕ is the electrostatic potential and μ_0 the magnetic permeability of free space.

The total heat flux in (3) describes the heat transported by conduction and the enthalpy transport by mass diffusion, and is defined as

$$\bar{q} \equiv -\kappa_e \nabla T + \bar{\Gamma}_e h_e, \tag{7}$$

where κ_e is the effective thermal conductivity and $\bar{\Gamma}_e$ is the electron mass diffusion that can be approximated by

$$\bar{\Gamma}_e \approx -\frac{m}{e'} \bar{J}, \tag{8}$$

where e' is the elementary electric charge and m is the electron mass. Equation (8) neglects the charge transported by ions. In (7) $h_e = 5 k_B T / (2m)$ represents the specific electron enthalpy (k_B is the Boltzmann's constant).

The effective viscosity is

$$\mu_e = \mu_l + \mu_t, \tag{9}$$

and the effective thermal conductivity is

$$\kappa_e = \kappa + \frac{\mu_t C_p}{P_r}, \tag{10}$$

where C_p, μ_l and κ are the plasma specific heat at constant pressure, viscosity and thermal conductivity, respectively. The turbulent Prandtl number P_r and the turbulent viscosity μ_t are given in the next subsection.

The source terms in (3) account for the Joule effect, the compression work, and the radiation loses $4\pi\varepsilon_N$, where ε_N was taken for a plasma radius of 0.5 mm [3]. The NEC of pure Oxygen for one atmosphere [9] has been multiplied by the factor p/p_{ATM} for other pressures [5].

2.4 Turbulent modeling

The closure of the system equations requires extra relationships to calculate the turbulent enhanced viscosity and thermal conductivity. The simple Prandtl mixing length model was chosen. Such length is given as:

$$l_m \equiv c\lambda, \tag{11}$$

where c is an adjustable parameter and λ is a local thermal radius defined as the radial distance from the axis to the point at 2000 K [10]. It has been found that for transferred arcs the turbulent Prandtl number can be approximated by unity ($P_r \equiv 1$) thus only the parameter c in (11) needs to be adjusted by comparing the numerical results with the experiment. The turbulent viscosity for isotropic turbulence was calculated taking into account the effect of the vortex injection [11]

$$\mu_t = \rho l_m^2 \left((\partial u_x / \partial y)^2 + (y \partial (u_z / y) / \partial y)^2 \right)^{1/2}. \tag{12}$$

2.5 Boundary conditions

Table 1 summarizes the prescribed values of the physical quantities (or of their spatial derivatives) on the boundaries shown in figure 1. In addition, the voltage drop between the cathode AC and the anode FG was adjusted in order that the integrated value of the axial current density on a given section corresponds to the value of the electric current of the torch. An external source term to increase the temperature [1] was applied at the axis of the torch AG to initiate the current. A current value of 30 A was used in this study [7]. Also, at the hafnium insert AB the maximum value of the axial current density on the axis of the geometry was limited to $\leq 170\ \mathrm{A\,mm^{-2}}$ [1]. Besides, the electrostatic potential value of the nozzle DE was calculated so as to preserve the zero current balance at its surface (i.e., the nozzle is electrically floating).

Table 1. Boundary conditions.

	p	u	T	ϕ
AB	–	0	3500 K	–
BC	–	0	500 K	–
CD	–	mass flow 0.71 g s^{-1}	300 K	$\frac{\partial \phi}{\partial x} = 0$
DE	–	0	500 K	–
EF	1 atm	$\frac{\partial u_x}{\partial y} = \frac{\partial u_y}{\partial y} = \frac{\partial u_z}{\partial y} = 0$	300 K	$\frac{\partial \phi}{\partial y} = 0$
FG	–	$\frac{\partial u_x}{\partial x} = \frac{\partial u_y}{\partial x} = \frac{\partial u_z}{\partial x} = 0$	$\frac{\partial T}{\partial x} = 0$	$\phi = 0$
GA	–	$\frac{\partial u_x}{\partial y} = 0, u_y = u_z = 0$	$\frac{\partial T}{\partial y} = 0$	$\frac{\partial \phi}{\partial y} = 0$

2.6 Numerical aspects

The unsteady form of the model equations was solved using a time-marching method [12]. The specific values used for the initial guesses did not impact on the final converged results. The set of governing equation was discretized in time using a Taylor series first-order accurate, in space using the finite volume method, and solved with the given boundary conditions on a 81 × 15 non uniform internal grid-points and 39 × 47 non uniform external grid-points, by using the predictor-corrector algorithm [12]. The time-step used in the time-marching algorithm was chosen so that the CFL criterion was fulfilled [12]. The calculation was stopped when the relative variation of the plasma variables between two consecutive time iterations was $< 10^{-3}$. The accuracy of the calculations was tested by repeating them with a 38 × 15 internal grid-points and 19 × 47 external grid-points. The change in the plasma temperature was everywhere less than 15 %, while the changes in the axial velocity were less than 20 %. The finer grid was then used for generating the results to be presented in the following Section.

3. Validation of the model

This section is devoted to the torch model validation for temperature and velocity by comparing the model results with experimental data. For the temperature, its radial profile at 3.5 mm from the nozzle exit was used. This profile was derived from electrostatic probe [13,14] and schlieren [15] measurements. For the axial velocity, the axial distribution derived from a time-of-flight technique, corresponds to light emitted from the arc central core [7].

Figure 2a) presents the radial profiles of the calculated temperature at 3.5 mm from the nozzle exit for $c = 0$ (i.e., laminar flow), $c = 0.08$ and $c = 0.20$. As shown, all the temperature profiles are similar (their differences being smaller than the experimental temperature uncertainty ≈ 10 %). Figure 2b) shows the comparison among the theoretical profile corresponding to $c = 0.08$ and the experimentally derived temperature profiles. It can be seen from this figure that the model results are in good agreement with the experimental data [13-15] for $c = 0.08$.

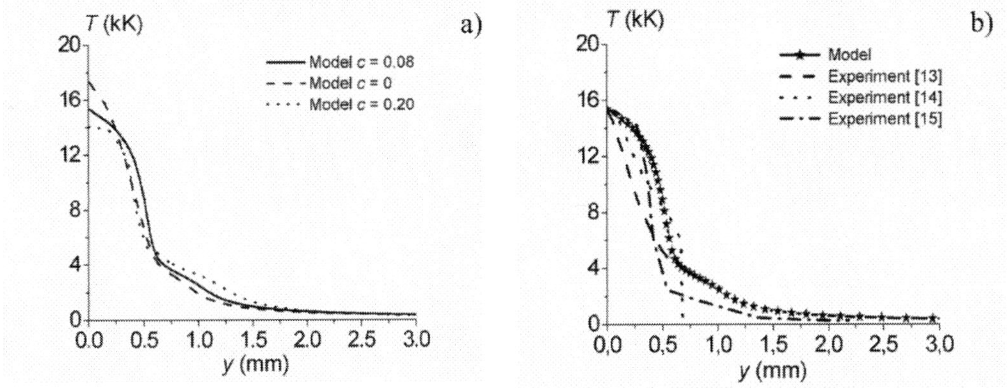

Figure 2. a) Radial profile of the calculated plasma temperature for $c = 0$, $c = 0.08$ and $c = 0.20$. b) Radial profile of the plasma temperature for $c = 0.08$ together with the experimental data.

The theoretical distributions of the axial velocity on the axis for the same c values presented in figure 2a) are shown in figure 3. For comparison purposes, the measured values of the axial velocity corresponding to light emitted from the arc central core [7] are also included in figure 3. It can be seen that the theoretical profiles are close among them at the vicinities of the nozzle exit (reflecting the well known fact of the little importance of the turbulence inside the nozzle [1]) but soon after the scatter in the u_x values is larger than those found for the temperature, reaching about 100 % at the middle of the gap. Hence, it can be concluded that the fluid velocity strongly depends on the particular value of the turbulent parameter c. On the other hand, the theoretical profile presenting the best matching with the experimental data is that corresponding to $c = 0.08$.

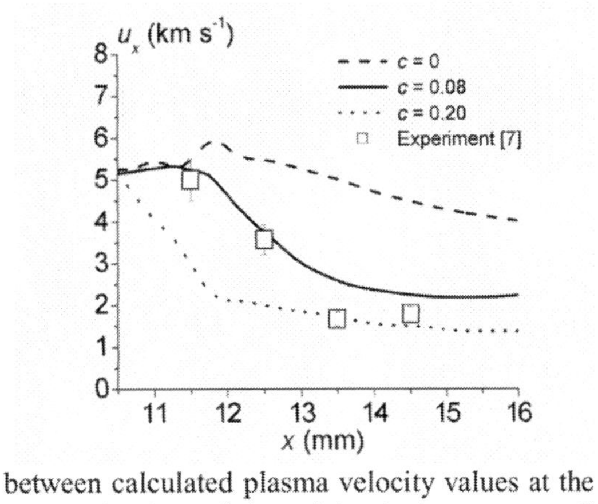

Figure 3. Comparison between calculated plasma velocity values at the axis for $c = 0$, $c = 0.08$ and $c = 0.20$, and the measured values of the axial velocity corresponding to light emitted from the arc central core.

4. Conclusions

Up to now, turbulent models for the plasma generated in cutting torches published during the last ten years have been validated using temperature data derived from spectroscopic measurements in

Journal of Physics: Conference Series **296** (2011) 012025 doi:10.1088/1742-6596/296/1/012025

the nozzle-anode gap. It has been shown in this work that the plasma temperature is not the most appropriate quantity to validate numerical codes since it is not quite sensitive to changes in the model numerical parameters. Instead, it has been shown that the plasma velocity appears to be a more adequate quantity to perform such validation. In order to realize this validation to such a sensitive variable as the plasma velocity, a 2-D model similar to those proposed in the literature was developed and applied to the same 30 A high-energy density cutting torch that was used in the velocity measurements recently published by some of the authors [7]. Within the experimental uncertainties, it was found that a Prandtl mixing length turbulent parameter $c = 0.08$ allows to reproduce both the experimental data of velocity and temperature.

Acknowledgements
This work was supported by grants from the Universidad de Buenos Aires (PID X108), CONICET (PIP 5378) and Universidad Tecnológica Nacional (PID Z 012). F. O. M. and H. K. are members of the CONICET.

References
[1] Gleizes A, Gonzalez J J and Freton P 2005 *J. Phys. D: Appl. Phys.* **38** R153
[2] Nemchinsky V A and Severance W S 2006 *J. Phys. D: Appl. Phys.* **39** R423
[3] Freton P, Gonzalez J J, Gleizes A, Camy Peyret F, Caillibotte G and Delzenne M 2002 *J. Phys. D: Appl. Phys.* **35** 115
[4] Ghorui S, Heberlein J V R and Pfender E 2007 *J. Phys. D: Appl. Phys.* **40** 1966
[5] Zhou Q, Li H, Xu X, Liu F, Guo S, Chang X, Guo W and Xu P 2009 *J. Phys. D: Appl. Phys.* **42** 015210
[6] González-Aguilar J, Pardo Sanjurjo C, Rodríguez-Yunta A and Calderón M A G 1999 *IEEE Trans. Plasma Sci.* **27** 264
[7] Prevosto L, Kelly H and Mancinelli B 2009 *J. Appl. Phys.* **106** 053308
[8] Murphy A B and Arundell C J 1994 *Plasma Chem. Plasma Process.* **14** 451.
[9] Naghizadeh-Kashani Y, Cressault Y and Gleizes A 2002 *J. Phys. D: Appl. Phys.* **35** 2925
[10] Yan J D, Nuttall K I and Fang M T C 1999 *J. Phys. D: Appl. Phys.* **32** 1401
[11] Yan J D, Zhang J L, Liau V K, Dixon C M and Fang M T C 2003 *XV Symposium on Physics of switching arc*, Czech Republic Brno, 278
[12] Ferziger J H and Períc M 2002 *Computational Methods for Fluid Dynamics*, (Springer-Verlag)
[13] Prevosto L, Kelly H and Mancinelli B 2009 *IEEE Trans. Plasma Sci.* **37** 1092
[14] Prevosto L, Kelly H and Mancinelli B 2008 *IEEE Trans. Plasma Sci.* **36** 271
[15] Prevosto L, Kelly H and Mancinelli B 2010 *J. Appl. Phys.* **107** 023304

AUTHOR INDEX

Institute of Physics
Dirac House, Temple Back
Bristol BS1 6BE UK

CURRAN ASSOCIATES INC
proceedings.com